ECOPIETY

HUDSON
PUBLIC LIBRARY
WITHDRAWN
FROM OUR
COLLECTION

D1235974

RELIGION AND SOCIAL TRANSFORMATION

General Editors: Anthony B. Pinn and Stacey M. Floyd-Thomas

Prophetic Activism: Progressive Religious Justice Movements in Contemporary America
Helene Slessarev-Jamir

All You That Labor: Religion and Ethics in the Living Wage Movement
C. Melissa Snarr

Blacks and Whites in Christian America: How Racial Discrimination Shapes Religious Convictions
James E. Shelton and Michael O. Emerson

Pillars of Cloud and Fire: The Politics of Exodus in African American Biblical Interpretation
Herbert Robinson Marbury

American Secularism: Cultural Contours of Nonreligious Belief Systems
Joseph O. Baker and Buster G. Smith

Religion and Progressive Activism: New Stories about Faith and Politics
Edited by Ruth Braunstein, Todd Nicholas Fuist, and Rhys H. Williams

"Jesus Saved an Ex-Con": Political Activism and Redemption after Incarceration
Edward Orozco Flores

Solidarity and Defiant Spirituality: Africana Lessons on Religion, Racism, and Ending Gender Violence
Traci C. West

After the Protests Are Heard: Enacting Civic Engagement and Social Transformation
Sharon D. Welch

Ecopiety: Green Media and the Dilemma of Environmental Virtue
Sarah McFarland Taylor

Ecopiety

Green Media and the Dilemma of Environmental Virtue

Sarah McFarland Taylor

NEW YORK UNIVERSITY PRESS
New York
www.nyupress.org

Library of Congress Cataloging-in-Publication Data
Names: Taylor, Sarah McFarland, 1968– author.
Title: Ecopiety : green media and the dilemma of environmental virtue /
Sarah McFarland Taylor.
Description: New York : NYU Press, 2019. | Series: Religion and social transformation |
Includes bibliographical references and index.
Identifiers: LCCN 2019001449 | ISBN 9781479810765 (cl : alk. paper) |
ISBN 9781479891313 (pb : alk. paper)
Subjects: LCSH: Ecology—Religious aspects. | Mass media—Religious aspects. | Ecotheology.
Classification: LCC BL65.E36 T39 2019 | DDC 205/.691—dc23
LC record available at https://lccn.loc.gov/2019001449

New York University Press books are printed on acid-free paper, and their binding materials are chosen for strength and durability. We strive to use environmentally responsible suppliers and materials to the greatest extent possible in publishing our books.
Manufactured in the United States of America

10 9 8 7 6 5 4 3 2 1

Also available as an ebook

For my beautiful mother,

Sandy McFarland Taylor

CONTENTS

Introduction

"It's 3:23 in the morning and I'm *awake*." A man sits up on the side of his bed with his head in his hands, running his fingers through his thick brown hair in frustration, and we as the viewers get the sense that he is being haunted by some unseen force. The eerie sounds of Icelandic rock band Sigur Rós front man Josí playing guitar strings with a rosined cello bow evoke the creaks and groans of calving glaciers. The scene shifts to the twinkling lights of the Golden Gate Bridge in the San Francisco night sky, and the man is now up and dressed, soberly staring unblinkingly at us. He explains: "I'm awake because my great-great-grandchildren won't let me sleep." Images of children playing on merry-go-rounds and swinging on swings sequence one after another. "My great-great-grandchildren ask me in dreams," he continues, "what did you *do* while the planet was plundered? What did you *do* when the earth was unraveling?" Images flash of power plants, polluted skies, melting arctic ice, seals searching in vain for land, sea birds blackened in crude oil, choked highways, burning oil fields, and bomber jets headed off to war in what appears to be the Middle East. The great-great-grandchildren's questions fill the man's head and are unrelenting. They want to know of him, "Surely you did *something* as the seasons started failing? As the mammals, reptiles, and birds were all dying?" The children press further and demand, "Did you fill the streets with protest when democracy was stolen?" Their final question hangs heavy in the air as a haunting refrain of accusation: "What did you do *once you knew*?"[1]

How many of us imagine that we might adequately answer that weighty last question one day by responding, "Well, I made sure to shop for the right stuff. I bought green products, I shopped organic, I supported green capitalism, purchased ecologically mindful cell phone apps, a nice recycled beverage container, a hybrid vehicle, and generally consumed my way to helping solve global climate change and save the earth." Would this satisfy the specters of our future offspring who wish to hold us morally accountable for the state of the earth they have inherited? And yet, much of the environmental messaging we encounter on a daily basis through marketing, advertising, and mediated popular culture is simplistically reassuring and absolving. Shop for the right stuff, and it will all be okay.

The voiceover in the video recounted above, which was created by Bay Area–based spoken-word poet and filmmaker Drew Dellinger, features an excerpt from his longer poem, "Hieroglyphic Stairway," drawn from his award-winning book of environmentally themed poetry, *Love Letter to the Milky Way*. Dellinger, who also holds a PhD in philosophy and religion, jolts us awake from the complacent fantasy that we can sit back, make minimal changes, and it will all be okay. His video intervenes in these reassurances, telling a different story—one of late-night hauntings and the horror of deep regret as we are called on the carpet by our suffering descendants for failing to act when we might have. Dellinger's video goes on to show hopeful images of protests filling the streets, of student activists braving pepper spray and police, and of arts and culture leading the way to a vibrant and sustaining ecological human culture. But the haunting question hovers, "What did you do *once you knew*?" What is it going to take to ensure we never face the haunting questions this video conjures? Consumer culture sets our minds at ease that individual acts of environmental virtue will do the trick. Dellinger's answer is broader ranging and calls for *more*—more imagination, more collaboration, more connection, and more coordinated acts of conscious collective social transformation. Borrowing a phrase from the work of civil rights activist Martin Luther King Jr., the final words of the video appear as striking white letters glowing against a black screen: "Planetize the movement."[2]

Piety, Virtue, and Saving the Planet

"Eco-friendly," "green," "environmentally sustainable," "ecologically mindful," "earth-conscious," "environmentally responsible," "planet-friendly," "eco-smart," and even "eco-elegant" are just some of the sobriquets used in popular parlance and consumer marketing to signal products, services, acts, behaviors, or lifestyles associated with the practice of a kind of environmental virtue, or "ecopiety." Piety, a complex social, civic, philosophical, and religious designation, is classically considered to be a virtue associated with the practice of appropriate duties or obligations. The Greek term "*eusebeia*" ("*εὐσέβεια*") connotes reverence and respect in engaging in right behavior or service that pleases the gods. In the ancient world, the Roman ideal of *pietas* was associated with "dutifulness to parents, homeland, and emperor," and with maintaining good relations with the gods—acts that if performed well were often credited for Roman empiric success. The proper performance and practice of *pietas*, in Roman terms, also suggestively correlated with the maintenance of order in the universe and the sustaining of the world.[3] Some scholarship on religion in the ancient world has probed the complex sociopolitical and economic transactional value of *pietas* in the Greco-Roman Mediterranean "ideological marketplace."[4] This book explores the marketing, representation, and popular mediation of environmental piety in our own contemporary "ideological marketplace," its contested cultural meanings, and its complex transactional value in the lives of North American consumers.

"Ecopiety" is a shorthand term I use to refer to contemporary practices of environmental (or "green") virtue, through daily, voluntary works of duty and obligation—from recycling drink containers and reducing packaging to taking shorter showers and purchasing green products. Practices of ecopiety evoke an idyllic harmonial model of proper relations cultivated between humans and the more-than-human earth. Much of the language used in the mediasphere to motivate practices of ecopiety speaks of "doing our part," "saving the earth," "shopping for change," "buying green," "going green," "practicing earth care," and "becoming more sustainable"—with exhortations often promising at the same time that "every little act," no matter how seemingly insignificant, "all adds up."[5] Works of popular culture and environmental marketing campaigns are rife with messages that reassure publics that personal, individual, private acts of ecopiety are efficacious remedies to address an overwhelming array of environmental ailments, if only everyone would virtuously *do his or her bit*.

Like so many other aspects of contemporary social and political activism, stories of ecopiety and how to practice it reflect and perpetuate the logics of global capitalism and market ideology. As marketed in the contemporary US context, the devotional practice of ecopiety most often requires the performance of a correlative "consumopiety," or acts of "virtuous consumption." This book is highly attentive to correspondences drawn between ecopiety and consumopiety and probes in depth what role media and culture play in their making. As media scholars Roopali Mukherjee and Sarah Banet-Weiser point out in their research on the mediation of commodity activism, "within contemporary culture it is utterly unsurprising to participate in social activism by buying something."[6] Ecopiety thus thrives, this book argues, within the hospitable conditions of a depoliticized marketplace environmentalism and mediasphere that generate story after story of privatized, small-scale, voluntary, individualized acts of "green virtue" as being utterly adequate to dealing with our monumental planetary challenges. Whether circulated via the adscape, reality television, popular books, films, games, or social media, these stories explicitly or implicitly promise publics that the practice of an individualized, consumer-based ecopiety holds the key to making things on earth right again. In so doing, these stories market an imagined moral economy, in which tiny acts of voluntary personal piety, such as recycling a coffee cup, or purchasing a green consumer item, can be exchanged as an offset to justify the continuance of current consumption patterns and volume.[7] No need to make any fundamental structural changes, implement public policy or legislation, or enforce stricter, much less existing, regulations. The trick is simply for the consumer to buy the right things, the eco-piously green things—to engage in individual simple acts to save the earth—and all will be well.[8]

Plato's dialogue *Euthyphro* presents a number of contested meanings of "piety," but many of these reference acts or works associated with caring for the gods, performing devotional sacrifice and service to them, and doing what is right by honoring and pleasing them.[9] Having the gods actually agree with one another as to *what* they love, and to achieve consensus as to what is pleasing to them, turns out to be a moving target. The figure of Socrates in Plato's dialogue implicates an absolute definition of piety as being broadly disputed.[10] Ecopiety, as represented in and through contemporary popular culture, is about cultivating a proper and respectful relationship, an ecologically responsible connection, between individual citizen consumers and the more-than-human earth, but on *whose* terms and by whose definition is similarly a moving target.[11] Performing the *right* kind of devotional service is key to the notion of ecopiety, and green consumer marketing is quick to offer a template for which acts are most pleasing and pious, virtuous, and efficacious. Engendering what is arguably an unmarked "Protestant ethos of capitalist individualism," green consumer marketing proclaims that each citizen-consumer is empowered to effect the healing and restoration of the planet directly through the swipe of a personal credit card.[12] All this "priesthood of all credit-card holders" needs to do is "shop our way to a better world."[13]

Piety as a concept is complicated, means different things in different cultural contexts, and can be both personal, individual, and private, as well as collective, public, and communal. It can be ascetic, ecstatic, or both.[14] Though they do so with different approaches and in different geographic contexts, anthropologist Saba Mahmood's *Politics of Piety* and sociologist Rachel Rinaldo's *Mobilizing Piety* both investigate Islamic women's piety in relationship to feminist theories and do so focusing on piety as devout religious *practice* in everyday life.[15] In her study of transnational devotion to the Mexican Virgin of Guadalupe, Elaine Peña, too, sees piety as something fundamentally practiced and performed.[16] Peña emphasizes that, from an early age, her elders taught her that piety "is not something talked about, it is something you *do*."[17]

Within the milieu of contemporary global capitalist culture, or what a number of critics now name in geologic temporal terms, "the Capitalocene," *doing* ecopiety, to a large extent, has become a devotional practice performed within the "sacred buyosphere."[18] Writing about American consumer culture, design and culture critic Thomas Hine has identified the "buyosphere," the sphere of purchase, as simultaneously a set of physical and virtual spaces, as well as a state of mind. Hine asserts that, today, we come to know ourselves through our acts of consumption within this mystical realm of promise.[19] Practices of ecopiety take up the promises of the "sacred buyosphere" and extend them to the promised restoration of the planet, one organic-cotton-sweater or electric-hybrid-vehicle purchase at a time.

In archiving and analyzing sightings of ecopiety, as observed mostly in and through North American consumer marketing and mediated popular culture, this book argues that fundamentally individualized, free-market, privatized, voluntary approaches—promoted as addressing the monumental environmental challenges facing us—are not simply inadequate to the task but in some cases are *counterproductive* in the worst possible ways. Ecopiety, as marketed, is both too dourly restrictive in some ways and grossly facile in others. It simultaneously asks too much and expects too little, making pious actions taken on behalf of the environment grim, unappealing, onerous duties or obligations, on one hand, while they constitute superficial, perfunctory modes of practice that are by and large insignificant in terms of scale and scope of impact, on the other. In studying the concurrent dynamics of both too strict and too lenient jointly at work within certain religious institutions, sociologist Laurence Iannaccone famously labeled this combination the "worst-of-both-worlds position."[20] Messaging ecopiety as fun, playful, hip, sexy, and appealing, while also making it more rigorous, potent, systemic, policy linked, and effectual, is challenging but not impossible, as certain ecopiety sightings in this book illustrate. The contradictions and tensions between ideals and practices of ecopiety are explored in each chapter of this book, which also considers proposed alternatives, challenges, and creative cultural paths into the future, as conjured by various media works, practices, and narratives.

Since this book appears in a series on Religion and Social Transformation, readers may legitimately wish to know what notions, stories, practices, and the marketing of ecopiety and/or consumopiety have to do with religion. Religion appears in this book as it does in my other work on religion, media, and consumption—not as a solid, circumscribed, institutional, organized entity but more as (to borrow a phrase from author Anne Lamott) "the water at the edge of things."[21] Depending on context, contours, and conditions, this "water" shapes what cultural theorist Stuart Hall calls our "life-worlds," how we know them, and how we make sense of them.[22] Religion, like language, is a fluid representational medium and set of dynamic sociocultural practices in which contextually defined meanings are produced and exchanged according to a variety of vested interests.[23] Fluid semantic, rhetorical, and historically inherited themes, stories, symbols, sensibilities, moral codes, and conceptual currents from a variety of systems/streams of knowing and ways of being in the world flow both at the edges of, and subterraneously underneath, the processes and practices that frequently get labeled as "culture," or as "not religion."[24]

This water at the edge of and flowing underneath things both sculpts and is sculpted by its areas of contact—lapping at and remaking the edges, flowing under and eroding what we often assume to be terra firma, carving channels and rivulets, making deposits, sweeping materials along, and carrying all sorts of sediments that we may not explicitly be aware of or immediately recognize. "Piety" and "devotion" are themselves religiously inflected terms, and ecopiety may indeed be practiced by religiously devout persons as an extension of self-identified religious commitments to God/gods, nature, Creation, a love of humanity, an observed monistic sacred unity of all things, and so forth. As a more broadly cultural phenomenon, however, and as a celebrated demonstration of civic virtue, practices of ecopiety and their mediated popular representations challenge artless binaries drawn between religion and secular culture, or between religion and not religion, as conventionally socially identified.[25]

Representations of ecopiety in a variety of popular cultural forms tap into a media maelstrom of values and meanings: the logics of a global consumer capitalist economy; the popular production of environmental (green) virtue as a commodity; ideals of piety, impiety, and their practice; and conceptions of personhood and identity defined by "sovereign individualism"; and yet also ideals of collective civic engagement.[26] That is, in the sphere of ecopiety, all is not simply opportunistic consumer marketing, depoliticized marketplace environmentalism, and late capitalism's digestion of all things that would challenge its ubiquity and dominance. Even as this book teases out the all-absorbing economic forces of peristalsis that would digest ecopiety, regurgitate it, and market it as a product, it also spotlights how media making, especially grassroots and DIY media making, imperfectly but potently crafts moral interventions. These intervening stories, images, and performances consciously pursue a turn toward a more just, sustainable, compassionate, and even joyful future for human/planetary relations.[27]

The intent of morally engaged environmental media interventions and their makers can be to edge humans in different existential directions by providing a kind of rhythm and set of community movements with which to align. Sometimes eco-pious media interventions are contradictory, and more often than not they are messy and experimental.

Nonetheless, they contribute to emergent social energetics that can shift already existent and even entrenched social energetics. Climate change philosopher Roy Scranton characterizes social energetics as involving "how bodies harvest, produce, organize, and distribute energy," which in turn "determines how power flows, shaping political arrangements."[28] To visualize this phenomenon, Scranton uses the analogy of honeybees doing storied "hive dances." To create a livable, compassionate, and more sustainable Anthropocene—the age, already upon us according to Scranton, in which humans have become the defining agent affecting life systems on the planet—he argues that we need new ideas, myths, stories, and visions of who "we" are that change our relationship to our world. We need a "new way of thinking our collective existence."[29] Communities of honeybees, explains Scranton, send out a series of scouts to figure out their next destination—that is, how the hive should get from point A to point B. The scouts then return from their explorations and dance becomes the *medium* by which they communicate or tell the story of various options and directional plans to the rest of the hive. Different scouts dance various options or visions for the future, and each scout's dance itself composes a story of how to get there. These multiple dances engage in contrapuntal motion, at once distinct from one another and moving in different directions, and yet danced parallel to each other, and simultaneously. Bees start aligning themselves with one dance or direction, based upon the scouts' communications, and eventually one dance story receives more critical mass of those aligning with it than another.[30]

Media making, systems, practices, uses, and designs all have a critical role to play as directional *agents* in the social energetics of our storied human "hive dances" into earth's future. Media are how we *do* these dances. Examining the dynamics of representation, resistance, power, control, intervention, and cooption of stories, symbols, expressions, and practices related to ecopiety provides pivotal insights into our agential ongoing media "hive dances" at a time of environmental crisis.

Sightings of Ecopiety in Practice

This book explores the conflicting and yet often symbiotic narrative relationship between representations of ecopiety and consumopiety through a series of environmentally themed media case studies or sightings drawn from popular culture. It argues that rather than functioning as a mere hegemonic delivery system for dominant cultural narratives, mediated popular culture performs a crucial role in dynamics of environmental moral engagement and processes of civic social transformation. Porous and adaptive, especially as it is remixed, repurposed, and shared through the tools and technologies of participatory digital new media and social networking, mediated popular culture has become a prime location for contemporary moral/civic engagement and public debate. Prime examples of this can be seen in Henry Jenkins's analysis of activist political protest actions that strategically deploy many of the thematic tropes, symbolic resources, and aesthetics of resistance drawn from film franchises such as *The Matrix*, *Harry Potter*, *Avatar*, and *The Hunger Games*.[31]

Jenkins and his Civic Imagination Project research team at the University of Southern California have documented the "appropriation and remixing" of popular culture among two hundred youth social activists, "including the use of Superman as an 'illegal alien' by the [undocumented youth] Dreamer movement."[32] In 2018, Jenkins and the team turned their attention to Marvel Comics' *Black Panther* as a catalyst for civic engagement. "Before we can build a better world, we need to imagine what one looks like," asserted Jenkins, pointing out that "*Black Panther*'s fictional [African city of] Wakanda provides a vivid contrast to the poverty and hopelessness depicted in Oakland [California] in the film's opening and closing scenes."[33] Germane to *Ecopiety*'s examination of popular culture and its engagement with environmental themes are the ways in which *Black Panther*'s depictions of the ecotopian city of Wakanda present not a model of individual ecopiety but one of collective societal public investment in green infrastructure and planning. The film's rendering of Wakanda has in turn sparked policy and planning discussions online among engineers for sustainability, urban policy wonks, and green city planners—all discussing what it would take to construct a Wakanda, or at least the fictional city's high-tech transportation system, in the United States.[34] What it would take is a reorientation of environmental messaging away from the notion that every tiny act counts and toward an unapologetic emphasis on broad-scale policy enactments and serious public investment. Imaginative planners and engineers in the United States looking for models could easily study Germany's successful and subsidized EnergieWende program and public transportation system, or Copenhagen's real-life urban cycling infrastructure—both discussed later in this book. But the visually rich ecotopian scenes of vibrantly green Wakanda and, more importantly, the compelling narrative that drives our experience of sustainability in *Black Panther* are what truly engage public imagination. It is in the realm of mediated popular culture that a complex and contested moral discourse of ecological citizenship, green civic engagement, ecopiety, ecopolicy, and their envisioned impact on our planetary future, emerges.[35]

Popular narratives delivered through storied media can intervene, disrupt, and agitate. They can reveal new angles through which to view our world in expansive ways. At the same time, the wave-like contracting muscles of capitalist commodification and assimilation are powerful in their alimentary propulsion of narratives in a particular direction. Media theorists such as Nick Couldry are correct to point to the repressive power dynamics involved in the political economies of new media.[36] Active modes of storying and *restorying* the world are always vulnerable to the swallowing forces of directional assimilation that ingest resistant voices and activist narratives of social change. Neutralizing such resistance can be effectively achieved by channeling it toward pacifying consumer actions within a depoliticized marketplace.[37] The ways in which some counternarratives resist market-driven logics and defy directional forces, however, speak not only to the social function of popular culture but arguably to the role it can play in shaping the kind of world in which we might live. Pivotal to the dynamics of both resistance and assimilation (and it is imperative that we pay critical attention to *both* in their simultaneity), the digital world of communication is squarely at the center of this larger story of global transformation and its narrative outcomes.

This book focuses primarily on the North American cultural landscape, but in a globalized world, these borders are increasingly artificial, if not misleading, especially when one is studying the circulation of media content. In turn, this book crosses borders: geographical, technological, as well as disciplinary. Informed by study, research, scholarly formation, and teaching in multiple fields (religious studies, media studies, and environmental studies), I acknowledge what media and cultural studies scholar Michael Pickering refers to as "the need to negotiate the limitation of academic specialisms when investigating the multiple convergences and flows in cultures of modernity."[38] Boundary-crossing convergences and flows can thus be found in each of this book's chapters.

The first chapter introduces the reader to a number of theoretical tools and analytical lenses for understanding ecopiety and consumopiety, defining more closely the meaning and use of those terms in this book and their relationship to one another. In this chapter, I also explain the concept of what I call in shorthand "restorying the earth"—the ongoing processes of mediated moral engagement in recrafting or remaking stories of earth and our place in it in an age of environmental crisis. Two conflicting but inexorably related stories take center stage in this discussion: (1) what economist David Korten calls the "sacred money and markets" story; and (2) what Korten identifies as the "sacred life and living earth" story.[39] We will explore why tracking these stories through mediated popular culture in particular provides an effective approach to getting at the kind of agential cultural work in which they are involved. This chapter also engages theories of media intervention, considering how media interventions become moral interventions in popular narratives of ecopiety. This theoretical discussion in turn plays upon therapeutic uses of the term "intervention" in popular discourse in order to think through associations between practices of environmental virtue and notions of addiction and recovery.

Chapter 2, entitled "Fifty Shades of Green," attends to the role played by "moral offsets" and what social psychologists term "moral self-licensing" in intertwined stories of ecopiety and consumopiety in the not totally unrelated realms of both popular erotic fiction and corporate public relations messaging. Reading across platforms, this chapter teases out various portrayals of environmental sin and virtue, juxtaposing the corporate public relations practice of greenwashing with the eco-pious storying of CEO and philanthropist protagonist Christian Grey in the popular mass-market romance *Fifty Shades of Grey*. As critics/activists use social media to organize and voice objections both to the corporate practice of public relations greenwashing and to the romanticized representations of abusive power in *Fifty Shades*, these protesters wield digital technologies as tools of narrative interruption and contestation. Their citizen interventions and transformative works of media offer insight into the participatory dynamics of what I argue is an emergent environmental economy of virtue as mediated through popular culture.

Chapter 3 explores Toyota Corporation's hybrid automobile, the Prius, as a consumer icon of green virtue, charting media representations that portray the Prius in the US market as a vehicle of *ecopiety* as practiced through acts of *consumopiety*. This chapter sharply contrasts eco-pious cultural readings of the Prius with ones that are intensely hostile and resistant. Drawing insight from moral foundations theory and theorizing its relevance to political disparities among environmental attitudes, this chapter's media analysis of the Prius provides an opening into the complex incongruities between media encoding and media decoding. As messages about the Prius as icon of environmental piety get filtered through different power dimensions of class, gender, sexuality, and race, oppositional narrative decodings give rise to the new and fascinating subgenre of pornography called "pollution porn." These works of pollution porn constitute seemingly unlikely, but nonetheless operative, class-based media interventions into a dominant environmental discourse often perceived to be elitist, self-righteous, and smug.[40]

Chapter 4, "Green Is the New Black," delves into more entanglements of ecopiety and consumopiety in the narratives of green capitalism as found in the world of mobile-device carbon sin–tracking software applications, reality TV programs, and popular fashion manuals. Each of these products models and markets an ideal of a depoliticized, individualized, privatized, stylish, consumer-based environmental practice that reinscribes the virtues of capitalist consumption as an effective solution to pressing environmental problems. These messages championing individualized ecopiety either explicitly eschew or more subtly obscure and distract from collective solutions to environmental crisis that entail organizing for broad policy initiatives, increased government regulation, and substantial public funding to address environmental problems.[41] This chapter, however, explores two sides of the same coin and their respective merits. On one side is Theodor Adorno and Max Horkheimer's sharp critique of an American media and culture industry that has a vested interest in promoting small, individual, private acts as wholly suitable and sufficient responses to address broader social problems. On the other side is an examination of how moral modeling gets actively challenged in the digital mediasphere through works of media prosumership and via critical fans and viewers engaging in "public spheres of the imagination."[42] To absorb fully the contested multiplicities of popular moral engagement in environmental issues and their functions variously as both obstacles and catalysts to social transformation, this chapter argues that we need to take cognizance of both of these sides, leaving neither ignored nor dismissed. What is more, at a time of environmental crisis, functionality and utility of theoretical tools to get critical work done take priority, even when placed in unconventional combination, and outweigh hereditary loyalties to particular schools or to the orthodoxy of ideological silos.

Chapter 5 explores conflicting expressions of ecopiety and consumopiety as worked out in and through the contemporary remixing and remaking of vampire narratives for an age of environmental crisis. Today's socially conscious, self-identified "vegetarian vampires" virtuously subsist on synthetic blood replacement or on animal blood and are analogous to humans who for ethical reasons abstain from consuming animals. This self-restraint in the face of powerful supernatural desire makes ecopiety surprisingly hot and sexy when practiced by its vampire ecovirtuosi. In this chapter, I tease out how, as one reads across these popular narratives of piously abstaining vampires, a moral sensibility emerges that equates the deep, monstrous desire to consume and deplete the earth's resources with the vampire's voracious hunger to consume and drain the life of its host. Intermingled with messages of virtuous environmental temperance, however, vampire media franchises market endless tie-in consumer products to enthralled fans, in turn whetting a voracious appetite to consume. These counterpoints of pious consumer restraint and yet enthusiastic fan participation in vampire narratives via the virtues of consumer capitalism, ironically and strategically, tap into and feed off one of America's most powerful consumer demographics: *teenagers*. Fans engage environmental virtues through popular cultural texts such as *The Vampire Diaries* (The CW, 2009–2017), the *Twilight* film saga series (2008–2012), *Dracula* (NBC series, 2013–2014), and *True Blood* (HBO, 2008–2014). In so doing, they explore moral and ethical subjects in communal online discussions, strategically and purposefully invoking the iconic image of the green or vegetarian vampire in blogs, Twitter feeds, and other forms of shared digital media to discuss and debate monstrous environmental threats, such as global climate change. Fans' online comments, commentaries, and cultural creations (memes, jokes, videos, visual cultural productions, etc.) suggest a hunger for more meaningful participation and involvement in their favorite vampire story worlds. I approach their participatory work in vampire fan culture using archival methodologies and engage their commentaries through discourse and content analysis.[43] Fan practices of online interactive cultural analysis and debate, as well as fans' material consumption, implicate both ecopiety and consumopiety as being bundled together with what media theorist Nick Couldry explores as "the *social process* enacted *through* the varieties of media-related practices."[44]

Chapter 6 analyzes the online marketing and media representations of green burials and eco-funerals. As the market for green burial grows, conflict ensues between green burial movement advocates and messages of consumopiety promoted by eco-friendly funerary product corporate marketers. Whereas green burial movement activists stress that the most eco-friendly products are the ones you *don't* buy, marketers swallow this message and coopt movement ideals of eco-piously giving back to the earth through eco-pious purchases. Ecopiety in turn becomes the basis for more and more elaborate lines of imported and expensive "green" consumer products. Green burial activists concertedly build bridges between personal green burial planning and collective civic engagement to effect policy making. Mega marketers, by contrast, portray the virtuous purchase of eco-friendly funerary goods as an end in itself. In both marketing and funeral practices on the ground, the bodies of eco-pious corpses have a story to tell about humans' relationship to stuff and to their own mortality in an age of environmental crisis. "The body," observes media theorist John Durham Peters, "is the most basic of all media, and the richest with meaning."[45] With this understanding of the body in mind, this chapter examines corpses as media that signify and are signified in contested ways. Beyond merely identifying the relentless influence of capitalist logics and forces of assimilation, this chapter makes a concerted analytical pivot. That is, it probes what kind of potential green burial practices might hold for prompting Americans to face their own mortality as they begin restorying death in more positive ways that could in turn induce limits on death-denying consumption.

Chapter 7 is a tale of two environmental organizations that each deploy popular short-form media as tools to engage younger cohorts, moving them from environmental ideals into environmental action. While both organizations' media making challenges the consumer capitalist myth of disposability, they intervene in this myth with very different priorities and environmental urgencies in mind. The first organization's orientation is largely biocentric and uses endangered-species tattooing to extend the moral sphere of who "counts" to include vulnerable and threatened nonhuman populations. The second organization's mission is squarely human-focused, attending to environmental justice issues and the plight of poisoned minority communities that have been morally excluded from what we explore in the chapter as the "sphere of justice."[46] Too often low-income urban minority neighborhoods are rendered by corporate, industrial, and city planning interests as expendable or "sacrifice" populations. Mediations of ecopiety, suffering, and sacrifice infuse the skin media of tattooing and the storied media messaging of green hip hop music, respectively, as each medium is used strategically to make a case for greater moral inclusion of those vulnerable, invisible, and left out.

The first half of the chapter takes a close look at the practice of endangered-species environmental tattooing, the consumption of these images through digital media sharing, and the use of skin media for activist goals. As rituals of sacred ordeal, identity, protection, devotion, and self-discipline, environmental tattoos epidermally testify to the uninitiated about what it means to embody (literally) a moral commitment to practicing green virtue. I analyze these storied tattoos in contrapuntal relationship to the commercial corporate storying of bodies, in which companies employ body branding as a marketing tool. Corporately branded bodies are used as paid human billboards. This economy of branded bodies (referred to as "skinvertising") somewhat ironically has created a space for environmental activists to use ink and skin to bear witness to such transgressions as habitat destruction and species endangerment, while extolling the virtues of veganism, bicycling, anticonsumption, and renewable energy technologies. Environmental tattoos, of which endangered-species tattooing is a subset, subversively insert messages of ecopiety and environmental virtue into a corporate-message-dominated mediasphere.

The second half of chapter 7 focuses on urban environmental justice concerns and the use of green hip hop also as a tool of media resistance. Suffering is a major theme in green hip hop and eco-rap rhymes, but in contrast to the pain and suffering self-imposed by tattoo activists, the suffering attended to in green hip hop is imposed by external sources. This section of the chapter looks at the way the moral narrative interventions of eco-rap artists, through the genre of green hip hop, are deeply involved in the cultural work of extending the sphere of justice to include human populations dismissed as disposable or expendable. In this context, hip hop lyrics, videos, circulated online poetry, and visual cultural aesthetics become activist tools to witness against the devastating economic and environmental health conditions perpetuated by environmental racism. Hip hop media also become prophetic tools to craft a more sustainable future for low-income minority communities and, by extension, the planet. This chapter explores what kind of restorying hip hop artists are doing in the growing genre of eco-rap and how their prophetic stories of a great future, in concert with proposed communal solutions for getting there, sharply contest many of the standard images associated with the environmental movement. The prophetic visions of eco-rappers reveal a distinctive and broader vision of environmental values, the moral community, priorities, and practices from those that primarily dominate national discourse. Rather than fixating on individual acts of consumopiety, eco-rappers call for *collective* moral engagement, structural change, grassroots activism ("people power"), and a new living economy based upon inclusivity rather than exclusion, extraction, and exploitation.[47] Finally, this chapter argues that the mediated differences in biocentric approaches by endangered-species tattoo activists, on one hand, and the human-focused efforts of eco-justice urban activists, on the other, have much insight to offer us into what has been called the "eco-divide" between predominantly white and minority communities.

The book's concluding chapter focuses on storied visions of the future, examining the speculative storying of earth's fate embedded in space-exploration company SpaceX's marketing of Mars colonization, contrasting it with the earth-reinhabiting storied play enacted in environmentally themed alternate reality gaming. Both of these mediations of the future—planetary and extra-planetary—situate environmental action, not in a framework of grim duty or obligation but in the inviting sphere of *play* and delight, though with vastly different frameworks, goals, and outcomes in mind. This final chapter argues that ecoplay, as an inviting conduit into the work of ecopolicy, provides a far more effective strategic approach than ecopiety for moving environmental ideals into substantive action. With such an approach in mind, this chapter makes the case that delight, not duty, will prove to be a compelling motivator for catalyzing social change as we experiment with more life-sustaining ways to live into the future.

Rescripting the Future

Endemic to the cultural work of restorying the earth are fundamental questions that redefine our world and our place in it: "Who are we? Where are we going? Can we even agree on who 'we' includes?"[48] And finally, "What are we to do?" Conflicting narrative responses to these questions take up the challenge of renegotiating the borders and boundaries of "us," not just as members of nations, or self-identified communities, but as planetary denizens. The "art of world making" and its *remaking* is a profoundly prophetic one, and one that is intensely mediated.[49] Using phrasing very similar to Elaine Peña's descriptor of piety as "something you *do*," Nick Couldry reminds us that "media are something we *do*."[50] Both are practiced, performed, enacted, and engaged. The realm of mediated popular culture can act as both a powerful mirror and an engine for moral engagement, as once again, we tell our stories, and they in turn tell us.[51] As this book demonstrates, both of these dynamics—reflective and generative—are at work in the ongoing vibrant processes of remaking our world and rescripting the future.

1

Restorying the Earth

Media Interventions as Moral Interventions

We do not have an environmental problem so much as we have a *story* problem. Or so says Stanford Business School PhD and former Harvard Business School professor David Korten, who argues that the grave environmental challenges we face today have been set in motion by the predominating cultural lens of what he calls the "sacred money and markets story." This has been a story with catastrophic outcomes, contends Korten. Is there a way to fix it? Korten thinks there is: intervene. *Change the Story, Change the Future*, Korten proclaims in his book on creating more sustainable and just earth economies.[1] Intervening in the "sacred money and markets story" and supplanting it with the "sacred life and living earth story" as the prevailing story of our time, he argues, will result in changing the human cognition and behavior that got us into our current environmental mess. Change the story, and we will climb our way out of destruction and despair and begin the repair and recovery of life systems on the planet. Permit the dominant "sacred money and markets story" instead to proceed on its current trajectory uninterrupted, with *no* effective intervention, and we seal our own doom.

The narrative struggle between these two powerful world-shaping stories, each vying for control, Korten asserts, is the epic struggle we face on earth today.[2] The "sacred life and living earth story" is, in climate change philosopher Roy Scranton's terms, the "hive dance" with which humans need to align themselves—a dance that tells the story of moving in a new cultural and planetary direction. This chapter explores why, methodologically, analyzing data drawn from works of mediated popular culture provides an ideal approach for tracking and understanding stories and dynamics of social transformation. In doing so, it also introduces key media theories that, like extensions of one's hands, can act as *tools* to help us analytically "dig into" representations and meanings of ecopiety and consumopiety. These theoretical tools concomitantly offer leverage with both perspective and framing, and include expanded theories of media interventions, theoretical understandings of media as *action*, theories of elemental media, and theories of transmediation and storytelling.

Competing Stories and the "Magic Lens"

David Korten is not alone in his prescribed narrative solution for addressing our very real environmental ills. More than a few contemporary environmental philosophers, theologians, artists, and ethicists advocate a similar solution: change the story, change the future. In fact, Korten's narrative framing of competing "sacred money" and "sacred life" stories, and his discussions of our collective future hanging in the balance of this global narrative struggle, encapsulate the central message communicated in a number of works by public intellectuals and futurists writing about the environment.[3] Korten condenses the views expressed in these works, which articulate a vision of human partnership and kinship with earth, while identifying the power of story as a strategy to bring about a greener and more just future. Both reflective and generative, Korten's oppositional narrative schema offers a valuable resource that is "good to think with."[4] His work helpfully identifies, organizes, and gives name to salient competing narratives at work in the mediasphere. The telling of this story of opposing but critically defining cultural stories, the notion of the power of story to forge a new future, and the spread of this story through and across multiple media channels all signal an important conversation taking place about ongoing processes of global moral and social transformation.

But just *how* do defining cultural stories, paradigms, or "shared cultural lenses," as Korten calls them, get swapped out and replaced? Unlike *The Hunger Games* popular teen novel and film series, in which a game master can easily manipulate terrarium-confined players into a new story world, *deus ex machina* style, global narrative transformation would seem a tall order. In the competition and contestation over narrative control, which rival story or stories prevail? And is a shared cultural lens even possible anymore, if indeed it *ever* was?[5] Sociologist Anthony Giddens contends that in modernity, it becomes specious to think of societies as bounded, unified wholes—the kinds that would share a cultural lens. The romantic illusion of society as a whole, sociologist Michael Mann similarly asserts, communicates the way we never were: "Societies are not unitary. They are not social systems. They are not totalities. We can never find a single, bounded society in geographical or social space." Instead, societies are overlapping, intersecting, extending, and internally as well as externally contested social-spatial networks—promiscuous in their multiplicities.[6] That is to say, social energetics do shift and move in different directions, but there is unlikely ever to be a unified, ubiquitous *human* hive dance. That does not mean, however, that some stories danced might not become more popular and pronounced than others, gaining influence and circulation momentum, and in the process, drawing a critical mass of participants who align themselves with the dance's movements and direction.

Whether or not Korten and similarly minded contemporary thinkers are correct in their proposed narrative solutions to our environmental problems, their sense for the potency and alchemical properties of stories in our lives is indeed quite insightful. We know ourselves and our world through stories. Engaging in story, participating in its making and telling, helps us to make meaning in our lives, to make sense of our place in the universe, to make and remake our social worlds, and to discover and define what is important to us and why.[7] Stories move us and change us. They also entertain and enliven us, provide solace, and let us know, as *Shadowlands* author William Nicholson's C. S. Lewis character famously observed, that "we are not alone."[8] Depending on how captivating a story is, we can enter it imaginatively, live among its characters, immerse ourselves in its settings, and become completely consumed by the story world. Arguably, what makes religions persistently compelling to so many in our time, when many theorists have long predicted religion's demise, is the power and persuasion of a good story.[9] Whether it be the biblical Yahweh speaking the cosmos into being or, as in Laguna Pueblo culture, Thought-Woman telling stories that transmogrify into the beingness of the world, the generative powers of story and the generative powers of creation are feraciously linked. Laguna Pueblo storyteller Leslie Marmon Silko observes that it is story and its very telling that create the world. Cosmic creatrix Thought-Woman, the Grandmother Spider, spins a yarn, and as she tells the story, it begins to appear before her. Silko depicts story as alive and moving, like an unborn child in the storyteller's belly.[10] It is through stories, she says, that we hear who we are.[11]

What is more, stories are not simply told orally or through written texts but are performed and enacted through cultural practices and technologies, inscribed into bodies, architecture, art, rugs, pots, fashion, rituals, ceremonies, games, computer software applications, and myriad cultural works that do the storytelling for us. We use multimedia to tell story.[12] Historian of religion Mircea Eliade famously made a career of drawing attention to the embedded cosmological stories in the human design of earthly cities, temples, shrines, and countless artifacts.[13] He argued that these material inscriptions of stories of origin and cosmic ordering, in their telling and retelling, served to collapse symbolically the time dividing our own time from the primordial time, *in illo tempore*—the distant time when the universe first began.[14]

Although scholars have enthusiastically critiqued Eliade for his ahistorical romp through a wide variety of religious expressions and his cavalier cross-cultural leaps that asserted ubiquitous symbolic patterns of human meaning, he was spot on when it came to the penetrating power of stories in people's lives across space and time.[15] Arguably, it is telling stories that makes us human. We are what evolutionary psychology writer Jonathan Gottschall terms "Homo fictus" (fiction man) or "the great ape with the storytelling mind." Long before *Hamlet* or *Harry Potter*, we "thronged around hearth fires trading wild lies about brave tricksters and young lovers, selfless heroes and shrewd hunters, sad chiefs and wise crones, the origin of the sun and the stars, the nature of gods and spirits, and all the rest of it." As a species, observes Gotschall, "we are addicted to story. Even when the body goes to sleep, the mind stays up all night telling stories."[16]

Plotting the stories we are telling ourselves today about who we are as humans in relationship to the earth at a time of environmental crisis enables us both to mark and to probe this ongoing making and remaking of our world for its critical implications.[17] The content and dimensions of these stories are themselves revealing—after all, as we tell our stories, they in turn tell us.[18] But perhaps more significant is the process of *how* these stories get told and how people *engage* with them. That is, how do people become involved and attached to them, participate in them, discuss and deal with them, respond to and remake them, and, in some cases, even act upon them? And how might this storying process involve certain moral dimensions—mediated assemblages, representations, and interpretations of what is morally right and wrong in both our individual and collective relationships and responsibilities to the world at a time when we face momentous environmental problems?[19]

As cultural *restorying* of earth to respond to environmental crisis unfolds across multiple media platforms, from myriad sources, and in myriad ways, what meanings do we see being made of the extreme planetary challenges we face? And how are these stories told, performed, enacted, played, and even physically embodied, more often than not, in and through our enmeshment with contemporary popular culture? If, as philosopher James Conlon contends, story acts not as a map but as a compass that points us in a particular direction, then in which directions (if not perhaps opposite ones) are environmentally themed narratives now pointing us?[20] Most interestingly, how do these stories compete and conflict, vying for space as contested narratives in public imagination? How might a story actually contain multiple stories, sometimes at odds with one another? What stories dominate and how might the tools for newer DIY (do it yourself) modes of digital media practice (blogging, vlogging, Internet discussion and social media, amateur video making, meme making, and the like) open a space for media *interventions* into dominant stories of the earth and its relationship to its inhabitants?[21] How might narrative interventions that seemingly subvert what Korten has thematically encapsulated as the prevailing "sacred money and markets story" ironically contain powerful reinscriptions of the same? Far from achieving a shared cultural lens, perhaps restorying the earth yields something more akin to the product that Adobe software systems company has dubbed the "Magic Lens." This is a compound camera lens made up of nineteen separate sub-lenses that shoot the object from multiple angles simultaneously, resulting in numerous and different focal points.[22]

Remaking the story of earth and our place in it constitutes an ongoing project and process of moral engagement. This restorying process produces what can be viewed as something along the lines of the "*Rashomon* effect." That is, it works much like the 1950 classic Japanese

Figure 1.1. "Adobe Magic Lens." Adobe's prototype "Magic Lens" is a compound lens consisting of nineteen sub-lenses that generate nineteen different but simultaneous focal points. Credit: LightFieldForum.com.

period film, *Rashomon*, famous for its plot device by which characters who are witnesses to the same incident produce multiple conflicting accounts of what has transpired.[23] Like Adobe's Magic Lens, multiple eyes, in this case eyewitnesses, provide the audience with multiple angles or views on the events surrounding a murder. Ecological imagination as mediated through popular culture similarly affords us myriad, and often disparate, focal points. As with *Rashomon*, multiple eyewitnesses to the state of the earth and our human impact on its current condition tell competing stories.

Grassroots digital participants in restorying the earth, for instance, challenge or "hack" corporate-generated stories that reassure us and soothe our anxieties about our destructive human impact on the biosphere and its sobering consequences. A prominent example of this phenomenon is the 2006 Chevrolet Tahoe SUV campaign, in which General Motors provided easily downloaded slick video and music clips, inviting consumers to produce their own DIY online commercials to promote the company's low-fuel-efficiency vehicle ("gas-guzzler" in popular parlance). The best ad maker would win a prize. It was Chevy instead that in turn received a whole host of unexpected gifts. Activists hijacked the campaign and posted critical ads with slogans such as "Global warming is not a pretty SUV ad—it's a frightening reality," "Chevy Tahoe: Larger than any normal human needs," and "Our planet's oil is almost gone," set against a Middle Eastern desert scene and accompanied by messages protesting America's launch of war in Iraq.[24] A similar tactic triggered an unpredictable activist response in the summer of 2017, when Republican legislators in Indiana asked Facebook users the question, "What's your Obamacare horror story? Let us know." Instead of eliciting the kind of public relations ammunition the legislators wanted, the page was flooded with more than fifteen hundred responses, the majority of which were laudatory and offered comments such as, "My sister finally has access to affordable care and treatment for her diabetes"; "My father's small business was able to insure its employees for the first time ever. #ThanksObama"; and "The only horror story is that Republicans might take it away."[25]

Digital "prosumers" (simultaneous producers and consumers of media), in their various creations and participatory digital practices, can engage in disruptions or in-breaks into a lulled consumer conscience. Multiple voices and viewpoints that thrive in online contexts carve out spaces to challenge and try to walk back the *inevitability* of the Anthropocene.[26] But it is far from simply those involved in overtly activist or alternative media who play a role in disrupting pervasive capitalist complacency about such human activities as deforestation, fossil fuel consumption, habitat destruction, mining, industrial groundwater contamination, and first-world overconsumption. Mainstream media players (corporate institutional media), as we will see, also generate narrative interventions into the dominant "sacred money and markets story," producing cultural works that participate in restorying the earth. There is often more to these environmentally themed stories, however, than meets the eye. Korten casts the "sacred money and markets story" and the "sacred life and living earth story" as fundamental rivals. And they are. These stories *are* oppositional and competing, but they are also fundamentally intertwined and in a number of instances even co-constituting. This book takes a critical look at multiple agents of restorying, examining in part how fans of corporate-produced media narratives participate in those stories of ecopiety or consumopiety, and at times do so by remixing or challenging the moral basis of environmental messaging as authoritatively presented.

Adjusting Our Viewfinder

Research that examines both institutional religion and more diffuse, informal religio-cultural phenomena, such as popular representations of moral narratives, commitments, and sensibilities, gains perspectival depth when we recalibrate the spectrum of source materials and points of access to include the rich data provided by cross-platform cultural analysis. Media scholars' once-standard approach to studying "discrete media and textual forms," Henry Jenkins reminds us, is no longer adequate to our understanding of storytelling, content, and characters that develop across and in between interconnected media platforms. Instead, we need to look at the convergences of media and the increased prosumer participation in media production made possible by media's more decentralized and now digitally accessible points.[27]

More than a decade ago, Barbara Klinger notably advocated a more expansive conceptual approach to media scholarship that would shift attention away from the large institutional "Cineplex" (what religion scholar Elijah Siegler has called "cathedrals of the image") to the handheld "Gigaplex," a diffusion that signaled that moving pictures are no longer singular and discrete but transmitted in and through a rapid multiplication of channels.[28] That is, in the wake of proliferating cross-media phenomena, scholars in media studies have made the realization that they could no longer ignore the fracture of what was once a celluloid monopoly on moving-image technologies. Where once film, as delivered in the Cineplex, dominated the study of media, now the delivering of content through cross-platform entities (cell phones, computers, advertising, gaming, Broadband, etc.), has increased the various points of access for the cultural consumer who now is to a greater degree also active producer and interactive participant.[29] Influenced by scholarship in media studies, environmental studies, and cultural studies, this book seeks to broaden the "viewfinder" in religious studies to incorporate more research that purposefully and passionately engages multiple angles of cross-platform research. What is more, there are pronounced parallels between the way media scholars study mediated popular culture and the expanding lens of religious studies scholarship, which has widened from a tight focus on more vertical institutional and formal expressions of religion to include more horizontal, lateral, informally practiced, DIY, "on the ground" expressions.[30]

The study of media, popular culture, and moral engagement also brings to the fore a too-often-overlooked dynamic of reciprocity in cultural production. That is, we as humans *make* things: television shows, films, web sites, fan gear, caskets, urns, video games, music, tattoos, sci-fi novels, and so forth. But then, these things make *us*, too. And through new media technologies, we in turn *re*make them, renegotiating their meanings, and then we share those meanings through myriad channels.[31] The contemporary values discourse about the relationships and responsibilities of humans with regard to their more-than-human environment is not relegated to the domain of elite broadcasters transmitting environmental information to audiences/consumers in traditional one-directional media forms. New media technologies create increased and more accessible political opportunities for what feminist theorist bell hooks has called "talking back," the bold move of "speaking as an equal to an authority figure." As hooks explains, in the southern African American community where she grew up as a young girl, "to back talk" meant "to disagree and sometimes it just meant having an opinion." Either way, it was an "act of risk or daring" that had real consequences in terms of disrupting the established hierarchies of power.[32] What is new about participatory culture and mediated back talk? Very little, as the remnants of casual graffiti still visible on the seats in the Roman Colosseum well illustrate—commentaries and critiques left by participatory audiences and publics of long ago.[33] As my doctoral students and I discuss in the seminar I teach on religion, media, and digital culture, it is the "3 Ss" (*speed*, *scope*, and *scale)* of digital technologies, and the increased access to now relatively inexpensive and easy-to-use tools of mass cultural production and circulation, that intensify media/cultural patterns and practices of the past.[34]

In the 1930s, cultural theorist Walter Benjamin famously brought to light the prospects and promise that recent or emerging media technologies (photography, film, sound reproduction) held in the age of mechanical reproduction for broader public access to the tools of production. This greater access would narrow the gap between producer and consumer. He, too, spoke to the then-rapid shifts in speed, scope, and scale, reminding us that "in principle a work of art has always been reproducible," but the sheer scale of reproduction, the accuracy, standardization, and speed of reproducibility in photographic prints, for example, held radical implications for art and culture.[35] Before the advent of personal computers, user-friendly software, multipurpose mobile devices, and social media, Benjamin drew attention to the impact of letters to the editor, in which media consumers engaged in criticism and "talked back" about the issues of their day. Readers became writers, and Benjamin predicted that "the distinction between author and public is about to lose its basic character."[36] Now, in an age when, as new media scholar Clay Shirky has argued, "everyone is a media outlet," this narrowed gap between consumption and production, and audience and authority, has profound implications for the engagement of moral issues, including that of environmental crisis.[37]

Today's digital culture offers a stark contrast to the top-down model of circumscribed elite dissemination, in which a single authority—be it ethicist, philosopher, theologian, curate, scholarly expert, etc.—merely hands down authoritative moral prescriptions. As Nicco Mele, director of Harvard University's Center on Media, Politics, and Public Policy, observes, in some ways, it is "the end of BIG."[38] In other ways, especially through the mechanisms of horizontal and vertical media integration, it is the apotheosis of BIG. The digital mediascape, in theory at least, allows for collective learning and debate through horizontal mutual engagement in shared content via activating "pass-along power."[39] Social media, including major social networks, media sharing sites, blogs, and DIY publishing sites, discussion forums, social shopping/hobby/sharing-economy sites, and readers' comments sections in both blogs and other online publishing outlets, together, virtually efface Benjamin's already fuzzy distinction between author and audience. This upturning of vertical structures of media authority unmistakably has its upsides and alarming downsides, as Americans witnessed in the 2016 presidential election. The horizontal dynamics of digital media culture, and the ease of strategic and camouflaged media buys, cleared space for a hostile foreign power to manipulate easily fabricated and targeted media messaging about the presidential candidates, specifically designed to manufacture confabulations in the minds of voters.[40]

And yet, digital technologies and the evolving milieu of digital culture make possible an expanding community of interactive media and socialities, in which environmental questions of moral concern can be worked out in what cultural historian Michael Saler terms "public spheres of the imagination"—arenas of invested ideological and textual discussion, debate, and exchange that can operate as vehicles for moral change.[41] It is not simply that audiences encounter moral dilemmas through characters in their favorite television series, popular novels, music, or games; it is that the web *presence* of those works provides spaces for fan (or simply "reader") engagement with those moral issues and indeed their interactive and community theorizing of them.[42] In the digital age, a do-it-yourself culture of talking back permeates the interlaced realms of media practice, moral engagement, and popular culture, and if we have not yet adjusted our viewfinders to include such realms, we are missing out on what is arguably one of the most powerful and pervasive dimensions of our contemporary world.[43]

Transforming Culture by Invading Dreams

Producer/director Joss Whedon, famous for his television series *Buffy the Vampire Slayer* (1997–2003) and *Angel* (1999–2004), and with screenwriting credits ranging from *Toy Story* (1995) to *The Avengers* (2012), keenly recognizes the potential and potency of narrative intervention. Whedon proclaims, "The idea of changing culture is important to me, and it can only be done in a popular medium. . . . I don't want to create responsible shows with lawyers in them. I want to invade people's dreams."[44] The realm of popular culture is the forum where we engage and work out the most pressing moral issues of the day. Invading people's dreams includes the environmental imagination and our visions for the future of our world. To get at environmentally themed narrative interventions, this book focuses on where Americans most often encounter stories that concern moral issues—not in formal religious institutions but in the realm of mediated popular culture.

For more than two decades, ABC News' familiar tagline proudly stated, "More Americans get their news from ABC News than from any other source." When first broadcast, that tagline mainly referred to the sum of ABC's news programs, which then included a large number of

ABC-owned or ABC-affiliated radio stations. Toward the end of the slogan's run, ABC News could still technically make this claim to being a dominant source but only on the basis of a much more diverse network of delivery channels or platforms, which included television, radio, Internet news sites, Facebook and other social media pages, ABC-affiliated Twitter accounts, an *ABC News Now* application for mobile phones, and a special deal ABC made with Walmart to broadcast *ABC News* on its in-store television network. Over time, technology changed to provide both more diverse platforms for encountering information and more participatory ways for people to talk back to that information through various Internet blogs, social networks, and video- and audio-sharing formats. What is more, the very tools used to convey information have changed—so that phones, for instance, once simply used for making phone calls, became cameras, video and audio recorders, Internet access sites, GPS mapping devices, movieplexes, gaming devices, and conduits to any number of software applications—from guided meditations to sonic mosquito repellents to fitness competitions.[45] In other words, there has transpired a convergence or coming together of more than one (previously separate) technology into now-multipurpose devices; at the same time, there has been a simultaneous proliferation of communication channels and an intensification of corporate media ownership consolidation and conglomeration.[46] These fundamental shifts, arguably in contrapuntal directions, necessitate new tools for studying media and culture, as they are *cultural* shifts as well as *economic* and *technological* ones.[47]

As the study of religion and culture, including the cultural study of shifts in developing moral sensibilities and engagement, faces a similar kind of protean dynamic, scholarship needs to adjust accordingly. The convergence of religious and environmental idioms in the productions of American popular culture provides a matrix for exploring new ways of seeing and representing cultural narratives that flow across and through many different channels. The study of religion and ecology has directed scholarly attention primarily to the institutional manifestations of religious environmentalism, focusing on such phenomena as eco-congregations, green seminary programs, green sanctuary projects, green scriptural exegeses, interfaith environmental programming, experiments in eco-liturgy, institutional and clergy environmental statements, and the production of academic texts by eco-theologians, eco-ethicists, and eco-philosophers. This emphasis on the institutional "greening of religion" has addressed a notable emergent dimension of restorying the earth. Scholarship (mine included) on the greening of major religious institutions, the eco-friendly reinterpretation of scriptural texts and theologies, and the ecologizing of institutional liturgy and doctrine can help tease out the various ways in which institutional religious communities are working to restory moral commitments in more ecologically minded ways.[48] However, research also points to the efficacy and actual impact of such greening-of-religion efforts being negligible at best, and at worst, losing ground.

A 2018 research study, conducted by public policy researcher David Konisky, analyzed longitudinal, cross-sectional data drawn from repeated nationwide Gallup annual surveys on environmental attitudes. Konisky's results not only point to US Christians *not* getting any greener over time but indeed provide evidence to the contrary. The annual surveys suggest that concern for the environment has actually declined among US Christians since the 1990s. Konisky's study analyzes data on Christians, and it may be that the greening of other religions is taking place with greater impact on those affiliated, but those who self-identify as some variety of Christian make up the bulk of the US population—70.6 percent, according to Pew Research Center polling.[49] The effort to green religion, or at least to green a Christian majority, has been a major faith-based strategy for getting a broader swath of Americans to care about the environment and to support the implementation of policy shifts critically needed to address climate crisis. Konisky's research suggests that this strategy is not working. Across multiple measures, Konisky finds that "Christians tend to show less concern about the environment. This pattern generally holds across Catholic, Protestant, and other Christian denominations and does not vary depending on levels of religiosity. These findings lead to a conclusion that there is little evidence of a 'greening of Christianity' among the American public."[50] If not via the context and messaging of institutional religion, especially for the growing number of the population nonaffiliated with religious institutions, how else then might Americans be morally engaged and moved to respond to our pressing environmental challenges?

Moment to moment, the immense rhetorical power of mediated popular culture imbues our everyday lives, morally affirming or discouraging its every aspect. Especially when delivered in storied entertainment form, mediated popular culture texts have a particular and pervasive power to, as strategic communications scholar Deanna Sellnow describes, "[influence] our taken-for-granted beliefs and behaviors about how things ought to be or, perhaps, just are, as well as what is normal or abnormal, desirable or undesirable."[51] We now clock an average of eleven hours of screen time a day on our smartphones and other electronic devices alone, not even counting printed books, magazines, comics, radio, billboard advertisements, flyers, nonelectronic games, or even films viewed outside of home at an actual cinema. In terms of narrative encounter and critical mass of messaging hits, mediated popular culture is where most Americans live.[52] Practically speaking, most of us encounter messages about moral concern for the environment not through reading eco-theologians or eco-philosophers, or listening to green clergy, but as viewers/users of and, indeed, participants in the stories of popular culture. It is these mass-mediated stories of popular culture that not only "invade people's dreams" but accompany us in our daily lives.[53]

Pope Francis's 2015 environmental encyclical, *Laudato Si'*, achieved the public reach that it did precisely because of Francis's savvy in using digital new-media technologies and social media. Francis had already been an avid Twitter user and by 2014 was named Twitter's most influential tweeter. The pope's unique mass recognizability as a popular cultural figure, frequently cast in the US press narratives as an underdog unafraid to challenge the status quo, combined with the Vatican's use of younger and hipper tools of communication to get Francis's message out, created a partnership between pontiff and mediated popular culture.[54] Strategically, the Vatican did not release Pope Francis's much-awaited environmental encyclical in one copious and laboriously read document but instead released it via (140 characters or less) tweets, in small doses that quickly went viral on the Internet. Within a short period of time, sociologist Christopher Helland recounts, the pope's "tweet was shared more than 30,000 times and it was quoted and referenced in more than 430,000 news articles. Throughout the day, the Pope continued to tweet short statements from his 183-page text, savvily inundating the online world, to the point that almost everyone on the web that day was aware of the event."[55] *And yet*, in terms of impact, a poll conducted by Yale researchers, the Associated Press, and the NORC Center for Public Affairs Research found that a month after the environmental encyclical's big debut, only 31 percent of respondents in general and only 40 percent of Catholics polled had ever heard of it. Fewer than two in five Catholics had heard anything about it from their priests.[56]

In covering the hoopla around the encyclical and the (apparently) nonmaterializing "Francis Effect" roughly two months after its release, *Religion Dispatches* senior correspondent Patricia Miller wrote, "Catholics aren't getting the pope's message that climate change, environmental degradation and global poverty are intimately linked to fundamental Catholic social justice teaching."[57] By 2015, the number of those who followed Pope Francis on Twitter had reached twenty million, and in 2018 Francis was the most followed world leader on the social networking site (with US president Donald Trump just behind him in second place).[58] These statistics speak to Francis's acumen for mass communication and his skillful embrace of new media. Still, in the same year as the encyclical, pop singer Katy Perry's following of highly devoted Twitter fans reached over seventy-four million, and Korean entertainer Psy's music video, "Gangham Style" (which had originally gone viral in 2012 and became the first YouTube video to surpass one billion hits), reached 2.3 billion hits. Using the powerful tools of new media and its knack for manifold multiplication, vectoring content through myriad popular cultural channels, the Vatican may have sought to turn the environmental encyclical into a new kind of digital loaves and fishes, but reception, sharing, and circulation dynamics of media content are unpredictable.[59] This may be especially so when many US parish priests simply ignored the fact that the pope had said anything on climate change and proceeded as though the encyclical never happened.

Grassroots environmentally activist media prosumers, however, may make more headway than top-down Vatican communications strategists, as they work to restory the encyclical for popular audiences. Not long after Francis tweeted out his bite-sized encyclical portions, environmental activist groups began telling and retelling a story of a superhero pope doing battle with the evil forces of corporate industrial greed—grim men in suits who would shortsightedly perpetuate the carbon-producing culprits of climate change. One of the most notable examples of these secondary viral storied cultural productions, and picked up by entertainment weeklies like *People* magazine, was a mock trailer called "The Encyclical—the Movie," produced by the Brazilian climate-change NGO Observatório do Clima and circulated in multiple languages via YouTube.[60] In the trailer, the character of the pope takes on sinister fossil-fuel industry executives, Rocky Balboa style, assisted by Jesus himself, who serves as the pope's Burgess Meredith–like boxing coach and spiritual motivator. The trailer not only shows the pope using his smartphone as a weapon, tweeting out ring-side salvific messages of environmental goodness and virtue, but also captures His Holiness in the act of blessing solar panels, using ninja moves to out-maneuver polluting corporate villains, and issuing trailer taglines from the actual encyclical, such as the direction, "To change everything, we need everyone."[61] The story culminates in the on-screen hashtag, #PopeForThePlanet.[62] In cultural economies of scale, the power, saturation, and impact of mediated popular culture in people's everyday lives, as singer/storytellers Katy Perry's or Lady Gaga's devotees can attest, are immense, as is the potential for what Henry Jenkins and collaborators Sam Ford and Joshua Green term the "spreadability"—or the rapid, unpredictable "remixing" and circulation—of media content.[63]

Popular culture, as folklorist and literary scholar Ray Browne defines it, is "the force that has overwhelming impact on shaping our lives." Browne's wording, sounding Vatican II–like, speaks of culture as "the everyday world around us" and the "culture of the times." Browne famously (and passionately) declared, "Popular culture studies *is* the new humanities,"[64] arguing that, far from being trivial, analysis of popular culture is critical both to fundamental and to deeper understandings of society and culture. What makes popular culture such a powerful force? Browne points out that it does not have to be made accessible or translated in order to make it applicable to people's lives. Popular culture, writes Browne, "is already *with* the people, a part of their everyday lives, speaking their language. It is therefore irresistibly influential. What it is, the way it works and its relation to the other humanities need to be understood if we are to appreciate its overwhelming influence on our lives."[65] But popular culture not only matters to our everyday lives, reflecting our cultural moment, generating meaning and situating the "*current* state of society." Perhaps most critically, stresses historian of film and culture Marcelline Block, popular culture "helps to shape and inform [society's] *future*."[66]

Mediated popular culture in particular (i.e., what we experience through television, radio, sports, music, comics, advertisements, film, games, the Internet, etc.) merits greater attention by scholars as a source for understanding the moral dilemmas of our time.[67] "Popular media, for good or ill," Phyllis Japp and her colleagues declare, have become "our cultural thesaurus of everyday life, often the only common frame of reference across race, gender, class, and other social divides. Popular culture appears to have replaced religious texts, literary classics, history, ritual, and oral tradition as the source for immediately recognizable examples. Because we understand new life situations by their connections to old one, plots, analogies, phrases, proverbs, and metaphors from familiar sources provide a 'shorthand' that allows us to fill in the particulars and quickly categorize immediate happenings."[68]

"Replaced" religious texts is perhaps too absolute an assessment here; more appropriate would be a "convergence" model. Traditional religious texts are clearly still influential in the lives of those who are more orthodoxly religious, while more broadly, popular culture concomitantly provides mass-accessible reference points in terms of meaningful, inspirational, identifiable, and guiding narratives. In the act of reading popular culture (and, I would add, in the act of "remixing" it as well), Japp and colleagues find that prosumers come to "recognize 'ethical flashpoints' in mediated texts, those moments when ethical awareness is engaged, when action moves one way or another, when we sense—however dimly—the implications that hang on that moment, when we realize there is 'more to the story' than the simple problem/solution of surface scripting."[69] That moment of realization is key to restorying the earth as an act of moral engagement, as it generates meaning, critique, reflection, and debate.

Earth as Transmediated Core Text

In probing the ongoing process of restorying the earth in its various forms, interventions into the prevailing "sacred money and markets story," and into ongoing competition for narrative control, Henry Jenkins's work on transmedia storytelling, new media, and fan culture proves a fertile analytical resource.[70] Jenkins draws attention to the nonnormative but potent and fascinating phenomenon of transmedia storytelling as fueled by expanding digital outlets for participatory culture, in which fans play a substantive role in the co-creation of narratives.[71] Transmedia storytelling, or multiplatform storytelling, involves telling a story across multiple media forms, in which rather than simply repeating the story in different forms, producers develop and extend narrative content over myriad outlets (film, television, gaming, comic books, interactive web sites, online videos, etc.). Classic examples of transmedia storytelling that Henry Jenkins analyzes in his work are *The Matrix*, *Star Trek*, *Star Wars*, *Harry Potter*, and Frank Baum's *Oz* series.[72] As fans buy into the story world created in these media franchises, they take media into their own hands, actively participating as simultaneous producer/consumers in their favorite narratives and taking on self-appointed roles as story co-creators.

One of the prime illustrations of this is the composition and circulation of fan fiction (often dubbed "fanfic"), in which fans rewrite shared texts, authoring new sub- and side plots that franchise producers never wrote into the core text.[73] For *Star Trek* fans, this might include supplementary plotlines like gay love affairs between otherwise straight characters or crew adventures that never took place in the actual series.[74] The site *HarryPotterFanFic.com* alone boasts over eighty thousand stories of fan-generated content. Other Harry Potter fan sites, like the active fan fiction site at MuggleNet.com, provide supportive spaces where fans can construct elaborate background stories for minor characters or generate new plotlines that take characters well into the post-Hogwarts era.[75] Much to the consternation of media franchise copyright lawyers, fans assert a kind of *co*-ownership of a story and its telling, playing with it, remixing it, and subverting what is often termed "the canon" or "core text" in favor of grassroots-generated narrative arcs, tangents, twists, character makeovers, alternative endings, or other acts of *restorying*.[76]

Imaginatively extending Jenkins's illuminations of transmedia storytelling provides us useful ways to think generatively about contemporary restorying of the earth across multiple media platforms.[77] That is, narrowly conceived, transmedia storytelling is, for the most part, a strategic story-based marketing strategy implemented by for-profit media franchises. Writers/producers tell stories, extending and developing compelling story worlds across multiple platforms, and media companies then sell those stories to consumers and make money. When fans become so invested in a story world as to desire deeper participation in it and agency in the direction of its telling, the storytelling itself becomes a more free-ranging, and sometimes wildly collective, enterprise of contest, negotiation, and coauthorship. When this happens, fans also become culture makers.[78] If, as Jenkins identifies it, "Transmedia storytelling is the art of world making," then restorying the earth is the art of world *remaking*—rewriting the story of who we are as humans in relationship to our story world with an intent of changing its course.[79] We are both transmedia storytellers and consumers, and this restorying process is a collective enterprise—contested, negotiated, infused with power struggles, in an ongoing process. Interested parties, stakeholders with different aims, may vie for narrative control, but, as storyteller Leslie Marmon Silko keenly observes, once a story is out there, it cannot be called back. It is already set in motion.[80] As Chevrolet discovered in its DIY Tahoe commercial debacle, story has a life of its own and is unpredictable in the directions it takes.

With an expanded notion of collaborative transmedia storytelling in mind, *what if*, in restorying the earth, instead of a core film or video game providing the driving platform (what drives the story forward), the driving platform were ongoing environmental devastation—fracked groundwater, poisoned marine mammals, burst oil pipelines, fish-depleted oceans, nuclear reactor radioactive oceanic drift, toxic chemical spills, mountaintop removal, encroaching desertification, skyrocketing urban asthma rates, vanishing species, and the emergence of global-climate-change-intensified "Frankenstorms"? Can the earth itself, as it has been shaped by anthropogenic (human-based) activity, serve as a "core text," perhaps even *the* core text? Could the story of earth's systems in crisis, as humans are telling and retelling it, constitute a large textual system, not simply repeated and mediated across multiple platforms but codeveloped and extended with multiple variations and subplots? The fields of biology and earth science, most expressly, remind us that humans are an embedded and integral part of the earth system.[81] Gottschall refers to our species as "the storytelling animal," but who are we as human beings, if not part of the earth system, telling stories across varied earth media? The second half of this book, drawing on media theorist John Durham Peters's theorizing of the earth and its natural phenomena (clouds, wind, fire, water, etc.) as "media," highlights the transmedia nature of restorying the earth across such media as corpses, graves, caskets, tattooed skin, and the real-life backyards of alternate reality gamers.

Moral Engagement, Media Interventions, and Media as Action

Until relatively recently, the concept of "media interventions" has been understood fairly narrowly. This term has been used to refer to the use of media power to influence behavior and cognition through activist measures such as citizen journalism, public health campaigns, or newspaper advocacy campaigns.[82] Attention to media intervention has also primarily focused on alternative media activism. More recently, media scholar Kevin Howley has pressed for opening up the concept of media interventions, defined as "activities and projects that secure, exercise, challenge, or acquire media power for tactical and strategic action," to include, in activist media researcher Mitzi Waltz's term, the "mass media monolith," or concentrations of power in mass media institutions. This more capacious understanding of media interventions makes room for the impact of both mass-media-produced and mass-media-delivered political narratives, in addition to the more informal talking back to mass media that takes place in less structured and more free-wheeling social media contexts and fan discussions. As a specialist in US and global media activism, Victor Pickard has created a model of media interventions that depicts multiple and often overlapping types of media intervention, including not only "contesting representations in mainstream media" and "creating alternative media" but, significantly, "using mainstream media to advance political messages."[83] In examining the role of media in the ongoing restorying of earth, this expansive typology of media intervention, or what Nick Couldry characterizes as a "wider angle lens," is useful for getting at the scope and scale of media power and its practice. The sheer "range of who can intervene in and through media—and how," observes Couldry, "has been massively expanded by the connective hypertext of the Internet, meaning that bundles of practices comprising any particular media intervention are becoming more diverse."[84] These diverse acts of media intervention square with Couldry's theorizing of media as something we *do*. "Understanding 'media' as action, as part of the wide set of practices in which each of us is open-endedly engaged (rather than something confined within boxes of 'text,' 'production,' and 'audience')," explains Couldry, "helps expand what we pay attention to in media."[85]

In analyzing the *restorying* of earth as morally engaged media practice, I intentionally go one step further in the theorizing of media intervention by playing with the associations the term "intervention" takes on in a therapeutic context. In this milieu, the term signals a concerted *intervention* action taken by family members when they witness a loved one spiraling into a self-destructive crisis (alcohol addiction, drug abuse, gambling problems, etc.).[86] In this sense, an "intervention" is a break or disruption into the narrative that the addict tells himself or herself ("*What* problem?"), confronting the addict with a community-endorsed counternarrative that challenges the addict's view of the current state and well-being of his or her life. Media interventions into the storying of earth in an age of environmental crisis ("*What* global climate change?") are simultaneously narrative interventions and moral interventions. Many examples of media intervention have historically included cases such as smoking-cessation campaigns, HIV/AIDS-prevention campaigns, or prenatal-maternal-health campaigns, as "common forms of media intervention designed to alter individual and public health attitudes and behaviors."[87] The restorying of earth through popular culture narratives similarly takes up a major public health problem, indeed a fundamentally global one.

A number of moral themes dominate these narrative interventions. A close examination of Korten's two major competing stories of earth as writ through mediated popular culture reveals that a recurrent theme governs each. "Sacred life and earth" stories tend to be stories of environmental virtue and fidelity to pious environmental practices that aim at sustainability by tempering human activities and harmonizing with the ways of nature. These are, as we saw earlier, what I term stories of "ecopiety." Stories of "sacred money and markets" also convey a prevalent theme, one that encourages and validates material consumption as both a "path to personal happiness" and a social good, as the drive for corporate profit "maximizes economic growth and thereby the wealth and wellbeing of all."[88] These are stories of consumopiety, if you will. Consumer spending is lauded as pious practice and one that is key to ideals of freedom, good citizenship, and, ultimately, a kind of "trickle-down" planetary well-being. Like contrapuntal lines in music, or contrapuntal contrary motion in dance, in which independent melodic lines or choreography move in different directions and yet play and are performed together, stories of ecopiety and consumopiety, though often seemingly disparate, paradoxically and powerfully "play" together in various registers of the contemporary mediasphere.[89] This contrapuntal contrary motion, a favorite approach employed by choreographer Alvin Ailey in dance storytelling, also characterizes the story of new media. That is, paradoxically, as the tools of media making become more accessible and widely democratized among grassroots media makers, so too the forces intensify of media consolidation, manipulation, and control, in conjunction with the power of corporate and state surveillance, or what might be regarded in Foucauldian terms, today's digital "panopticon").[90] Thus, that which can appear at first encounter to be media interventions deployed as moral interventions, consciously working toward the social energetics of a greener, more sustainable world, can turn out to be advancing a contrary agenda of reinscription and regression. The chapters that follow deploy these theoretical insights to delve into contrary, at times oppositional, but frequently co-constituting, dynamics within popularly mediated moral discourses on the environment.

2

Fifty Shades of Green

Moral Licensing, Offsets, and Transformative Works

In just three years (2009 to 2012), Erika Leonard morphed from a British television producer and part-time amateur online fan fiction writer into *New York Times* best-selling author "E. L. James," winner of the 2012 United Kingdom Book of the Year National Book Award, winner of *Publishers Weekly*'s Publishing Person of the Year award, and one of the "100 Most Influential People in the World," as named by *Time* magazine.[1] James has now sold over 125 million copies worldwide of the novels in her *Fifty Shades of Grey* (*FSOG*) series—selling 20 million copies in just four months alone, selling two books every second at the series' sales peak, and outselling the *Harry Potter* series on Amazon.co.uk.[2] James secured major motion picture deals for the film versions of *FSOG*, three of which were released between 2015 and 2018, grossing more than $1.3 billion—making it the fourth-highest-grossing R-rated film franchise in history.[3] And James did all of this by *rewriting* and *remaking* someone else's story.

James began as a participant in the online site FanFiction.net, a closely networked community of fans who creatively and collaboratively rewrite, supplement, and otherwise "mod" what are referred to as "source" or "core" texts. Critics and copyright lawyers alike may charge that these are mere rip-offs. Others regard fan fiction as remixes that provide an expressive outlet for fans to participate in "their" story in a deeper way—not simply as passive consumers of content but as would-be co-creators and extenders of content.[4] Fan studies scholars Karen Hellekson and Kristina Busse refer to fan fiction as "transformative works," which can be used as pedagogical tools for writing and yet significantly transmit reinterpretations of the source text that are resistant and can engender sociopolitical critical arguments.[5] Participants in online fan fiction writing communities do this kind of experimental "messing around" with their favorite stories in the company of a collaborative community of fellow and sister fans, who mutually comment on and edit each other's work, often rating it or endorsing it in some public way.

After Mormon author Stephenie Meyer's popular *Twilight* young adult vampire romance novel series captured James's imagination, she began participating in Internet collective outlets for the story's fandom, including its very active fan fiction community, by participating in the "All-Human Alternative Universe (AH-AU)" section of FanFiction.net. In this community, fan writers would not only extend the *Twilight* story and invent new plotlines but would also make supernatural characters into human ones. Then they would write those characters into different settings and genres—a pirate ship, a western, a mystery, an office romance, and so forth. James (known online as "Snowqueen's Icedragon") followed the lead of others before her in the FanFiction.net community and began writing a popular BDSM (bondage/discipline/sadomasochism)-themed version of *Twilight.* Her story was one of a number of BDSM remakes that fan writers were sharing via postings to the site.[6] In James's version, Stephenie Meyer's sexually chaste teenaged characters, Bella Swan and Edward Cullen, morph into Anastasia ("Ana") Steele and Christian Grey—legally consenting (human) adults who ultimately explore kinky sex in what Ana refers to as Christian's "red room of pain." The result was wildly popular. James's online installments went viral, and she built an active and devoted online fan base, ultimately "pulling to publish" (or "P2P") onto her own web site. She then self-published *FSOG* as an e-book in 2011 and put out print-on-demand books through the Writer's Coffee Shop virtual publishing house. In 2012 Vintage Books re-released the series to meet increased demand. *Fifty Shades* became the number one best-selling genre fiction novel on Amazon.com, broke sales records as the fastest-selling novel in the United Kingdom, and now has been translated into fifty-two languages. One might say that *Fifty Shades* is a supreme example of a product designed to maximize its broad appeal for mass consumption. But in rewriting virtuous vampire Edward into BDSM-practicing billionaire CEO Christian, James performed a provocative transmogrification, one that traded in an economy of environmental virtue and its cultural meanings in order to make Christian Grey's violent behavior more palatable to a mainstream, largely female, audience.

Christian Grey's performance of ecopiety throughout the text offsets the behavior of a sadistic, stalking, violent, psychologically manipulative, and emotionally unavailable character, and renders him less sinister, more attractive, and promisingly redeemable by signaling his underlying virtue. The very fact that James can bank on a shared cultural understanding—that a mass audience will read green pious acts and a heartfelt concern for the environment as the stuff of good guys—tells us something instructive about contemporary embedded cultural notions of the environment as a site of moral engagement. To use literary theorist Northrop Frye's term, Christian Grey's ecopiety taps into cultural "resonances" that go beyond James's series and open into a larger cultural conversation.[7] The seemingly contrary and contradictory oppositionality of Grey's *ecopiety* and *consumopiety*, and yet their intimate interrelationality and codetermination within the narrative, bear resonances that extend beyond James's lucrative romance franchise and into a larger production and reproduction of meanings connecting capitalist consumption and popular environmental moral engagement.

James's BDSM-practicing billionaire stalks the heroine, tracks her cell phone, confines her movements, spies on her, polices with whom she is allowed to socialize, and dictates the acceptable maintenance of her body and appearance, procuring her an approved hair stylist and gynecologist—both on his payroll to implement his instructions and tastes. At one point, Christian sadistically and savagely beats Ana with his belt, later observing, "I like to whip little brown-haired girls like you because you all look like the crack whore—my birth mother." And yet, this is a romance.[8]

Fifty Shades has been thoroughly criticized by a wide spectrum of BDSM community and lifestyle advocates, who decry the depiction of BDSM in the novels as both inaccurate and dangerous, especially to uneducated newcomers and inexperienced BDSM lifestyle experimenters. Emily Prior, a BDSM/kink/fetish teacher at Sex Coach University and director of the Los Angeles–based Center for Positive Sexuality, writes that Christian Grey's seduction of Ana breaks every rule in the BDSM rule book, and that BDSM contracts, which lay down agreed-upon BDSM rules, and which neither character in the book ultimately signs, are to be freely entered into, not emotionally or physically coerced.[9] For a community that espouses strict ethical rules about healthy/safe practices and play, the relationship depicted in the novel is both exploitative and abusive. These are points on which organizations that advocate for victims of sexual violence and BDSM-community representatives both agree. Both are troubled that the *Fifty Shades* series repackages a story of an abusive relationship into a love story, glamorizing abuse and validating it with the kind of fairytale happy ending that real victims of abuse rarely if ever receive.[10] Ana, like so many women, is convinced that she can change her abuser and heal his pathology with enough love and understanding, and miraculously, she does! The payoff is a billionaire husband, a series of luxury estates, a supercar of her own, an elite job (as an inexperienced twenty-something) heading her very own publishing house, beautiful, healthy children, lavish European vacations, and a life of privileged domestic bliss.

James assiduously offsets Grey's troubling and violent behavior with a strong specific emphasis on his credentials as an *eco-pious* philanthropist and, more broadly, as a socially responsible American businessman. Under his executive direction, Christian Grey's company avidly invests in renewable energy and works tirelessly to generate breakthroughs in environmental technologies to help the planet, while Grey himself stalwartly insists on doing an eco-friendly renovation of his colossal Puget Sound mansion. In James's fourth book installment in the series, she doubles down on Grey's eco-virtue, and we learn that his company is redeveloping "brown fields" (toxic urban areas) in Detroit in addition to preparing an air drop shipment of food to Darfur. We discover more about his substantial financial contributions to the Washington State University (Ana's school) Environmental Sciences Department. Grey's ecopiety seems to know no bounds as he discusses his considerable financial investment in his brother's construction company, which we learn is dedicated to green building projects and is in the midst of the Spokane Eden Project, an eco-friendly affordable housing project north of Seattle. As Christian Grey drives his Audi R8 luxury, gas-guzzling supercar down the highway, he and his brother chat in detail about installing domestic wastewater, or "graywater," recycling systems at the low-cost green housing project, and how this and other measures will reduce the residents' water usage and home bills by 25 percent. Those are some pretty sexy reductions, but it *does* make one pause and wonder. *This* is erotica? The discussion of graywater system technologies?

What is going on here? But all of Grey's demonstrations of ecopiety, I would contend, are doing not only important narrative work but narrative work that relies upon a kind of resonance, or culturally shared significance, that ties the practice of environmental virtue to upstanding moral character.

On balance, Christian Grey's beneficent acts on behalf of the environment function as reassurance to readers that he is really a good man underneath the acts of violence and control, and his acts of ecopiety signal the promise of redemption. Even as Christian spanks Ana for rolling her eyes at him or for daring to have an independent thought in her pretty little head, his eco-pious credentials and green corporate track record provide narrative evidence that he is fixable as a romantic partner. Universal Studios, the producer of the film versions of *Fifty Shades of Grey*, takes the green profile James creates for Christian Grey in the book series even further by maintaining a participatory alternative-reality film promotional web site (similar to the "real marketing" strategies used for HBO's *True Blood*) that depicts the character's fictional company, Grey Enterprises Holdings, Inc. The faux company's real-looking web site is illustrated with images of wind turbines and other alternative energy technologies, as the web site touts Grey's substantial financial investments in sustainable energy. Fans are greeted with the question, "Passionate about change for the greater good?" and then invited to apply for an internship in Grey's socially responsible green company, where they, too, can help change the world for the better.[11] (In reality, "applicants" get put on a fan list and receive information about upcoming movie openings and publicity events.)

Yes, Christian Grey is a fictional character in a fantasy romance novel, but the lines of fantasy and reality blur as romantic narratives of Grey's capitalistic virtuousness get mediated across various platforms. *Forbes* magazine published its own calculated estimation of Christian Grey's net worth to be at 2.2 billion dollars, extolling his talent for taking underperforming companies and "whipping" them into shape. *Fortune* magazine ran a special feature for CEOs on "7 Leadership Lessons from *Fifty Shades of Grey*," which prescribes "valuable lessons" from the boardroom to the bedroom of America's favorite BDSM CEO. This included advice such as recruiting passionate team members willing to put in long hours, tying down (retaining through incentives) your best employees, and devising ways to leverage risk and turn it into opportunity.[12] *Business Insider* featured a story on the "real Christian Grey," a tantalizing twenty-seven-year-old Seattle-based successful entrepreneur named Brayden Olsen, who drives a supercar similar to Christian's and runs a holding company strikingly similar to Grey Enterprises Holdings, Inc. Women have flown to Seattle from all over the world to meet Olsen in hopes of kindling a real *Fifty Shades* romance with him, and each time a new installment or film in the *FSOG* series is released, Olsen is flooded with Facebook messages and romantic propositions.[13] Dynamics of "role-play" and "real-play" intermingle as financial media outlets breathe heavily over the figure of Christian Grey and his dominant style of executive leadership. Even as these magazines play along with the blurred lines of reality and fantasy, their treatments of Grey seem to suggest a sense of familiarity and even admiration for the sadistic CEO, or figures like him. *Forbes* and *Fortune* seem to know and recognize Christian Grey as one of their own, and here, once again, Frye's principle of resonances comes into play.[14]

Greenwashing Grey

When oil companies and other polluting industries get a public relations makeover through strategically designed, slick, eco-friendly corporate rebranding advertising campaigns, it is commonly referred to by environmentalists as "greenwashing." After highly publicized industrial oil spills, British Petroleum ads, for instance, announced that "BP" now suddenly stands instead for "*Beyond* Petroleum," as the corporation began to surround its logo with images of wildlife and sunflowers, communicating messages of a company that reveres and protects the environment.[15] The company, which is now "formerly known as British Petroleum," simply identifies itself as "BP." Visitors to BP's official US web site are immediately greeted with a splash of green graphics and more of the company's ubiquitous green-rimmed upbeat sunflower logo. Site videos feature BP-funded examples of green virtue, like happy teenagers wearing t-shirts that proclaim, "Protect, Restore, Preserve," as the students make their way through forests and participate in habitat restoration projects. A video voiceover talks about just how much money BP has contributed to the Student Conservation Association in the Chicago and northwest Indiana area. Nowhere in this video is there mention of the sixteen hundred gallons of toxic heavy crude oil that BP's Whiting Refinery spilled into Lake Michigan in 2014, thus contaminating the very water that those students, their families, and their communities drink and rely on for their health and well-being.[16]

The web site for BP Global is somehow even greener, sporting several fonts and hues of green, so that it appears one has mistakenly stumbled upon the web site for an environmental activist organization, such as Greenpeace or the Sierra Club. Just after "About BP" and "Products and Services," "Sustainability" immediately follows on the menu priorities, where readers are assured that BP is meeting the "global energy challenge" and doing so "affordably, sustainably, and securely."[17] On the menu list appears a section called "Gulf of Mexico Environmental Restoration," a section illustrated by photos of healthy-looking blue herons and pelicans, beautiful sunsets over pristine waters, and placid beaches with happy beachgoers—images that communicate a very different story from the fiery devastation of BP's April 2010 Deepwater Horizon explosion that constituted the largest marine environmental disaster in US history. Instead, the focus in this section is on how the Gulf is "healing itself" through the ecosystem's "natural resilience," how the "environmental impact of the accident was [note past tense] of short duration," and how life has rebounded and flourished in all its splendor. The word "accident" is prominently highlighted in bright green font just above a report that never mentions BP as having been responsible for that accident, nor the estimated 10 million gallons of crude oil still sitting on the floor of the Gulf of Mexico.[18]

One of the successor companies to Standard Oil, Chevron Corporation, runs advertisements that similarly show the company saving sea turtles and tigers.[19] Clicking on Chevron's home page on the Internet, one is immediately greeted with the exuberant slogan "Chevron: Human Energy—Finding Newer, Cleaner Ways to Power the World."[20] Looking strikingly similar to Christian Grey's corporate web site, Chevron's web site features stories that are categorized under such headings as "Corporate Responsibility," "The Environment," and "Making a Difference." A section on "Human Power" features stories of how Chevron, through its use of "human power," is protecting marine algae in Angola, saving tortoises in the Mojave Desert, and even "addressing climate change." In clicking through all these upbeat, eco-friendly stories of environmental virtue, one quickly forgets that Chevron is the fourth largest petroleum company in the world, is part of an industry that spends hundreds of millions of dollars each year lobbying the US Congress *against* passing climate change legislation, and has a long list of oil spills, toxic sites, pollution violations, and habitat destruction to its dubious credit.[21]

E. L. James's dark, brooding, sadistic neo-Gothic hero, Christian Grey, gets a similar virtuous greenwashing to *offset* his other destructive and troublesome behavior.[22] Just imagine how differently we might feel about *Jane Eyre*'s Mr. Rochester or *Wuthering Heights*' Heathcliff if, from the get-go, they had diligently recycled, piously maintained an organic garden out back, supported the local farmer's' market, and donated to student conservation groups. In the realm of mass-consumed erotica, and what some have dubbed "mommy porn," *Fifty Shades* functions simultaneously as a reflection of and an engine driving contemporary currents in popular culture, including the phenomenon of ecopiety and popular perceptions of environmental virtue.[23] Both the moral scale in *Fifty Shades*, in which good deeds on behalf of the environment outweigh other serious and violent human transgressions, and the very casual way in which author James employs this scale, signal an increasingly mainstream embedded discourse of environmental virtue and ecopiety communicated in and through systems of cultural meaning. As Phyllis Japp and colleagues argue, "Popular culture provides a daily catalog of cultural attitudes, values, and practices. From television sitcoms to the daily news, from the theater to the sports stadium, we observe embodiments and enactments of character, virtue, honesty, and integrity (or lack thereof) in situations we find understandable, if not familiar."[24]

But more than simply *observing* these embodiments and enactments, in the age of new media (the digital world of online social media, mobile applications, chat rooms, message boards, e-mail, streaming video and audio, Internet telephony, and so forth), we are actively encountering and engaging these depictions of character and virtue in our everyday digital practice. As prosumers, we talk back, we write back, we tweet back, we vlog back, and we *interrupt*, *contest*, *rewrite*, and *reframe* as we engage in media as *practice*.

E. L. James came face-to-screen with this phenomenon, arguably getting a taste of her own "discipline and punish" thematic, when she hosted a social media event on Twitter (#AskELJames). This was a marketing event crafted to launch her latest book, *Grey*, a sequel to the original trilogy in which Christian Grey retells the story of *Fifty Shades* from his point of view. James's publicity team had framed this online exchange as an opportunity for an accomplished author's many adoring and devoted fans to gush excited questions to James about the recently revealed insights into the complex mind of heartthrob Christian Grey, as featured in her latest series installment. Instead, users of the popular social media platform unceremoniously interrupted and reframed this smoothly constructed PR narrative, supplanting it with contestation, resistance, and moral critique. Questions from participants included postings like the following: "Which do you hate more, women or the English language?"; "Does it make you happy to know young girls everywhere may think abuse disguised as love is part of a normal relationship?"; "Why did you ever think it was okay to write books that romanticize abuse and reinforce rape culture?"; "When humiliating my date to tears, how much money must I spend to be seen as sexy and eccentric?"; "If I stalk a girl and GPS her car, does that mean it's true love?"; "What's it like telling millions of women it's okay to be in an abusive relationship as long as he's rich?"; and "Do all these negative tweets sent to you seem abusive to you? I think it's romantic enough to be turned into a novel!"[25]

These scathing tweets are only a glimpse into a much larger and complex backlash against the *FSOG* series, which has included public outrage against a highly publicized incident in which a nineteen-year-old female student at the University of Illinois brought charges against a male student for allegedly tying her up for more than ninety minutes, sexually assaulting her, and beating her with a belt. The accused first-year student, also nineteen, defensively answered these charges, explaining that he was simply reenacting scenes from *Fifty Shades*, which he was under the impression young women like.[26] After all, Christian Grey assaults Ana, but his ecopiety and demonstrated consumopiety—heck, even his eco-friendly house renovation—absolve him, avowing that he is a good guy who cares about the environment and wants to save the planet.[27]

The very dynamics of digital culture that had given rise to James's soaring popular success—mass sharing and "spreadability," networked online communities, remix and prosumer culture, and the use of grassroots media channels—are the very same dynamics that resulted in her collective "spanking" from Twitter users. As James—of all people—should have known from her fan fic days, once content is set in motion in the digital mediaverse, creator control is a bit of a Sisyphean task. In the online mediaverse, there is *no version* of Walter Benjamin's "Finis," which he identifies in his classic essay, "The Storyteller," as one of the defining characteristics that mark the cultural shift away from oral storytelling to the ascent of the novel in modernity. Unlike the novel, in which an author makes a definitive and static cutting-off of the story at the end with the declarative finality of the word "Finis," the orally told story *continues* when the storyteller stops speaking, as the discussion carries on communally, face-to-face, opening the story to mutability and alternative narrative possibilities.[28] But Benjamin perhaps could not have foreseen the digital age, where there is no "Finis" in the mutability and spreadability of the Internet.[29] Although E. L. James's *FSOG* is a published novel, it was conceived and produced in a very different process from what Benjamin had in mind when he wrote about the static nature of the novel. In Benjamin's binary, it is the realm of the oral storyteller that constitutes what we might term now a "participatory space" of collective didactic instruction, learning, counsel, and shared social interaction.[30] But *FSOG*, for better or worse, was not simply highly derivative of previous storytelling but *collectively* constructed in a communal space, and gained popularity through social media word of mouth and digital retelling. What is more, the authority of the author, especially in the Twitterverse, becomes just one voice among many and not one that can silence all others with the conventional authorial pronouncement of "Finis." When E. L. James says "The End," no one listens, and the conversation continues with or without her, a conversation that neither she nor her public relations minions can control.[31]

As transformative works, the *Fifty Shades* series and the corporate public relations messaging of ecopiety-asserting greenwashing media machines share more affinity than we might imagine. BP's CEO Bob Dudley may well be able to commiserate with the BDSM erotica author's inability to "control message" since the multinational petrochemical corporation got a spanking of its own from Twitter-ignited activists upon the announcement that BP would become the official sponsor of the British Museum in London. Through its high-profile sponsorship of both the British Museum and the 2012 London Olympics, the company sought control over its image and corporate message in the wake of bad publicity from its associated refinery explosions and large-scale environmental disasters. Sponsoring the arts and goodwill global sporting events like the Olympics would strategically offset sullying news stories of the company's repeated and numerous environmental calamities.

Using tweets to organize, a group of environmental "*actor*-vists" who call themselves "BP or Not BP?" have launched a series of Shakespearean theatrical flashmob protests at the British Museum, circulating protest information and footage via the Internet. (Flashmobs are public gatherings called together through the use of social media.) In one of the group's many online videos publicizing their anti-BP actions, one of the members explains the reasons for her troupe, who call themselves "The RSC" (not Royal Shakespeare Company but *Reclaim* Shakespeare Company), having converged a flashmob to protest BP: "They are exploiting the natural resources of the world and also trying to give themselves a lovely greenwash by supporting the arts and the Olympics."[32] Through Internet-circulated videos of performance protests, activists work to interrupt BP's controlled public relations narrative of corporate virtue, unmasking and condemning this greenwash. In the tradition of *Macbeth*'s witches, one intervention staged inside the BP-sponsored Shakespeare museum exhibit involved protesters hexing BP by repeatedly chanting, "Double, Double, Oil and Trouble. Tar Sands Burn and Greenwash Bubbles!"

The RSC's protests against BP, which, incidentally, adhere to iambic pentameter poetic form, attract about two to three hundred flash-horde participants, but these performances extend well beyond the immediate on-site witnesses or even journalistic coverage. Circulation across multiple and varied media platforms creates, in effect, a digital loaves and fishes effect, multiplying the actor-vists' reach and impact. Use of social media and its networks has also facilitated the teaming up of the RSC and the BP or Not BP with other groups such as Liberate Tate (art gallery), UK Tar Sands Network, Art Not Oil, Rising Tide UK, and Arts Fossil Free Commitment, a movement to reject greenwash funding of the arts by fossil-fuel-purveying companies. Working with meager financial resources, but drawing from followers in the Twitterverse, Instagram and Facebook communities, and an extended and expanding online network of activist partners, RSC-led Lady Macbeth–like chants of "Out, damned sponsor! Out, BP!" grow louder.[33]

These groups talk back, write back, and literally *perform a contradictory story* of BP than the one suggested by their sponsorship logos on British Museum banners and brochures—a counternarrative that aims to dislodge and subvert the company's well-funded greenwashing strategies. Access to the tools of online networking, content production and circulation (protest scripts are circulated ahead of time), and inexpensive publicity and advertising, all at the click of a mouse, in addition to the reposting of action videos and post-action blogs, contribute to mobilizing media interventions and fostering environmental moral engagement. These storied interventions effectively resist and rewrite BP's carefully crafted corporate public relations messaging of good corporate citizenship and eco-pious "green" concern. Discussions of media have long split along binary lines of debate as to whether media constitute tools to control and dominate—fostering submission and complacency in a media-anesthetized populace—or whether media provide tools to liberate, empower, and organize resistance against dominant messages and sources of power and authority.[34] Yes to both. "We think through, with, and alongside media," writes media theorist Katherine Hayles, and I would add that we likewise enact dynamics of dominance, submission, as well as resistance (all of the above) through, with, and alongside media practice.[35]

Trickle-Down Virtue

Why look at the moral dimensions of greenwashing through the popular mass-market romance *Fifty Shades of Grey*? Narratology, the study of narratives, particularly mediated popular culture narratives, as they are told and engaged, tells us something important about the relationships among media, culture, and moral engagement. "Not only do we tell stories," point out literary theorists Andrew Bennett and Nicholas Royle, "but stories *tell us*."[36] In James's greenwashing of Christian Grey, replete with interjected environmental moral offsets, never once is the BDSM CEO's green virtue called into question vis-á-vis his unbridled capitalistic consumerism and its impact on the planet's dwindling resources. Earlier in the series, we learn of Grey's interests in developing environmentally sustainable agricultural practices as part of his agribusiness holdings around the world, and presumably this *offsets*—much as one might purchase carbon offsets—his conspicuous consumption of wildly expensive gas-guzzling supercars, his yacht, his personal helicopter, the vast square footage of his enormous Seattle penthouse, his other luxury properties, and his violent and controlling nature toward women.[37] Grey rails against wasting food of any kind (presumably a green virtue), so much so that he wants a contractual signed agreement that permits him to punish Ana physically for not eating everything on her plate at mealtimes. And yet, Christian Grey seems to have no remorse or even self-consciousness about the sheer volume of *waste* generated by his profligate consumer lifestyle, nor the planetary toll of doing what it takes in manufacturing and mergers and acquisition deals to make one's first billion dollars by the age of twenty-five.

When, as a student, Ana interviews Christian for the school newspaper and the discussion turns to his possessions, she observes, "You sound like the ultimate consumer." Christian agrees, replying, "I am." In the fourth book, told from Christian Grey's perspective, we hear the rest of Christian's internal reply to Ana's comment, in which he remarks to himself, "Nothing wrong with consumption—after all, it drives what's left of the American economy."[38] Intermingled in the character of Christian Grey are the fundamentally intertwined practices of *ecopiety* and *consumopiety*. Grey's sleek fleet of Audi luxury vehicles, his collection of elite real estate properties, his private plane, his helicopter, his impeccably tailored expensive clothing, all *drive* the American economy, and this *consumopiety* contributes to the good of the nation. There is an implicit "trickle-down goodness for all" activated by his opulent luxury spending, from which even starving people in Darfur appear to benefit. At the same time, Grey is also an *eco-pious* consumer, investing in renewables and green construction (buying the *right* stuff, purchasing ecovirtuous materials), while sparing no expense on the installation of environmentally conscious features to his mansion. These acts of ecopiety are fundamentally and simultaneously also acts of patriotic consumopiety, bolstering the US economy and thus increasing beneficent outcomes for the nation and beyond.

The term "moral licensing" refers to a phenomenon in which "past moral behavior makes people *more likely* to do potentially immoral things without worrying about feeling or appearing immoral."[39] Behavioral psychologists Benoit Monin and Dale Miller's seminal study of prejudice demonstrated how individuals who had previously established credentials for *not* being prejudiced were *more* likely to go on to express prejudiced attitudes.[40] More recently, Monin and co-investigators have described moral self-licensing as "when being good frees us to be bad" and describe two major models for this phenomenon. In the first model, there is a kind of banking system of moral credits, in which one deposits good deeds that will offset or balance out bad deeds (moral withdrawals). Monin and his colleagues explain that, in this first version of moral licensing, "[I]t feels fine to commit bad deeds as long as they are offset by prior good deeds of a similar magnitude (Nisan, 1991)." In effect, the researchers find, one "purchases the right to do bad deeds with impunity."[41] Think if you will of the preacher who has served his flock loyally and worked tirelessly to build a church community from the ground up that emphasizes purity and "traditional family values," only to be discovered later to be hiring male prostitutes to accompany him for his extracurricular activities. In the second model of moral licensing, the establishment of moral credentials (a demonstration of previous moral virtue or a known identity as a virtuous person) *changes* the meaning of bad behavior, or reframes bad deeds in the mind of the one who commits them as not being bad at all, since the person is already known to be a virtuous person. That is, "Past behavior serves as a lens through which one construes current behavior, and when the motivation for current behavior is ambiguous, it is disambiguated in line with past behavior."[42] One of a number of behavioral realms to which Monin and colleagues apply both the "credits" and the "credentials" model is the realm of questionable consumer choices. The researchers find that "one can self-license frivolous consumption by behaving in ways to establish one's morality"—for instance, self-licensing self-indulgent purchases by having banked prior frugal and sensible purchases.[43]

Does Absolution Promote Sin?

Duke University and University of California–Davis economists Matthew Harding and David Rapson's research specifically examines moral licensing and American consumer habits in the context of energy conservation and energy offsetting programs. In their study, "Does Absolution Promote Sin? The Conservationist's Dilemma," Harding and Rapson find that households that voluntarily subscribed to an energy carbon offsetting program to reduce their carbon footprint actually went on to increase their energy use once they committed to the program and enjoyed a feeling of virtue from having done so.[44] Change strategist Tom Crompton identifies a similar rebound effect, pointing out that consumers who feel that they have done their bit for the environment often go on to justify their larger excessive consumption patterns and political complacency.[45] This behavioral response is akin to the dieter who drinks a diet soda and then reasons that this justifies consuming a chocolate cake. Harding and Rapson's work, when read in the context of Monin and colleagues' research, suggests a kind of "economy of environmental virtue," in which pious acts of "going green" serve to legitimate or morally license other environmental sins.

We see this economy of virtue and the exchanging of "green credits" or "green identity credentials" reflected in the narratives of mediated popular culture. In the case of *Fifty Shades*, Christian Grey's personal ecopiety, his acts of environmental virtue, as performed throughout James's series, serves as moral licensing, offsetting and absolving his violent transgressions—his whipping little girls who look like his mother and liking it—thus reassuring *Fifty Shades* readers that appearances are deceiving, and Christian Grey is promisingly virtuous. If enough effort and understanding are applied, the right girl can get him to reveal his true self. In the process, James powerfully and repeatedly *reinscribes* the capitalistic virtues of consumption. James can employ ecopiety as a character trait in this economy of virtue and bank on her presumption that most readers of her (now more than 125 million copies sold) book will get that corporate ecophilanthropist Christian Grey has got to be more than the guy who stalks, beats, and abuses young women—that he is promisingly redeemable—and that his environmental piety signals this to a contemporary mass audience. Her using green credentials to signify or communicate underlying (and promising) moral credentials resonates and reflects what sociologists and psychologists have shown for decades in their quantitative research on both pro-environmental attitudes and pro-environmental values held in the United States and abroad. That is, pro-environmental attitudes around the world are generally high, and responsibility toward the environment is associated with good citizenship and civic virtue (although there is a gap between the scales of self-reported pro-environmentalism and actual behaviors that prioritize environmental concerns over economic ones, particularly when it comes to global climate change).[46]

In an economy of virtue, does eco-pious practice in one area truly offset violence and abuse toward *both* women and nature in other areas? Do Chevron's ads about supporting turtle conservation or BP's "going green" ads seduce us into giving them a free pass for oil spills and the violence they and other corporate fossil fuel purveyors inflict upon the planet? In both cases of greenwashing, how many of us are lured by the fantasy of somehow *changing* an abusive partner with whom we are nonetheless habitually linked and on whom we find ourselves dependent? In the United States, where one resident consumes as much as thirty-five people do in India and fifty-three times as much as one person in China does, do we convince ourselves that the strategically contrived stories of goodness and virtue underlying violence are true simply because we cannot imagine a life beyond our current levels of relative comfort and consumption?[47] Singer/songwriter Sheryl Crow's haunting lyrics of toxic-relationship dynamics come to mind here, as she soulfully croons, "Lie to me. I promise I'll believe. Lie to me, but please don't leave. Don't leave."[48]

Restorying Grey: Fifty Weeks of Green

Linda Watson, an environmental activist, sustainable agriculture advocate, and food blogger, was sufficiently disturbed by what she read as a consumer of *Fifty Shades of Grey* that she set out to *rewrite* and *retell* E. L. James's story, publishing her own authentically green version of the erotic romance. In Watson's transformative work, we see a prime example of the kinds of sociopolitical arguments and feminist critiques often interjected into fan fiction renderings. Unlike James, Watson, who is a former IBM projects manager, is not identified as a fan fic writer, but there is perhaps poetic justice in James's having her story rewritten, and remade, much in the manner that James did to *Twilight* author Stephenie Meyer. The chain of derivation, remix, and redux begins to gain more links here as Watson restories James's story, which had restoried

Meyer's story, whose story, some have argued, is essentially a mashup of vampire romance and the founding of the Church of Jesus Christ of Latter-Day Saints as told in the Book of Mormon.[49] Meyer's book series can be situated within the tradition of Mormon derivative literature, which itself is a kind of fan fic storying of the Book of Mormon. We could go back farther still, of course, and look at various theories about novel manuscripts or other books that Joseph Smith allegedly rewrote and restoried in order to create the Book of Mormon, and we could probably go even further beyond that.[50]

Remixes and remakes are certainly not a new phenomenon, as a number of cultural historians have pointed out.[51] In the cut-copy-and-paste culture of the digital media age, however, the intensified speed, scope, and scale of remixes and remakes make them easier, cheaper, and astronomically faster than ever to accomplish.[52] Filmmaker Kirby Ferguson has argued in his documentary series, *Everything Is Remix*, and now in subsequent TED talks, that "creation requires influence," and "combining and editing existing materials to produce something new" is simply the way culture works: "This is how we live and how we create." The dynamics of remix have always been with us, but we have reached a point, he argues, in which the sharing of information, the advancement and reach of sophisticated information search engines, the sheer volume of information exchanged, the readily available accessibility of unprecedented volumes of information, and the speed of communicating that information, tracking it, and retrieving it, are such that "technology is now exposing this connectedness" of remix and remake that were once more hidden and have made those connections easier to see.[53] What is of interest in the study of media, environment, and popular moral engagement is not the *fact* of remix or remake, but the *how* of it and what that interpretive process might suggest about restorying in an age of environmental crisis.

Linda Watson's *Fifty Weeks of Green* has been characterized as both "the sustainable antidote for *Fifty Shades*" and "a romance for the 99%." The book's title refers not to the green of money but to the green of vegetables. And these are not just *any* vegetables, but the kind of sustainably, organically, and locally grown vegetables that support local economies and encourage small-scale alternatives to industrial farming. Instead of Ana and Christian's BDSM contract of sexual do's and don't's, in the book's title, Watson plays with community-supported gardens' (CSGs') customary requirement of a fifty-week contract. In a CSG contract, the consumer partners with the farmer in the risks of growing food in return for weekly shares of fresh fruits and vegetables throughout the year. Watson refers to herself as a "food evangelist," and she is the founder of the organization Cook for Good in conjunction with its eponymous food blog at CookForGood.com. Watson also blogs for *Mother Earth News*, *Huffington Post*, and *Good Veg*, while maintaining a social media following on Twitter, Facebook, Pinterest, and YouTube. She began the Cook for Good project in 2007 after being galvanized into food activism by taking the "Food Stamp Challenge." This challenge involves individuals not requiring government assistance attempting to live on meals that cost no more than an average of one dollar per day. Watson describes her mission as helping people to "eat fabulous food, get healthy, and save the planet." The same year that James self-published *Fifty Shades of Grey*, Watson published *Wildly Affordable Organic* (2011), in which she promotes cooking with real, unprocessed ingredients, buying locally grown sustainable food, and eating low on the food chain.[54] Watson aims to help people to "vote with [their] forks to change the world" and "to reduce greenhouse gasses—one pot of beans at a time."[55]

Watson's restorying of *Fifty Shades* was her way of taking an active role as a cultural prosumer and talking back to the story as told—a case of change the story, change the future. "The more I read, the more Christian Grey resembled the villains I've fought all my life," explains Watson. "He isolates himself from the world using blindfolds, gliders, yachts, and private islands. Then he roughly takes what he wants, crushing lives and exhausting valuable resources. Would young women assume, given the blockbuster status of the books or movie, that Grey behaves in an admirable, manly way? This thought so haunted me that I wrote *Fifty Weeks of Green*."[56]

In Watson's remake, Christian Grey morphs into Roger Branch, an organic farmer who delivers boxes of produce to his CSG shareholders at the local farmer's market. Instead of schooling the heroine in the use of riding crops, he schools her in the usefulness of cover crops and how well they can fix nutrients in the soil. Subservient, submissive, twenty-something ingénue Ana from *Fifty Shades* morphs into Sophia Verde, a self-possessed, mature, ex–political strategist who designs a marketing campaign for Roger's CSA (community supported agricultural cooperative) in exchange for a weekly share of organic produce. Their romance blossoms at the local farmer's market, as Sophia wisely figures out ways for Roger's CSA to appeal to the growing demographic of "optivores"—people who want to optimize something through their consumer choices. Sophia explains to Roger that optivores have a variety of reasons for choosing what they choose to consume: "health, religion, or just good old do-gooding. These days folks want to know if something is organic, local, vegan, or vegetarian. Some want 'free' food: spray-free, sugar-free, oil-free, or gluten-free." She notes that one can be an optivore for selfish reasons (to look better, to enjoy better food taste) or for civic and social-activist reasons. An optivore, she explains, "votes with their fork." (And to Sophia's "optivore" explanation, we could now add the new designation of "climatarian.")[57]

In Watson's book, we see, once again, a narrative articulation of an intertwined ecopiety and consumopiety but framed in a different setting and within more egalitarian power dynamics. Optivores may have a variety of reasons for being choosy about what they consume, but by voting with their forks, they are also connecting consumption choices with moral engagement and, in some cases, the practice of "good old do-gooding."[58]

Sophia and Roger's contract is more formally known as a "mutual support agreement" and includes such commitments as Sophia's promise not to waste food she receives from her CSA share and Roger's vow to grow food "from heirloom or naturally selected seeds, without use of any seeds from Genetically Modified Organisms (GMOs)." These contractual bonds bring the couple together, and romance blooms. Instead of a fancy ride in his private helicopter on the first date to visit his impressive Seattle penthouse and red room of pain, Roger instead meets Sophia at REI (the outdoors outfitter co-op Recreational Equipment, Inc.) to help her pick out a comfy pair of hiking boots so that she can tour the farm. The REI episode involves an erotic scene in which Roger, massaging Sophia's sock-covered instep, slips the hiking boots onto her feet in Cinderella fashion and then firmly cross-laces them up—a scene that I wager will make *Fifty Weeks of Green* readers never see REI quite the same way again. The couple does not go out to eat at fancy restaurants, but instead supports the Conscious Café, a local farm-to-table restaurant where the new chef has just graduated from the nearby community college's Natural Chef program.[59] Rather than buying Sophia luxury consumer items and expensive clothing, as does Christian for Ana, Roger hand-carves Sophia a beautiful figurine from reclaimed walnut hardwood. And rather than his-and-hers Audi R8 supercars, Sophia drives a Prius and Roger a well-worn farm truck.

When things heat up after a summer solstice dance party and the couple romantically tangles under the stars in the flatbed of the truck, their lovemaking sessions are followed by suggestive seasonal "plant-powered" (vegetarian) recipes that, as Watson puts it, "will ignite your inner frisky while making a difference." One game of flirty chase ends up in Roger's bed, followed by a recipe for Sweet-and-Tart Collard Tangle.[60] What's so sexy about Roger, who is fifty-one and spends much of his time planting fields and educating people about the benefits of supporting local organic growers? Watson speaks of the magnetic appeal of a man "who loves to feel good soil and makes it better with sustainable practices. He delights in women who know their own minds, in his lively community, and in being part of the dance of nature."[61] Her narrative of optivore romance suggests that doing good for the earth can be sensual and fun, and that ecopiety can be less prim and more erotic, a theme we will explore further.

The story world that E. L. James created (or re-created) in *Fifty Shades* profoundly depressed Linda Watson, even "haunted" her, she recounts, and so using the same resources of digital new media and social media marketing that facilitated James's success, Watson *wrote back*, *remade*, and *restoried*, remaking the James story world of *Fifty Shades* into the kind of world she would like us all ideally to live in, a world where, as Watson avers, "Love doesn't have to hurt and healthy food can be delicious."[62] Watson seeks to create an authentically green romantic hero, in contrast to James's greenwashed Grey. Roger's demonstrated ecopiety is consistent with low consumption levels and an earnest repudiation of the patriarchal dominator role over *both* women and nature. However, the model of ecopiety in much of *Fifty Weeks of Green* is still very much a personal, individual, and privatized model of ecopiety, in which the practice of environmental virtue is about "voting with one's individual fork" and not about working to enact governmental policy initiatives, broad-based legislation, political solutions, regulatory changes, or other measures to implement significant structural shifts on a scale that would make a substantive dent in addressing the environmental problems we face.

The expression "vote with your fork" was popularized by *New York Times* writer Michael Pollan in his food articles and best-selling novel, *The Omnivore's Dilemma*. At the time the book was published in 2006, Pollan argued for changing our food system one fork at a time and one meal at a time.[63] The hero of Pollan's narrative is conservative Christian, Libertarian farmer Joel Salatin, who runs a nonindustrial, local, sustainable community farm in Virginia and argues evangelistically against what he calls "salvation through legislation."[64] Salatin might be doing good things for his local environment and for the integrity of animal husbandry via his sustainable polyculture farming methods, but his model is fundamentally an antigovernmental one that relies on the envisioned efficacy of localized individual forks and marginal microchanges of personalized conscious consumption. By the time the tenth anniversary edition of Pollan's book was released in 2016, he was singing a different tune. "It's important to vote with your fork. It's not trivial," he told an interviewer but then significantly qualified this by adding, "It's necessary but not sufficient. We also have to vote with our votes."[65]

Pollan now recognizes that without public/government policy and regulatory action, the earth does not really "notice" those individual forkfuls. His revelation over the course of a decade represents a critical narrative shift in the evolution of food movement activism and reflects a shift in messaging that is gaining ground in larger environmental discussions and debates. In fact, in a 2019 interview survey of sixteen of the world's most prominent sustainability leaders and environmental thinkers, fourteen out of the sixteen specifically cited collective action and organizing as the most effective strategy to prioritize for addressing global climate change. Far from Watson's "one pot of beans at a time" solution, Bill McKibben emphatically states in the survey, "You cannot do it [address climate change] anymore one lightbulb, one vegan dinner at a time." In the same survey, Paul Hawken likewise redirects and emphasizes that "[i]t is not about what 'I' can do. It is about what we can do."[66] This concerted collective approach to addressing environmental crisis is unfortunately mostly lost on the eco-pious characters in *Fifty Weeks of Green*.

After losing her job as a campaign strategist, Sophia Verde leaves the field of politics to do PR and marketing for local CSA farms, where she emphasizes that "trivial actions like saving salad scraps to compost can preserve the world."[67] Roger also champions small daily actions and personal choices of beneficent optivores as ones that will make a real difference. "We give people who want organic food and good conditions for farm workers and animals a chance to vote with their wallets," he explains. "No need to wait for elections and then pray that your representative represents *you*, not his biggest donor."[68] In the face of a broken political system and corrupt campaign financing that serves the welfare of corporate special interests to the exclusion of ordinary citizens and their needs, the primary option Walton presents is one on a personal micro-level of the conscious wallet or fork. The solution to planetary problems is one of eco-pious personal, voluntary consumption as civic virtue as practiced by those who presumably are already in an economic position to make those choices. Perhaps the story of privatized individual personal action, in which every little bit makes a difference, as an effective libertarian strategy for saving the planet is the real *romance* story in both James's and Watson's popular novels. Framing narratives, especially familiar ones that are firmly culturally embedded, are not easily interrupted or replaced. W. Somerset Maugham famously retells the Babylonian Talmudic story of a poor servant who upon encountering Death in a Baghdad market flees the city in a panic to seek safety in the town of Samarra, only to discover that all along Death has had an "appointment in Samarra" with him for that very evening. So, too, often our best efforts to leave stories behind end up with us smacking right back into them.[69]

The Fantasy of Fifty Simple Things

"Every little bit doesn't always help," declares Gernot Wagner, former lead senior economist at the Environmental Defense Fund and current director of Harvard's Solar Geoengineering Research Program. In his essay "Going Green but Getting Nowhere," Wagner, who has advanced degrees in political economy and government from Harvard University and in economics from Stanford University, instructs, "The changes needed are so large and profound that they are beyond the reach of environmental individual action."[70] He confesses, in his Amazon.com author's page description, that he does not eat meat, does not drive, and "knows full well the futility of his choices." It is changes in our *collective* way of life at a national and international scale, *major* structural changes, Wagner maintains, that will make enough of a difference that the "planet will actually notice."[71] What is more, although virtuous and commendable, the practice of individual small pious actions can "distract us from the need for collective action, and it does not add up to enough."[72] He points out that airlines now list voluntary carbon offsets on their booking web site pages, but these offsets are less than a drop in the bucket and serve as marketing ploys to make the eco-conscious feel better about flying and in turn book *more* flights. A real solution? "A heavy dose of government policy" to provide the right kinds of incentives would be a good start, contends Wagner. On both a national and an international level, governments can redirect market forces to make dirty energy exorbitantly expensive and clean energy very cheap.[73] Instead of "sweating the small stuff," as Wagner puts it, we should be focusing our collective energies on working to enact planetary-scale policy solutions.

Much as Monin and colleagues point to moral licensing behaviors, Wagner points to the problem of "single-action bias," or what is also sometimes called "crowding out bias." This occurs when people do one deed for the environment (shopping at the farmer's market for organic produce), or act in one area, then feel they have done their civic duty and move on. "If you catch yourself recycling that paper cup and thinking you've solved global warming for the day," warns Wagner, "think again."[74] Wagner does not dissuade anyone from recycling, using reusable water containers, going vegetarian, or consuming locally grown organic food. In fact, he recognizes that these can be personally mindful practices that keep people conscious of environmental issues. But people should engage in such practices with a very practical and *realistic* grasp of the fundamental planetary economics of scale needed to "move the needle" on climate change, while keeping in mind the very real counterproductive psychological dynamics at work in "single-action bias." Wagner's very first piece of advice to the civic green-minded who are committed to making a difference is, "Don't trust any list that gives you ten things you can do to stop global warming. . . . There is a fine line between simplistic single actions and doing what counts."[75]

Best-selling environmental how-to books, such as the now-classic and reissued *50 Simple Things You Can Do to Save the Earth* (1989, 1990, 2008), have stressed to audiences that it is *individual* acts of ecopiety that will ultimately yield efficacious and positive environmental results. Since its publication in the late 1980s and its multiple reissuings, including an additional version of *50 Simple Things Kids Can Do to Save the Earth* (1990, 2009), various environmental groups have challenged both the rigor and the framework of these handbooks that conveniently tout the salvific effects of simple eco-pious acts, which include actions like turning off the running water when you brush your teeth, recycling bottles and cans, and turning off lights when you leave a room. In response, *Earth Island Journal* printed environmentalist Gar Smith's "50 Difficult Things You Can Do to Save the Earth" (1997). Smith's list suggested alternative not-so-simple individually pious acts more in the hard-core realm of "Bury your car," "Don't have children," "Blockade a lumberjack truck carrying old growth trees," and "Have your power lines disconnected." But Smith also interspersed these suggestions of *personal* eco-pious acts with not-so-simple *collective* pious actions and policy changes: "Halt weapons productions and exports," "Replace majority rule with proportional representation," and "Pass a nature amendment to the U.S. Constitution." In that same year, J. Robert Hunter published *Simple Things Won't Save the Earth* (1997), in which he, like Smith, argues that real and significant environmental change involves not simple but complex and large-scale national and global policy changes, in conjunction with corporate policy changes, and massive collective action. Focusing on the case study of natural and synthetic latex rubber production, Hunter likens the advertising of eco-friendly products to the selling of indulgences that (falsely) reassure publics by reinforcing the narrative that *personal consumption choices* are all that is needed to solve the environmental crisis.[76]

Derrick Jensen's *As the World Burns: 50 Simple Things You Can Do to Stay in Denial* (2007) further takes to task the premise of the best-selling "50 Simple Things" books, pointing to the ways in which narratives of personal piety deceive us. Saving the earth, or at least achieving any sort of meaningful damage control at this advanced stage of the game, argues Jensen, necessitates serious collective sacrifice and discipline, enacted through broader civic action and collective commitment.[77] Or, as Barack Obama succinctly and bluntly put it to journalist Brian Williams after the cameras and microphones were turned off after a televised 2008 election debate, "Well the truth is, Brian, we can't solve global warming because I f——king changed the lightbulbs in my house. It's because of something collective."[78]

Books such as Derrick Jensen's interrupt the facile narratives of ecopiety as practiced through consumopiety and talk back to dominant narratives. These works are wake-up calls critical to advancing the conversation among the initiated—those already steeped in environmental issues and causes. Roy Scranton's discussion of the powerful impact of shifting "social energetics," however, prompts us to take seriously mediated popular cultural forms as those that actually engage publics more broadly and have the potential to move a critical mass of the uninitiated from mere feel-good environmental ideals into effective action. Environmental online discussion groups, as studied by ethicist Sarah Fredericks, reflect how group participants express personal feelings of guilt over negligent acts, such as forgetting to bring one's reusable bags to the grocery store and other personal ecological failures. These have become the subject of public online confessions and verbal self-flagellations. Not insignificantly, Fredericks found that the "everyday environmentalists" in these digital communities "almost never discuss governmental, social, or economic forces that contribute to environmental problems or could aid environmental actions." Although the environmental politics of her research subjects may be progressive, their solution model is a conservative one, or, perhaps more accurately, libertarian, in that it espouses a model of personal, privatized responsibility and individual culpability for an environmental crisis seemingly untouched by broader civic forces.[79] Linda Watson's *Fifty Weeks*, for all its efforts to restory and remake E. L. James's Christian Grey in a more authentically green way, which she arguably succeeds in doing in many respects, ultimately ends up focusing on a similar model of personal responsibility and private consumption choices. Still, Watson's book *does* show how an alternative micro-economic system—a cooperative system of community-supported agriculture and barter exchange—can work on the local level. With the marketing know-how of Sophia Verde, the local Raleigh farmer's market grows in leaps and bounds, its existing CSAs flourish, and new ones pop up. The growth in networked CSAs, in conjunction with an expanding number of local "farm-to-fork" restaurants in the book, other locally produced goods, and local college cooking and agriculture programs, meets economist Gernot Wagner's prescription for how to get from individual acts of "feel-good . . . free-lance environmental heroism" to a critical mass that makes a noticeable difference to the planet—that is, "start a movement."[80] *50 Weeks of Green's* epilogue further leads the reader toward increasingly expanding spheres of civic engagement, as Watson's couple, Sophia and Roger, rally the CSA network they have cultivated at the farmer's market, pooling their activist resources and energies to oppose the dangers presented by the incursion of the natural gas fracking industry into their local farmland communities.

The co-constitutive relationship between ecopiety and consumopiety highlights a concomitant and persistent tension between personal piety and collective civic action to influence policy change. As with ecopiety and consumopiety, personal and collective pieties are contrapuntal in relation to each other. By the 2018 release of the third film in James's trilogy, *Fifty Shades Freed*, the romanticized aspects of intimate partner violence and emotional abuse that run throughout the franchise played very differently in the #MeToo era and the launching of the "Time's Up" movement against sexual harassment. *Hollywood Reporter* film critic Kristen Lopez minced no words in designating the *Fifty Shades* 2018 installment a "relic of the past" when read in the context of the "real-world abuses of power in Hollywood" that had come to light via Miramax co-founder Harvey Weinstein's 2017–2018 sexual assault scandal.[81] Lopez highlights uncomfortable and considerable overlap between Weinstein and his abuse of his powerful executive position and CEO Christian Grey, who buys a publishing company in order to become Ana's boss so that he can have more control over her and pressure her into submission. "In a world where abuse against women is finally being discussed in-depth," declared Lopez, "it's nearly impossible to dissociate the pic's romantic fantasy about a controlling, wealthy man and his unbalanced relationship with a young, female subordinate from the current discourse about why such a setup is inherently problematic. The film takes a position that, as long as behavior is legal and consensual, it's also morally fine. It's a stance that feels particularly anachronistic in the midst of the productive national conversations about consent and abuse of power that have come with the current reckoning."[82]

James's eco-pious hero, Christian Grey, may also read differently in an era when prominent public discussions of the abuses of privilege and plutocrats have entered the mediasphere. James's individualistic, heroic, privatized, depoliticized, voluntary, libertarian, charity-based model of environmental virtue wears thin and conveniently challenges nothing structural. What appears at first to be a lauding of corporate responsibility and green virtue is instead a reinscription of capitalistic consumer norms and corporatized logics. An opportunity is missed, representationally, both to make more substantive environmental moral engagement fun, playful, and sexy and to extend that engagement beyond the salvific cliché of the lone green beneficent billionaire. Such a turn is of course not part of E. L. James's agenda. Grey's environmental virtue may provide a narrative device to offset morally his other acts of violence, effectively reassuring the reader of his redeemability, but he still presents an unappealing model of ecopiety that is dark, brooding, dour, painfully serious, emotionally abusive, and devoid of a sense of humor. As resistant readings that talk back to the series continue to emerge in collective participatory online settings, and as transformative works derivative of the series are produced that wield sociopolitical critique, we may see greener versions of Grey continue to interrupt and intervene in the series' trajectory of social energetics—calling "Time's Up" on the domination, manipulation, and control exerted in both partner and planetary abuse.

3

"I Can't! It's a Prius"

Purity, Piety, Pollution Porn, and Coal Rolling

Toyota Corporation press releases and marketing materials speak of their hybrid automobile, the Prius, as having become an "icon" in American culture in a fairly short amount of time since its introduction in 2000.[1] Multiple appearances of the Prius as a featured "character" in the narratives told on television series such as *Curb Your Enthusiasm*, *The Simpsons*, *Weeds*, *Six Feet Under*, and *South Park*, references and jokes made about the Prius on *Saturday Night Live*, and the officially inaugurated Prius game token in the "Here and Now" edition of the classic board game Monopoly, all seem to provide corroboration for the assertion of the Prius's cultural iconicity. But what does this icon symbolize and to whom? What moral meanings are associated with the Prius as a more "environmentally sustainable" consumer choice, and how do these meanings get mediated through popular culture? Indeed, how might disparate socioeconomic publics actively *contest* the moral meanings ascribed to the Prius in its consumer marketing? And what insight might these *oppositional* readings provide into the practice of ecopiety and its illegibility to white, working-class publics? This chapter argues that the production of "pollution porn" constitutes a class-based media intervention into moral narratives of ecopiety, or environmental virtue, as practiced through the consumopiety of green capitalist consumerism, within what has become a depoliticized marketplace-based environmentalism. Whereas the content of "pollution porn," which entails the performative practice of "rolling coal" through the exhaust systems of diesel trucks, may be repugnant to those who are environmentally concerned, analysis of this subgenre of media nonetheless provides a critical window into hostile readings of green capitalist marketing and points to concomitant class-based resistance to its perceived "elitist" messaging.

"I Can't! It's a Prius"

It is nighttime and the fans have all gone home after the game. Together in a car, a handsome high school football jock in his letter jacket and a cheerleader still in her uniform are parked in a dark spot next to the field for a little post-game "make out." It is a classic scene from American teen popular culture. Some heavy kissing is going on, when suddenly but reluctantly, the jock tells the cheerleader, "I *can't*." She doubts his sincerity and continues to kiss him. He is momentarily drawn back into passionate embrace but again, more forcefully this time, says, "I *really* can't." Still, his heart does not seem to be in his protestations, and the girl again continues to kiss him. The third time, he means business, pushes the pretty girl off him, and responds emphatically, "I *said I can't*!" Exasperated, she asks, "Why *not*?" The boy shakes his head regretfully and responds, "I can't! It's a *Prius*." Breezy music plays in the background as the camera pulls back to reveal the young couple sitting in a Prius. Titles on the screen sequentially fade in saying "Prius. Good"—followed by the Toyota "Hybrid Synergy Drive" emblem and the "Toyota Moving Forward" ad slogan.[2]

In a similar television commercial spot, entitled "Heist," an alarm sounds outside a bank as three men dressed in black and wearing face masks dash out of the building carrying sacks of loot. A fourth conspirator, who is in charge of securing a getaway car, is nowhere to be seen, and the men now search frantically for him. Finally, he screeches up in a black Prius and shouts to the men, "C'mon!" Two of the robbers jump in the car, but the third one stands frozen with a look of horror on his face. "What are you doing, man?" they shout. "Get in!" But he simply shakes his head and says, "I can't!" The men incredulous demand, "Why *not*?!" He replies, "It's a *Prius*." The stunned realization on the faces of the men inside the car register that he has a point, and they wait with solemn resignation as the police arrive. "Prius. Good.—Hybrid Synergy Drive—Toyota Moving Forward."[3]

Both of these spots are DIY Internet-distributed videos created by producer/director Randy Kent. "Lover's Lane" (2010) and "Heist" (2011) began as speculative "demos" on Kent's commercials reel on his production company's web site, but later, via postings on YouTube, Vimeo, and web sites such as FunnyOrDie.com, circulated through social media. I received links to the TV spots sent to me via Facebook, Flipboard, and e-mail, passed on from friends who know I study the intersections of media, environment, and moral engagement in popular culture. DIY spoof videos that adopt the Prius hybrid as their subject have become their own subgenre on the Internet. Another DIY commercial spot, produced by actor Josh Holt, called "Prius Lovers" (2012), features yuppies Beatrice and Brody, dressed in golfing clothes and driving their pristine white Prius around their upscale suburban neighborhood. As they drive, they trade remarks such as, "Do you know what I like about our Prius? It makes us *better* than everyone else." Later on, they piously intone, "People trust us because we drive a Prius. It shows we have high morals."[4]

This elitist, self-congratulatory note of moral superiority in Prius ownership echoes throughout DJ Dave's (Dave Wittman's) DIY video, "It's Getting Real in the Whole Foods Parking Lot," which depicts the rapper in his Prius, searching for a parking spot and losing patience with other Prius-driving, quinoa-purchasing shoppers at Whole Foods Market. In 2011, the video went viral upon its posting to YouTube, garnering more than five hundred thousand hits in just a few days as celebrities tweeted and retweeted it to their followers. The video was produced by the California-based creative collective Fog and Smog, who sought to satirize the kind of aggressive consumerism and moral superiority espoused by the natural foods, yoga-practicing, Prius-purchasing, LOHAS (Lifestyles of Health and Sustainability)-practicing Americans.[5]

Figurative painter and hip hop lyricist Delia Brown joined this ongoing media conversation about consumerism, LOHAS, and Prius ownership with her Prius-centered rap video, "Revenge of the Black Prius," a creative response to DJ Dave's video. Brown raps from inside a faux-"gangsta" black Prius in the Whole Foods parking lot, as she trash talks other drivers and invokes the trappings of LOHAS bourgeois entitlement—drinking kombucha, sauteeing kale, wearing hemp, doing tai chi, and meeting "a hottie" in the Whole Foods aisles who will share her "allergies to dairy and wheat."[6] Brown's Prius is paradoxically "bad ass" in its enforcement of socially conscious consumption, principles of sustainability, spiritually enlightened practices (yoga and tai chi are performed in the parking lot), and healthful eating (quinoa, wheatgrass shots, and goji berries all make video appearances).[7]

A precursor to these cultural works was the 2006 *South Park* episode "Smug Alert," which poked fun at the associations of the Prius with moral superiority, self-righteous environmental piety (ecopiety), and entitlement. The very presence of Prius drivers on the road gives rise in the *South Park* world, not to issued smog alerts but instead to "smug alerts."[8] Media scholars Mukherjee and Banet-Weiser highlight the mediation of consumption itself as a "platform from which to launch progressive political and cultural projects" and point to this "platform's" utility within the logics of neoliberalism to contain the potentialities of political activism.[9] Ecopiety, as mediated and practiced through the "consumopiety" urged by green consumer capitalist marketing, operates within a similar set of late-capitalist consumerist logics.

In Randy Kent's "I can't! It's a Prius" television spots, it is not a sense of smug virtue conveyed by the iconic Prius or a self-congratulatory practice of ecopiety by its drivers that is evoked; it is the premise of *moral purity* on which the narrative centers that makes the spot funny precisely because of its degree of resonance. In "Heist," the moral transgression of having committed a felony by robbing a bank pales before the prospect of sullying the *purity* of the Prius by somehow involving it in this act. To defile the Prius is a step too far. Similarly, in "Lover's Lane," we get the sense of the Prius as a consumer-consecrated sacred space—that premarital sex is okay in and of itself, but to "do it" in a Prius would be akin to desecrating a sacred building or sanctuary. As a consumer item, the Prius has become a multivalent and complex symbol, the recipient of admiration and laudation as much as derision and vitriol. However, in this pronounced statement—"I *can't*! It's a *Prius*"—there is a *resonance* that, again in Northrop Frye's sense, goes beyond a particular context and speaks to a broader cultural conversation about environmental ("green") virtue and its practice in an age of environmental crisis.[10]

In a series of studies on Americans' highly polarized attitudes toward the environment, social psychologist Matthew Feinberg and sociologist Robb Willer (2012) found that those at the liberal and conservative ends of the political spectrum respond very differently to pro-environmental rhetoric depending on how that rhetoric is framed in moral terms. Drawing from previous research that demonstrated the power of moral suasion as a motivating force for decision making, and that found "moral appeals about environmental issues tend to be more successful than non-moral appeals about environmental issues," the team set out to find out what kinds of messaging, appeals, and framing most resonated with moral commitments espoused by Americans identified with divergent political ideologies. Feinberg and Willer found that contemporary environmental discourse and messaging is largely framed in terms of "moral concerns related to *harm* and *care*." They also found that liberals tend to resonate more with the moral framing of "harm and care" and indeed are more likely in general to see environmental issues in moral terms, whereas conservatives do not.[11] However, the team also found that conservatives tend to respond to moral messages communicated in terms of "*purity* and *sanctity*," and so the researchers set out to investigate whether reframing environmental issues in terms of purity and sanctity would actually resonate positively with the values held by American conservatives and potentially sway their environmental attitudes.[12] Feinberg and Willer ultimately conclude that "reframing proenvironmental rhetoric in terms of purity, a moral value resonating primarily among conservatives, largely eliminated the difference between liberals' and conservatives' environmental attitudes," and their data suggest that "reframing environmental discourse in different moral terms can reduce the gap between liberals and conservatives in environmental concern."[13] If this is so, how might messages of the Prius's "purity" and "sanctity" engage diverse liberal and conservative publics in different ways?

The intention of the "I *can't*! It's a *Prius*" punchline in Kent's spots may be humorous and hyperbolic, but the joke works precisely because, as a consumer product, the Prius has become a signifier of virtue in the vernacular of popular culture. As an object, the Prius is what sociologist Pierre Bourdieu terms "a social marker" of aesthetic and cultural capital, and, I would add, a visible marker of moral commitment and ecopiety as practiced through green capitalist consumerism.[14] The focus in Kent's satirical ads on purity, sanctity, and potential defilement of the sanctified object ties in to a larger public conversation about personal responsibility, carbon footprints, and the moral offsetting of environmentally transgressive behavior. Built upon an earlier series of print-based speculative Prius advertisements that had gone viral on the Internet two years before, Kent's video ads are distinct (and funny) because, instead of the consumer purchase of a

Prius functioning to *offset* morally suspect behavior, it instead holds not only the Prius consumer but also Prius *passengers* to a higher standard of moral behavior.

"Well, at Least He Drives a Prius"

Former Ogilvy and Mather senior copywriter David Krulik's speculative Prius ad campaign created a sensation in 2008 when it went "viral," drawing widespread attention and commentary from advertising pundits and environmentalists alike.[15] In the first of the series, a slick print ad shows a Mafioso-style hit man in a dark suit dragging a body out of the back of his car in the dead of night, about to dump it in a nearby river. The tagline reads, "Well, at least he drives a Prius." A similar ad shows a man cruising the New York City waterfront in his hybrid, picking up a prostitute. Again, the tagline reads, "Well, at least he drives a Prius." And the final ad shows a suburban housewife and her landscaper *in flagrante* in the shrubbery as her husband obliviously drinks his morning coffee on the porch. To the side of the driveway, where her shiny little hybrid is parked, the ad copy proclaims, "Well, at least she drives a Prius." Krulik's campaign humorously but insightfully illustrates an economy of virtue that "trades" in the ecopiety of driving a Prius. As we saw previously articulated in behavior psychologists Monin and Miller's research on "moral self-licensing," reducing one's carbon imprint becomes moral "capital" that can then be "banked" to offset life's other sins. But driving a Prius is an act of ecopiety that is fundamentally tied to a capitalist consumopiety that promises redemptive solutions from buying the right stuff and making consciously green consumer purchases.

Here, it is fruitful to consider the territory of economist Gernot Wagner and his caution against "single-action bias"—that is, acting in one moral area and thinking that we have "done our bit" for the environment. In engineering these offsets, we falsely imagine, as Wagner says, "that my token gesture somehow makes a difference that compensates for other environmental sins throughout the day. That bias can result in phenomenal mass delusion."[16] Krulik's print advertisements take up the problematic phenomenon of moral self-licensing ("when being good frees us to be bad") and place it into pronounced comic relief, but they do so in a way that is especially telling.[17] Krulik plays with the notion of

Prius-owning piety, amplifying it to an absurd degree to get a laugh, but these ads only "work" to the extent that they tap into a cultural *resonance* about the virtues of consumer capitalism and commodity activism.[18] Notice that the ads do not show a man dumping a body in a river and then proclaim, "At least he *bikes* to work." Likewise, we do not excuse the unfaithful suburban housewife if the ad copy says, "Hey, but at least she takes *public transportation*." These actions—biking and taking public transportation—have greater efficacy in reducing one's carbon imprint than buying a Prius, but they involve no satisfying and economically generative purchase of a consumer product.

Gernot Wagner makes clear that adopting modes of greater energy efficiency is a good thing and that driving a Prius is preferable to driving a gas-guzzling Hummer, but he contends that "we need to be clear about what we get in return."[19] Even if the Prius is a more efficient car, Wagner unsettles the "feel-good" element of the Prius purchase, pointing to the counterproductive "rebound effect" in which Prius owners get twice as many miles per gallon as conventional sedan drivers but, as seen in Monin and Miller's "moral offsetting," this savings can actually prompt hybrid drivers to drive *more*.[20] Economic behavioral psychology is an important part of energy efficiency and complicates the way we think about ecopiety, and especially ecopiety derived from green consumerism.

The suggestion communicated by much of green marketing is that a virtuous product does not really count on a balance sheet of resource consumption. That is, a green product is a free pass—in some ways, like no product at all, or even a counterweight to other nonvirtuous products. Here is where both Wagner and consumer journalist Kendra Pierre-Louis engage in tactics similar to those used in family addiction interventions, like issuing "reality checks" and pointing out how consumers' behavior affects the lives of others. "Driving ten thousand miles in even the most fuel-efficient Prius," explains Wagner, "still produces four tons of carbon dioxide. That matches the annual [total] carbon emissions for the average human on the planet."[21] Wagner breaks off a piece of cold, hard truth that hybrid manufacturers would rather not have us hear—the Prius may be a better car, but it's *still a car* and is tied to a chain of fossil-fuel-generating production, delivery, and maintenance factors—raw materials extraction and transportation, manufacture and assembly, road building and maintenance. A 2018 report from Munich-based automotive consultancy Berylls Strategy Advisors goes one step further, challenging even the "may be a better car" premise, pointing to the high levels of carbon pollution generated to manufacture electric car lithium-ion batteries that are produced in countries such as China, Thailand, and Poland that rely on dirty coal-based power grids. The consultancy findings point to fuel-efficient conventional cars often producing lower emissions in total than vehicles marketed as "green."[22]

In her critique of hybrid cars and the illusions of green consumerism, Pierre-Louis similarly stresses that for every ton of steel that goes into manufacturing a car, two tons of greenhouse gases are released into the atmosphere. Like Wagner, she tackles mechanisms of consumer "denial" associated with the "green car myth"—the fantasy that if we all buy hybrids or electric plug-in vehicles, we will save the planet. Instead, she interjects some sobering "straight talk" to her readers, instructing, "[E]ven if we could get cars to run on hydrogen, or water, or pixie dust, the individual passenger car as the principle transport for the ten billion people who will call this planet home by 2100 can never be sustainable. And the emissions, or tail pipe, issue is only the tip of the iceberg of what makes the automobile unsustainable."[23]

Advertisements for the iconic Prius, whether commissioned or speculative, center on virtuous consumption and strategically tap into resonant and recognizable themes for American consumers.[24] Moralism in consumption, argue marketing professors Luedicke, Thompson, and Giesler, is key to the kind of stories consumers tell themselves as part of "identity work," enacted through consumer purchasing. Those who make moral distinctions in brand preference (e.g., choosing a local coffee shop over a mass corporate chain) cast themselves as the "moral protagonist" of their own story, and the way they consume and what they consume becomes a way of identifying and situating themselves within their social worlds. What is more, the research team contends that "consumers' moralistic identity work is structured by a variation of the classical morality play myth, in which a moral protagonist is called upon to defend sacrosanct virtues and ideals from the transgressive actions of an immoral adversary. By adapting this myth of the moral protagonist to their own life circumstances, consumers can "ascribe moral redemptive meanings to their consumer identities through implicit (and sometimes explicit) confrontations with other consumer groups that they ideologically construe as deviating from an inviolate normative order."[25] Consumers may, in turn, look down on or criticize other consumption practices for committing such transgressions as "wastefulness, personal irresponsibility, and selfish disregard for the collective good." In doing so, these consumers may also nostalgically lament the disappearance of traditional bedrock values in the marketplace.[26] But this "moral protagonism," its role in consumers' "identity work," and its oppositional positioning to an immoral adversary are *mediated*, multidirectional, and not simply confined to the purview of the environmentally minded and the socially progressive politically conscious.

Rollin' Coal and Producing Pollution Porn

One of many insights we glean from the late, great cultural theorist Stuart Hall is that the production and consumption of meaning is a complex, messy, nonlinear process. What producers intentionally *encode* into cultural communications is not necessarily what is *decoded* by consumers of those same communications. Unless "wildly aberrant," cautions Hall, messages are not open to *any* possible meaning, as there are certainly dominant or preferred meanings, but the "reader" of culture can decode those messages in a variety of ways, including engaging in negotiated and oppositional readings.[27] Modifying Hall's model of "oppositional readings," media scholar John Fiske argues instead for a model of multiple "resistive readings" in which readers read a cultural text in a variety of counterhegemonic ways but may even do so unconsciously.[28] In the case of the mediated responses of pollution porn producers to the icon of the Prius, Hall's original oppositional model applies, as these cultural productions are extraordinarily aggressive, conscious counterhegemonic readings. That is, they actively oppose and subvert an assumed preferred reading, in which the Prius is heavily encoded with messages of piety and virtue. Where Fiske's model is helpful in analyzing pollution porn, however, is in his discussion of the sheer pleasure and enjoyment often experienced by those who engage in "resistive readings." A kind of bad-boy transgressive ecstatic enjoyment is unmistakable in the performative practice of purposely expelling black exhaust from diesel truck engines (what is known as "rolling coal"), and the associated pollution porn media making that records this practice, circulates it, and celebrates it to the delight of fan viewers. Not only is this media practice infused with erotic pleasure and dimensions of recreational sport entertainment, but it engenders these alluring qualities while also engaging in sociopolitical messaging and class-based acts of cultural criticism and resistance.

Toyota Motor Corporation–commissioned television commercials (as opposed to DIY video commercial spots) tend to situate the Prius in an idealized, often storybook-like landscape, an illustrated world of pastoral perfection and bliss, accompanied by an upbeat folk song. The most memorable of these commercials, and the one that has received the most media attention and commentary, is the 2010 Prius "Harmony" commercial, created by Saatchi and Saatchi, Los Angeles, and directed by Hideaki Hosono. In the commercial, a white Prius initially emerges from a gray lifeless cityscape. As it drives along the road, it activates the surrounding inert and deadened landscape, which suddenly—Oz-like—comes alive in waves of vibrant color and motion. As we look more closely at the landscape, we see that what initially looked like animated waving grass, a flowing stream, and clouds in the sky are actually hundreds of people dressed in different brightly colored unitards all moving in unison. Nature is literally personified in the Prius world, as children dressed as butterflies and flowers sway to and fro in time to a folksy feminine voice singing a cover of the Bellany Brothers' 1970s feel-good hit, "Let Your Love Flow." When the Prius finally stops at the top of a grassy hill, we look back and see that the once gray and lifeless landscape, now vibrant and thriving, has been healed, if not redeemed. The tagline of the video is "Prius. Harmony between man, nature and machine."[29]

In a "How It Was Made" video on YouTube, director Hosono speaks about how he wanted the Prius world "all to look organic" and to show how "people make the world." Another production crew member in the same video talks about the production team's various techniques for "making the landscape come alive as people."[30] And a press release announcing the new "Harmony" campaign quoted Toyota marketing communications manager Kim McCulloch as saying, "Prius has emerged as a symbol of our time, empowering people to see how each individual can play a role in making the world a little bit better."[31] At this time of environmental trouble, the press release conveys, each individual can personally do *something*, and that something is to buy a car—the right car. In Toyota's words, we hear resonances of Murkherjee and Banet-Weiser's argument that in commercialized popular culture, social action itself shape-shifts into a "marketable commodity."[32] Environmental bloggers and web sites such as GreenWashingIndex.com had a field day with Toyota's "Harmony" campaign for what web commentators saw as camouflaging a vehicle that still, after all, burns fossil fuel in a gas engine, and does not "heal the world" in so doing.[33] In Stuart Hall–predicted fashion, others using the tools of new media, especially those expressing themselves through posted online videos, have asserted very *different* and *contrary* readings of the iconic Prius as "a symbol of our time."

Using aggressive media tactics, self-identified "coal rollers" communicate their oppositional reading of Toyota's advertising and marketing of the Prius with its accompanying harmonial environmental virtues. "Coal rollers" are diesel trucks whose owners have specially (and illegally) modified their exhaust systems to remove diesel particulate filters, while simultaneously disabling the "clean burn" programs in their trucks' computers. Truck owners then install after-market single- or double-barrel wide smoke stacks that belch out billowing clouds of black diesel smoke ("coal") when the truck accelerates. The preferred sport of coal rollers is to "blow up" Prius drivers. In this game, a modified truck will pull in front of a Prius driver, and then "roll coal" by accelerating in such a way that the Prius is enveloped in a thick, black cloud of diesel soot. Using a smartphone, a collaborator in the truck videos the Prius driver being engulfed in choking diesel pollution, followed by peals of laughter from the video producers, who then post and circulate these videos across what one pundit terms "red state social media."[34] Creators of photos and videos in this genre will identify and post these videos as "pollution porn" on a number of web sites specifically dedicated to the genre, or coal-rolling digital photos and videos will appear as features on more generalized diesel truck fan sites. The Wisconsin Coal Rollers group's public Facebook page alone has more than twenty-two thousand "followers" with active postings, but many states, such as Michigan, Oklahoma, and Texas, also have their own state-based sites, and national coal-rolling Facebook page "Let the Coal Roll" boasts more than four hundred thousand followers.[35]

"Rollin' coal," and video showcasing of its intentionally polluting practice through online-published self-described "pollution porn," first emerged as a phenomenon on the Internet in 2011, following the launch of the 2010 Prius "Harmony" campaign.[36] Since then, coal rolling and its pollution spectacles have been featured by major news outlets and have gained momentum with numerous YouTube channels that feature so-called pollution porn. A "Pollution Porn" site on Tumblr posts exemplary images of coal rolling, as do various coal-rolling Facebook pages ("Coal Rolling Rednecks," "Coal Roll'n Diesel," "Rolling Coal," and "Coal Roller") and web sites, such as PollutionPorn.com, which posts both photos and videos of polluting coal rollers.[37]

In the Internet-based self-organization of coal rollers and pollution porn amateur producers, we see what computer scientist Seb Paquet dubs "ridiculously easy group-forming" that the tools of digital new media and online sharing make possible at little or no cost. "Now that group-forming has gone from hard to ridiculously easy," further observes new media studies professor Clay Shirky, "we are seeing an explosion of experiments with new groups and new kinds of groups."[38] As Sarah Banet-Weiser has shown in her research on popular online misogyny, removal of barriers to information and media sharing has also facilitated the growth, recruitment, and efficient networking of alt-right groups.[39] Easy and relatively inexpensive access to wielding the tools of digital media production/circulation makes active social commentary "a breeze," albeit one, in this case, choked with diesel fumes. Like the "Harmony" commercial, coal rollers have a *story* to tell about man, nature, and machine, and as coal rollers tell their story, the story in turn tells them. These may not constitute media interventions in the more noble forms that media scholars commonly recognize and discuss—public health campaigns or consumer advocate reporting—but these are nonetheless grassroots media interventions and instances of moral critique.[40]

Coal rolling media producers concertedly work to dislodge dominant narratives of the virtuous hybrid consumer. They replace these with visual narratives that associate the Prius (more specifically) and environmental virtue (more generally) with a self-righteous moral superiority that masks the underlying inequities and hypocrisies inherent in class hierarchies that denigrate those who get their hands (and vehicles) dirty to make a living. Online postings by coal rollers and their fandoms express disgust that "yuppies" and "nature nuffies" (the entitled environmentally concerned) lack the skills to work on their own cars, and so must rely on mechanics to do their work for them. These "nuffies" then subsequently look down on manual laborers who perform these kinds of services. This "talking back" extends to construction workers and haulers who point out that they make possible the fancy houses of "self-righteous progressive f**ktards" who label them as ignorant and "inbred" for coal rolling.[41]

Class indignation and critique directed toward condescending "nuffie" elitism find popular expression in a number of cultural forms, but "Yuppie Folks Can't Survive," a song composed by Outlaw, a self-identified "hik-hop artist/vidja maker" on his *Backwoods Badass* album, appears to have achieved emblematic status as an anthem. Outlaw's song, which in music video form has received more than 2.4 million views on its official YouTube posting alone, receives enthusiastic affirmations in accompanying online video comments sections. As I visited postings and blogs linking to popular "Rolling Coal" video series—compilations such as "Best Burnouts/Rollin' Coal Compilations" (which go up to number 30), "Rolling Coal Trucks with Big Pipes," and "Best Badass Diesel Trucks" —Outlaw's "Yuppie Folks" music video repeatedly popped up in my Internet searches. In fact, once I began my research on coal rolling online videos, YouTube's algorithmic calculations helpfully placed Outlaw's "Yuppie Folks" video in my video feed, and his song repeatedly popped up in my suggested videos when I watched coal rolling DIY video. The song lyrics begin with Outlaw pulling into town only to find yuppies stranded with a dead battery, having no idea how to fix their own car and needing to find a mechanic who does. Outlaw, who in the video version is pictured wearing a Pabst Blue Ribbon–like logo t-shirt that proudly proclaims "Pure White Trash," derides these incompetents for not knowing basic skills like car repair, or how to hunt, fish, or plow a field. Yuppies would be incapable, he supposes, of even getting their car out of a ditch. He sings about the self-reliance and practical know-how of those who are "country born and country bred," who are in turn mocked by more urbanized yuppies who are unable to do for themselves.

> See them bitches relyin' on the mother truckin' government
> And I'm out in the woods working for the freakin' hell of it
> Huntin' all day just to feed my daggum family

> And y'all just at the store buying chips and unsweet tea
> World goes to shit and you're off the government's tit
> And you'll be looking at us hicks like we're God, sumbitch
> But go ahead, make fun of us now. I'll turn my head and go milk a cow
> One day it'll all come raining down, and you sumbitches gonna drown
> Yuppie folks can't survive
> Spit some Copenhagen [chewing tobacco] in your eye.[42]

Outlaw's song is actually a remake and riff on the 1982 Hank Williams Jr. hit song, "A Country Boy Can Survive." Williams reminds his audience that, as more Americans become dependent on consumer conveniences and apocalyptic times approach, existing class hierarchies will be inverted and those with practical labor skills, such as those accrued through "backwoods living" and rural farm life, will come out on top. Prius drivers may feel environmentally pious as good green denizens, choosing eco-friendly vehicles and purchasing organic milk at Whole Foods, but the class resentment expressed in coal rollers' digital online culture reminds their audiences that it is working-class people, often using dirty diesel engines, who do the physical labor that makes possible the self-congratulatory consumer lifestyles of more entitled Americans.

"To tell a story is to exercise power," contends literary theorist Ross Chambers, and whereas stories more often tell us about a society's dominant ideology, storytelling can be and is used as an oppositional practice.[43] What then do visually told stories of coal rolling and coal rollers tell us about forms of power and their enactment vis-à-vis already mainstreamed notions of environmental virtue and its eco-pious practice? Coal rollers post photos and videos of their trucks sporting popular anti-Prius decal stickers that say "Prius Repellent" on them and have arrows pointing to the truck's soot-blackened smoke stacks. The image of diesel trucks' black smoke as "Prius repellent" forms the basis of a whole host of Internet memes. Other memes show coal rollers gleefully playing a cat and mouse game of "hunting" Priuses, and one popular meme invokes an Elmer Fudd–style declaration: "Be vewy vewy quiet. We're huntin pwius."[44] Coal rollers' displayed Prius decals, positioned in between the trucks' gun-referentially named "double barrel" smoke stacks, which also frequently provide the location of a truck's mounted gun rack, add to the suggested predation and prey dynamic.[45] "Hunting" and "blowing up" videos show coal rollers targeting cyclists and pedestrians, "rollin' coal" directly into the faces of those not burning fossil fuel to transport themselves. Special vitriol, however, is reserved for the Prius as the preferred vehicle of "nature nuffies."[46] This sentiment is visually represented in the DIY video entitled, "F U Self Righteous Prius Lovers," in which the producer has edited together a pronounced response to Prius drivers, which consists of an extended montage of hard-working, dirty, four-wheel-drive pick-ups and monster trucks, plowing through mud, powerfully riding rough up off-road slopes, hauling equipment, and rolling coal, punctuated by the producer's descriptive commentary, "I am tired of many Prius owners thinking they better then [sic] the rest of us."[47] Viewers of the "F U" video cheer on this self-styled work of media criticism, and one automobile mechanic passionately remarks in the video comments section that each time he sees a Prius in his shop, he wants to "set it on fire and crucify the owner."[48] Perhaps when Toyota's Kim McCulloch called the Prius "the symbol of our time" and the ad team at Saatchi and Saatchi referred to the Prius as an "icon," this is not what they intended.

As Stuart Hall observes, in the extraordinarily complicated processes of communication, broadcasting structures "yield encoded messages in the form of meaningful discourse," but funny things happen when a message, via its decodings, enters into the structure of social practices. "We are now fully aware that this re-entry into the practices of audience reception and 'use' cannot be understood in simple behavioural terms," instructs Hall. Messages themselves, as they are crafted, are "framed by structures of understanding, as well as being produced by social and economic relations, which shape their 'realization' at the reception end of the chain [of communication] and which permit the meanings signified in the discourse to be transposed into practice or consciousness (to acquire social use value or political effectivity)."[49]

That is to say, coal rollers may have removed their smoke stack exhaust filters, but their socioeconomic, cultural, and political filters are still very much active and in place. These filters interact with the complex socioeconomic and political packaging of the Prius as formulated by Toyota's product marketers and advertising message makers. Coal rolling is not simply an entertaining diversion to its practitioners, although certainly it *is* that, judging from coal-rolling audience online comments and the peals of laughter recorded by giddy video producers. More than that, though, pollution-porn media producers and consumers engage in a political discourse of resistance through their pollution pageantry and its online sharing. One supplier of truck smoke stacks in Wisconsin explains the phenomenon, saying, "I run into a lot of people that really don't like [then President] Obama at all. If he's into the environment, if he's into this or that, we're not. I hear a lot of that. To get a single stack on my truck—that's my way of giving them the finger. You want clean air and a tiny carbon footprint? Well, screw you."[50]

A recurrent theme in coal-rolling culture is the high moral value placed upon "liberty," particularly liberty from the constraints of government regulations such as those that constrain carbon emissions and pollution produced by vehicles. In contrast to the "harm/care" foundation most often favored by American liberals as the basis for moral decision making, social psychologist Jonathan Haidt's research on moral foundations theory finds that American conservatives "sacralize the word liberty," lionizing libertarian ("hands off") values that morally "oppose domination and control by a secular government," particularly when it comes to economic, manufacturing, and environmental regulatory policies, although far less so when it comes to state moral interference in citizens' biological bodies.[51]

The "screw you" sentiment of the coal rollers presents a sharp contrast to the guilt-laden harm/care-based moral angst expressed by high-efficiency "clean diesel" Volkswagen vehicle owners, many of whom took to Twitter and confessional blogs after it became known in 2015 that the company had fixed its diesel cars' computers with a "defeat device" to cheat on emissions tests. Bloggers wrung their hands over their inadvertent culpability in pollution standards violations, and an assortment of VW owners who showed up in news coverage of the scandal confessed their guilt to reporters.[52] In her blog, *Sins of Emission*, Pastor Erica Schemper on ThinkChristian.com instructs her readers that the Volkswagen scandal "is a reminder of how our human sinfulness, in ways both individually and corporately, holds us back from shalom [peace]. It's a story of greed, pride, self-deception and outright lies, mostly by [VW] engineers and corporate officials. And even if I didn't know about it, on some level, no matter how clean my fossil-fueled vehicle seemed to be, I remained complicit in a world economy that is damaging creation."[53] No moral self-licensing or offsetting appears in the confession of guilt in this pastor's blog. Others confessed that on some level "they *knew*," admitting like a cuckolded spouse that there was something fishy about a "clean diesel" all along and that, in retrospect, they had deceived themselves about the virtuousness of their car purchase.[54]

The visible pleasure and delight of the diesel-belching coal roller is breathtaking in comparison to self-reproaching, angst-laden diesel VW–owner missives. The use of the self-identified term "pollution *porn*" within coal-roller culture is especially interesting, as it speaks to a pleasurable and gratifying play with moral and social taboos. It similarly connotes both obscenity (in this case, largely "environmental obscenity") and arousal as it suggests practitioner/audience sexual excitement not at the spectacle of human sex but at the spectacle of a kind of mechanical predatory intercourse that ends in a very visible and spectacular ejaculation of power through the machine as an extension of its driver. Reviewing the aesthetic and narrative content of coal rolling and pollution porn online videos, posted photographic stills, and coal-rolling Internet memes reveals dimensions of power, class, gender, sexuality, race, and taste in these media creations that are shared characteristics of both "porn proper" (if I may) and the subgenre of "pollution porn."[55]

The community's self-applied use of the term "porn" as a label to categorize coal-rolling media productions is apropos on a number of levels. The history of pornography is diverse, shifting, and situated. What constitutes "the erotic body," for instance, writes sociologist of culture Andrew Ross, is "not transhistorical."[56] Again, it is the "how" of coal rollers' cultural assemblages and interpretations of "obscenity" and "pornography," their chosen idioms and aesthetics, the political context of their emergence, and their ways of morally engaging themes of environmental virtue that most interest me about their media practices and what they might suggest about the making and remaking of social worlds at this point in time. In analyzing the history of pornography, film scholar Constance Penley highlights its kinship with "white trash" culture and its employment of both class resistance and class critique. Penley observes that popular forms of pornography are "based in a humor that features attacks on religion (and later, the professional classes), middle-class ideas about sexuality, trickster women with a hearty appetite for sex, and foolish men with their penises in a twist—when those penises work at all."[57] There is a lot of laughter and rebel hijinks performed in pollution porn videos that combine humor and play with a kind of "outlaw" culture, thumbing its nose both at the government and at a socially advantaged class of effete liberal hybrid-driving "nature nuffies." Coal rolling videos and images associate diesel trucks (as do television commercials for trucks more generally) with themes of a "working man's" demonstrated masculine virility and nature-conquering heterosexual prowess.[58] Videos and photos on PollutionPorn.com feature phallic shots of smoke stacks ejaculating black soot—a diesel-pollution "carbon cum shot"—accompanied by commentary that analogizes the power of (illegally) removing pipe filters from the trucks' exhaust systems to the power of men who perform bareback sex: "Just like in bed, a raw pipe spews the best exhaust."[59]

On the Pollution Porn Tumblr site, a user called "TheCruelCowboy" captions his own "carbon cum shot" (or "money shot") with the exclamation, "Fuck yeah, sexy truck pollution!"[60] A posting called "Smoke Stack Girls 2" shows a photo of scantily clad blonde women in Daisy Duke cutoff shorts and cowboy boots, standing in the bed of a diesel pick-up truck, where they straddle the smoke stacks. The caption details, "Chick in blue wants her pussy covered by some coal from the double barrel stacks."[61] Another posting from a different coal roller shows yet another blonde woman (also in cutoff shorts and cowboy boots) down on one knee and sticking her tongue out provocatively toward the diesel truck tail pipe. Here the caption reads, "Chick licking the tip."[62] True to the use of "porn" in the "pollution porn" label, the visual rhetoric of these images, accompanied by the verbal rhetoric of the captions, celebrates the male gaze, tropes of male control of production, and the male fantasy of the abject woman so characteristic of the aesthetics of heterosexual male–produced pornography.

The fact that diesel truck pollution has, in effect, become its own subgenre of hybridized sexual/commodity fetishism porn has a number of implications, but one of these is that notions of ecopiety and environmental virtue have become sufficiently mainstreamed within culture as to make the taboo of its "bad-boy" transgression erotic. That is, willfully and flagrantly polluting the environment constitutes "obscenity." Studying the contrasts between edgy coal-roller porn images and the images of Toyota's television commercials, where a pristine but bland, friendly, neutered Prius harmonizes with nature, makes one wonder whether in pollution porn media creations, it is Prius's associated purity and piety that feed the "dirty diesel's" naughty and "manly" sexual allure. Whereas Prius automobile reviews often complain of the car's "lack of power" but praise the Prius's decorous quiet engine and cleanliness, the diesel truck, as presented in pollution porn, is about power, noise, and stink. After all, a working-class man who labors, like his working truck, stinks at the end of the day. He does not have the cleanliness of a desk-bound professional. One posting to the Pollution Porn Tumblr site contributes a photo of trucks parked tandem in a driveway and in full "rollin'" mode as they crank out "the coal smoke" in unison. The caption offers the directive, "Your diesel should be able to scare children, small animals, and wake the neighbors."[63] Pollution porn thus embraces a kind of power and carnality not modeled by the well-behaved, good "green denizen" Prius drivers as portrayed in Toyota's commercial spots.

At the turn of the twentieth century, one of Henry Ford's great marketing coups was to cast electric cars, which early on competed for market share with his combustion engine–driven vehicles, as being "ladies' cars," and derisively so. Painting electric cars as effete, mundanely "simple," and fussily clean, Ford's marketing strategically stoked and fed men's anxieties about masculinity and a suggested homosexuality. As a consequence, "The speedier, more dangerous, complicated, and more 'masculine' gasoline-powered car took over the market and never looked back."[64]

The aesthetic differences between decorous silence and raucous noise, polite understated transportation and machine power, come into play in more areas of opposition, of course, than simply electric and/or hybrids cars and diesel coal rollers. We might point to similar conflicts engendered over mechanical speed, noise levels, cleanliness, and smell, between snowmobilers and cross-country skiers, powerboat drivers and sailboat owners, all-terrain-vehicle drivers and hikers—disparate recreational modes all influenced by class, taste, and the degree of physical labor one's job requires during the work week.[65] Cultural attitudes toward the consumption of porn, as opposed to the more tasteful "erotica," render it a class marker. As cultural theorist Laura Kipnis points out in her study of *Hustler* magazine, aesthetics are "historically specific" and always "completely class-bound." Class distinctions, for instance, classify "art" as being understated and implied versus pornography, which is deemed too "blatant," unartfully obvious, and thus, as Kipnis puts it, "too potent for art."[66] Pornography is arguably one of the most, if not *the* most, class-antagonistic expressive forms of mediated American popular culture.

Pornography scholars have also drawn attention to and analyzed the prevalent racist dimensions of "porn," and in this regard, coal rolling as pollution porn follows suit.[67] The "actors" in coal rolling videos and photographs all appear to be light-skinned, and yet a favorite stunt to pull, in the video, is for coal rollers and their friends to stick their faces in front of exhaust pipes and smoke stacks in order to put on an instant diesel "blackface." Routinely there is the "before shot" that shows a white boy or man, who is subsequently turned "black" after his immersion in "coal," emerging with a darkened mask of a face, eyes blinking, and his face sporting a wide grin of idiocy, which then sets off peals of laughter from his recording collaborators.

The history and politics of blackface performance are complex and tangled up with all sorts of cultural dimensions of race, class, gender, and sexuality, but social historian Eric Lott, in studying the relationship between blackface performance and the American working class, points not only to the racist nature of the practice but to the subversive humor involved in putting on the mask of the blackface "trickster." He details the thrill and transgressive pleasure white working-class men experienced in these "small but significant crimes against settled ideas of racial demarcation."[68] Once again, to tell a story is to exert power, and these diesel blackface videos exert power in telling a particular story of power. In the wake of the Black Lives Matter movement, coal rollers have coopted the protest slogan on countless t-shirts, memes, bumper stickers, and coffee mugs to read, instead, "Black Smoke Matters," again casting the coal rolling community as "oppressed" by government regulation, interference, and overreach, as they struggle for "liberty." A series of videos on YouTube features compilations of rolling coal on "Black Lives Matter, Trump Haters, and Tree Huggers." As "Show Me Diesels," a pollution porn producer whose compilation has over three million views, engulfs peaceful protesters in smoke, to laughter by the videog-

Figure 3.1. Photo of man wearing a "Black Smoke Matters" t-shirt. A whole host of tie-in consumer products, including t-shirts, bumper stickers, and travel mugs, celebrate "pollution porn" and market the coal rolling fan phenomenon. Credit: author's photo.

rapher, the driver sardonically remarks to the protesters, "Tastes like America, right?"[69]

Beyond coal burners' performances for the camera of a derogatory African American–referential minstrelsy and targeting of Black Lives Matter protesters, race continues as a theme throughout coal rolling videos and coal rolling Internet commentary in other ways. Coal rollers repeatedly speak about hunting and preying upon despised "rice burners"—the derogatory and racialized vernacular term used within the subculture for Asian-manufactured vehicles, most specifically the Prius. Rebellious counterimages of American-made trucks—Fords, Dodges, and "Chevies"—with their American-built Cummins diesel engines abound. "Liberated" from their oppressive environmental regulatory constraints, these trucks roam the highways in search of foreign vehicles to hunt and "blow up," and in so doing both endorse and enforce the performance of a kind of patriotic *consumopiety*. Mukherjee and Banet-Weiser observe that "citizenship in the U.S. has historically been understood and fashioned *through* consumption practices."[70] For coal rollers, "buying American," driving "American-made," and performing a kind of outlaw patriotism (*Mad Max* meets Patrick Henry) of working-class, smoke-belching truck owners assert *virility* while also suggesting an intense *vulnerability*. Financially injured by jobs sent overseas (the source of pejoratively cast "rice burners"); failed by a crumbling American public educational system with little opportunity for vocational training; angered by a growing disparity between more affluent liberal professionals (such as Prius owners) and a truck-driving working class struggling to make ends meet; strategically pitted politically and economically against immigrants and people of color in xenophobic, race-baiting rhetoric that blames economic troubles on immigrants and urban minority populations; and roused by right-wing talk-show hosts and conservative-news-outlet claims that environmentalists want to take away their jobs, coal rollers have a very different story to tell in their media productions about moral sensibilities, the environment, and the relationship of "man, nature, and machine."[71] The aforementioned list of contributing factors to coal rollers' anger and frustration does not even count the working-class stressors of dealing with exorbitant health care costs for treating such things as the estimated twenty-one thousand premature deaths caused each year in the United States by exposure to *diesel exhaust* inhalation, to which working-class members are exposed disproportionately through their jobs as construction workers, heavy-equipment operators, landscapers, factory workers, truckers, bus drivers, and now, of course, "recreationally" through the practice of coal rolling.[72]

Contrary to popular misconception, *story* is important in pornography. Pornography, Andrew Ross corrects, "is not at all inattentive to narrative, especially when considered within the context of its markets of consumption."[73] The stories that coal rollers tell through the use of new media are stories dominated by black smoke and angry white men who are literally "blowing their stacks." Visually and verbally, they expressly counter stories characterized by the green, clean, quiet "harmony" of hybrid-transformed and collectively personified utopian landscapes. As they do so, coal rollers' stories tap into much larger conflictual, class, and cultural differences than those simply present between hybrids and diesels. Both official corporate and unofficial DIY-produced media that take the Prius as their subject, whether seriously or in satire, morally engage the concepts of environmental virtue and its practice. Ironically, critiques by environmentalists that Priuses are not "pure enough" (better but still burning fossil fuel) and critiques by coal rollers that challenge the virtue of Prius ownership according to different moral standards, find ironic common cause. Coal-roller-produced media engage the Prius, but do so from moral foundations associated with US "political conservatives" who value principles of "liberty" and "loyalty" to one's country (patriotism) over and against the "harm/care" moral principle that Jonathan Haidt finds US liberals more likely to embrace.

Just to be clear, this is in no way to downplay or excuse the pronounced racist, sexist, homophobic, and environmentally repugnant dimensions of coal-rolling pollution and media, but close analysis of these media productions does provide important insights into the moral framing of coal rolling and perhaps into its appeal as a strategic oppositional media practice and narrative intervention. Indeed, varied and disparate media practices engaging both hybrid ownership and coal rolling suggest historian William Chafe's two conflicting American paradigms—one that prioritizes individual freedoms or liberties above all else, and one that champions the communal good of the whole (in this case the planetary biosphere).[74] In the circulation of mediated narratives and meanings of environmental virtue and its practice through acts of both ecopiety and consumopiety, and through the mediation of the Prius and the coal roller, these two paradigms occupy a relationship that is not simply oppositional but correlative and co-constituting. Where the Prius is quiet, reserved, tempered, and tasteful in its piety, coal rollers make predation a "game" and are explicitly ecstatic, explosive, and raunchy. In the mediated visual rhetoric of both the hybrid and the coal roller, *how* each vehicle means ends up being far more powerful and captivating than *what* each vehicle might mean. In restorying the earth in the Anthropocene, as we will continue to see, dimensions of *how* media mean, particularly aesthetics of play and delight, factor in shifting social energetics.

4

Green Is the New Black

Carbon-Sin Trackers, Reality TV, and Green Modeling

While coal rollers generate, post, and digitally circulate their pollution porn, users of carbon "sin-tracking" software busily enumerate, confess, and record their daily environmental transgressions. Carbon sin–tracking is just one subgenre in a larger body of sin-tracking software applications designers have developed for mobile devices such as smartphones and tablets. The range of sin-tracking applications more generally includes tracking programs for human infractions related to such self-improvement projects as dieting and cessation of drinking, smoking, drug abuse, and chronic lying. Got a vice? As the saying goes, "There's an app for that." But how might tracking software for environmental sins in turn effectively track its users along a narrow path of conformity, offering in their designs limited options for acts of piety and redemption to offset those environmental sins? "Design is always political," design author and Mule Design interactive studio cofounder Mike Monteiro reminds us.[1] Do media representations that model ecopiety in the form of green capitalism (i.e., buy the right stuff and everything will be okay) end up delivering consumers exemplary stories of exactly *who we need to be* in order for the "sacred money and markets" story to continue largely unchallenged? As we will see in the "sightings" examined in this chapter, pro-environmental messaging of green virtue—as delivered through mobile software applications, reality TV programming, and contemporary fashion guides—by design champion the purported power of negligible personal choices and reinscribe directives to consumers to "shop on" as being the most effective way to bring about a sustainable planetary future.

The famous feminist *cri de coeur* of the 1960s and '70s may have been that "the personal is political," but at what point and in what circumstances is the personal actually fairly apolitical?[2] In their encounters with and observations of entertainment industries during their period of exile in the United States in the 1940s, German cultural theorists Theodor Adorno and Max Horkheimer implicated Hollywood cinema especially, but also radio, popular music, and, in later work, television, as effective vehicles for communicating capitalist distortions to mass audiences. For Adorno and Horkheimer, key among these distortions was the mythic notion of the empowered individual consumer and the liberty of "consumer choice" in what is in actuality a narrow range of homogenized and homogenizing commodified cultural products targeted toward what the theorists assumed to be passive, easily manipulated consumers.[3] In an essay he wrote in the 1950s, analyzing the astrological horoscope column of the *Los Angeles Times*, Adorno further identified what he saw as a willing mass deception. This deception rests upon the narrative that the very real social and economic struggles people face are precipitated not by unjust systemic and structural realities but rather by one's personal makeup—the result of a person's individual fate as charted and determined by vague cosmic (or market, as the case may be) forces. The trick is simply to adjust one's behavior to these abstract forces—to modify one's behavior and outlook in keeping with the system, not to challenge or to change it.[4]

As sociologist Robert Witkin observes, for Adorno, "[T]he continuous exhortation to behave and act in accordance with certain private and individual interests generates the impression that those problems that are truly objective, that derive from the economy and from objective circumstances, are all somehow solvable by private individual behavior or by psychological insight into oneself and others. This [false impression] is not confined to the advice tended by astrology columns. It is a characteristic of film, television, and of radio and the culture industry generally to attribute both the cause and solution of objective problems to personal or private capacities."[5] But whereas Adorno warned of a culture industry that manufactures and models precisely the kind of consumer that laissez-faire monopoly capitalism needs in order to thrive, less visible to him at the time were elements of ideological and practical challenge to these models—oppositional and resistive readings and interventions into distorted narratives. This lesser visibility was perhaps especially so for foreign exiles who were in the United States for a limited period of time, and for those who had experienced first-hand, and been targeted by, the sinister power of Nazi-controlled media propaganda.

Media scholarship since Adorno—most notably in the theoretical work of Stuart Hall, the Michel de Certeau–influenced work of John Fiske, seminal reception studies conducted since the 1970s, and, more recently, media ethnographies—has challenged the very idea of "audiences" and passive consumption, conceiving instead of people as active "readers" of culture with different "filters" and situated modes of meaning making.[6] And yet, with the very real and pronounced economic shift in media ownership toward greatly intensified concentration and conglomeration, and vertical as well as horizontal integration, we see the narrowing dynamics of what Pulitzer Prize–winning journalist and media critic Ben Bagdikian once warned of as a "new media monopoly." In today's "digital panopticon," with its increased consumer surveillance capacities and corporate "data mining," further enabled by smartphones and our ever-expanding "digital footprints" or Internet "cyber shadows," elements of Adorno and Horkheimer's caveats in the twentieth century are arguably more salient than ever.[7]

All that was old is new again as Silicon Valley technology insiders, sounding in a series of articles and public talks a bit like Victor Frankenstein, have expressed tortured moral regrets over the various social-media-platform and cell-phone design features that they themselves created. Justin Rosenstein, the creator of Facebook's famous "Like button," has his assistant lock him out of his own cell phone during certain hours of the day and laments his role in creating a kind of dystopian existence in which "everyone is distracted all the time." He and other digital media technology executives worry that addiction to social media, combined with its unbridled power to manipulate, easily and quickly, the very moods and thoughts of its users, could actually dismantle democracy as we know it.[8] Tristan Harris, Google's former in-house design ethicist, very well might be channeling Adorno, Horkheimer, Noam Chomsky, and *Neuromancer* cyberpunk author William Gibson, when he warns, "All of us are jacked into this system. . . . All of our minds can be hijacked. Our choices are not as free as we think they are."[9] Harris's 2017 TED talk, "How a Handful of Tech Companies Control Billions of Minds Every Day," divulges to his audience the unprecedented power wielded in social media "newsfeed control rooms" to "schedule thoughts" into the minds of billions of people. Harris emphasizes that he knows of no more "urgent problem" than this invisible problem of hijacking minds, as it currently undergirds all of our other urgent problems.[10] In his talk, this urgency list includes environmental crisis and what billions of minds around the planet do or do not think about things like climate change—that is, if they are not too distracted by their control-room-manipulated news feeds to focus on such concerns.

The framing of a digital technology–induced "attention crisis," and concomitant abuses in "persuasive technologies," as somehow a *personal* or *individual* problem to solve—simply "sign off," turn it off, or keep it in your pocket—belies the research-based, calculated addictive mechanisms concertedly built into these technological designs.[11] Regulation is an unpopular topic in Silicon Valley, if not a heresy, but policy changes could establish ethical standards vis-à-vis exploiting user distraction and the manipulation of feeds. Sensible regulations, for instance, could require that certain technology features be integrated into designs that protect the consumer from these manipulations, guard privacy, and also mandate ethical accountability from technology companies that violate these basic consumer protection standards. Instead, in response to the identified problems of the attention crisis, as with the environmental crisis, industries repeatedly market messages that point to individual piety and responsibility as simple, adequate solutions.

With these considerations in mind, what makes sense is a *tertium quid.* We need a mixed model that both recognizes the profound social, psychological, political, and economic influence of media, its ubiquity, and its often hidden mechanisms of surveillance and control, *while* paying careful and close attention to active meaning making, resistive readings, and purposeful interventions into media narratives and processes.[12] Conversations and moral debates that are made more visible through interactive participatory digital media and ongoing online exchanges reflect the contested nature of manufactured media goods. Consequently, criticisms and counternarratives that intervene in these ideologically driven models have also in some ways become somewhat easier to identify and consider.

Green Penitents and Environmental Indulgences

The Roman Catholic Church attracted global press coverage in 2011 for its official (though qualified) approval of the iPhone application

Confession: A Roman Catholic App.[13] Marketed as "the perfect aid for every penitent" and providing a "sin summary" at day's end, the application provides a helpful "check-off list" of sins (as stipulated by Catholic Church doctrine) so that users can both clearly identify what constitutes sin in their lives and keep track of their lapses. Shortly after the Roman Catholic confession iPhone application release, a design firm called Rebel Box released *Sin Tracker: Confession* for Android, a more generic sin-tracking app in which, instead of following institutional definitions of sin (such as those stipulated by religious organizations), users can "customize sins" and add their own checklist of password-protected sins and good deeds. The app then provides progress graphs and Lifetime Deeds Breakdown Charts.[14] Unlike the Catholic iPhone app user, the user of the more customizable *Sin Tracker* can choose various levels of privacy, opting if desired to post both sins and good deeds to social media for more public accountability.[15]

Much like the Catholic confession app, environmental "carbon sin–tracking software" enables the penitent to log sins throughout the day and provides a summary at the day's or week's end that conveniently displays transgressions and their impact. Unlike the Catholic confession application, which requires a real-live priest to provide absolution, many carbon sin–tracking applications have absolution functions built in, allowing penitents to purchase "carbon credits" to offset their carbon consumption, a practice likened in some environmental circles to the historical practice of buying "indulgences" from the Catholic Church. "Rejoice! We have a way out," exclaimed environmental journalist George Monbiot in his influential 2006 essay, "Paying for Our Sins," in the *Guardian*.

> Our guilty consciences appeased, we can continue to fill up our SUVs and fly round the world without the least concern about our impact on the planet. How has this magic been arranged? By something called "carbon offsets." You buy yourself a clean conscience by paying someone else to undo the harm you are causing. . . . [Y]ou need have no further worries about what you and BP are doing to the atmosphere. . . . Without requiring any social or political change, and at a tiny cost to the consumer, the problem of climate change is solved. Having handed over a few quid, we can all sleep easy again.[16]

Surely, these carbon sin–trackers are media interventions in the more traditional sense, akin to public health campaigns, yes? They encourage better, more conscious, and responsible planetary denizens through measured, trackable, accountable "eco-pious" practice. But what story is really being sold to digital consumers via these apps, and what is the structural framework of their messaging? In different ways, consumers are encountering neither solely a "sacred life and living earth" story nor a "sacred money and markets" story but a mutually constituting convergence of both.

Pioneers in the area of digital carbon sin–tracking include smartphone green-friendly applications such as *Carbon Calc*, *Carbon Pulse*, and *iCarbon*, all designed to calculate the individual's carbon imprint. There is no need to guess at the degree of environmental sin associated with driving, flying, home energy use, and food consumption when applications now offer a precise, calculable measure. The tech-assisted specificity and quantifiable nature of one's carbon-generating turpitude create a structure for the practice of acts of carbon reduction or contrition as a daily *discipline* of self-monitoring and surveillance.[17] If, as media and religion scholar Rachel Wagner has argued in her theorizing of cell phone use and practice, "We *are* what we install," the downloaded carbon sin–tracking app, with its features of self-surveillance, confession, contrition, and even absolution, becomes a way of curating one's identity, making the practice of ecopiety and the daily digital confession of environmental transgression a part of the stories of who we are.[18]

The application *Carbon Pulse* not only tracks environmental sin but also helps its users make and uphold "pledges" to renounce current carbon production and to live more sustainably (i.e., "go and sin no more"). Another carbon-use-tracking program, *Green You*, helps users devise and stick to a plan for reducing their carbon footprints, offering links to additional virtuous eco-friendly goods and services. An application called *iamgreen* enables the user to compete in degrees of environmental virtue with other users around the world and plants a tree any time a new penitent purchases the app.[19] But a number of these carbon-sin applications redirect the user to opportunities for online "green shopping" to offset their sins via eco-friendly, green-sanctioned (as vetted by the app) sites of virtuous consumption.[20] *Green You*, for example, encourages steps such as purchasing a hybrid car and all-new home appliances (refrigerator, dishwasher, dryer, hot water heater, etc.) with all brand-new, more energy-efficient Energy Star ratings. Practicing *ecopiety* becomes a matter of practicing *consumopiety* through the purchase of new durable goods, reinforcing the notion that we can "buy our way" out of environmental crisis by purchasing the right sorts of things.

ShopGreen is a carbon-offset software application that gets environmentally conscious online shoppers together with retailers, who agree to invest 1 percent of the consumer item's purchase price into carbon offsets for which the app handles all the details. In this way, the consumer can feel free to shop unfettered by offset logistics. "You as an online shopper don't have to do anything at all," says *ShopGreen*'s explanation. "The web app goes on the retailer's end and helps them easily incorporate the [offset] program into their store."[21] *ShopGreen* can help you find local farmer's markets in your area or the closest recycling center to your house, and it gives you rewards such as shopping coupons for eating nonprocessed foods that are organically produced and locally grown. Users score extra points on the app for being vegan. *ShopGreen* also has a FitBit-like fitness tracker feature in which it tracks how much you walk, bicycle, or drive your car, rewarding you with more shopping coupons for choosing lower-carbon-producing options. An app called *Daily Green* gives you a checklist for daily acts of ecopiety: "Ran the dishwasher with a FULL load"; "Made an effort to take a shorter shower"; "Air-dried clothes instead of using the dryer"; and "Used a ceiling fan instead of AC." The app then rewards user piety through "earned" coupons to be used for purchases at tie-in eco-friendly restaurants and entertainment venues, and for eco-friendly grocery items. There is a seamless continuum in the design of the application between practicing acts of ecopiety and practicing acts of consumopiety.

The *Oroeco* app turns carbon reduction and ecopiety into a game in which the app user can progress to a status designated as "Climate Hero." Once you calculate your carbon footprint, you can "then assuage your guilt by purchasing carbon credits."[22] Partnering with various businesses and providing the user sustainability data on products, *Oroeco* states this description on its web site: "Our mission is to harness technology to empower us all to live our values, solving some of the world's biggest problems along the way. From climate change to living wages, your daily choices really do matter, and we think you deserve to be rewarded for creating a better future for everyone."[23] Here, the "mission" language emphasizes a model in which *individual* "choices" are what will create a realization of sustainability and solve the world's problems. Those choices "deserve" rewards. The app provides the categories Move, Live, Eat, Shop, Play, Work, and Invest as ways that our daily individual choices "make a difference"—the kind of difference that will solve world problems. Like the other carbon sin–tracking apps, the checklist of sins and good deeds is all about individual, daily actions—small pious deeds that will make a significant difference to the planet.

Nowhere in the software design of these applications do they advocate taking political action, protesting, organizing, or lobbying for broader-scale collective changes than those along the lines of running the dishwasher when it is full and "trying" to take a shorter shower. Nowhere are there checkboxes and rewards for activities such as teaming up with a local or national environmental group; joining a class action suit to sue oil companies for climate damage to fund critically needed climate-change-transition infrastructure; lobbying your representative to tax carbon at the point of extraction so as to provide real incentive for energy plant conversion and new infrastructure for renewables; or pressuring legislative representatives to fund a high-energy-efficient, low-cost national transportation system, and/or to triple the government's investment in renewable energy research, design, implementation, and subsidy—along the lines of social democracies around the globe. Instead, these software applications, despite their seemingly progressive intent to care for the earth and "solve" climate change, are framed in more laissez-faire ideological moral terms, in which governments do not provide real and powerful solutions. Instead, the small, more or less private, voluntary acts of virtuous individual consumers *do*.

European readers and others outside the United States will readily identify this privatizing ideological bent with the economic philosophies and policies of "neoliberalism." Although US academics routinely employ this term, it is often confusingly misread more broadly by US audiences to mean its ideological opposite. "Neoliberalism" is a political-economic theory that advocates *deregulated* capitalism and *unregulated* free markets as the best models for economic prosperity. In turn, this theory eschews governmental controls on manufacturing, including environmental controls and labor regulations. It promotes reduced social spending on collective systems that provide education, health care, and affordable housing, while in practice it often supports *increased* subsidies and tax breaks for businesses. Proponents of neoliberalism lobby for the privatization of the public sector and stress individual responsibility and culpability for economic circumstances. Media ecologist Mara Einstein's useful shorthand for defining "neoliberalism" is "believing in the magic of the market."[24] Espousing such "magic," mobile software carbon sin–tracking apps promote a kind of "One Thousand Points of Light" voluntary model of carbon reduction approaches.[25] In contrast, what might models of serious *public* infrastructural support add to the ongoing restorying of earth in the Anthropocene?

Green "Supermodels"

During a very windy July day in summer of 2015, Germany announced a new record—78 percent of the country's electricity had, for a period of time that day, been derived from renewable energy sources. At the time, the country's stated target was to hit a supply ratio of 100 percent from "renewables," even if only briefly, by 2020. On January 1, 2018, Germany met that 100 percent target a full two years ahead of schedule.[26] Such encouraging progress toward reliance on renewable energy (especially for a country with a large manufacturing base such as Germany and the world's fourth-largest economy) did *not* occur as a result of a thousand-plus privatized acts of individual volunteerism and carbon-sin confession, although such acts may certainly have been taking place simultaneously. Momentous energy changes materialized in Germany after the German government implemented the substantial, extensive, comprehensive, nationwide renewable energy plan called "Energiewende," or "Energy Turn/Change." This comprehensive energy "turn" included making substantial investments in clean-energy transportation, implementing high-efficiency energy measures and updates to technologies and manufacturing, building clean energy–generating infrastructure, and paying renewable energy producers for their energy at a guaranteed price, among other strategic measures. Germany's approach is often referred to as "the model" for the future.[27]

Another green "supermodel" brought up in discussions of climate change has been the city of Copenhagen. Just what exactly the case of

Copenhagen exemplifies or "models" depends on one's perspective. Gernot Wagner's book *But Will the Planet Notice?* received criticism from some environmentalists who worried that his emphasis on the impact of enacting environmental action on a planetary scale, in conjunction with his warnings about the feel-good environmental delusions perpetuated by "single-action bias," would merely perpetuate "climate fatalism"—people doing *nothing* because they feel individual action has no impact and makes no difference.[28] Wagner, in fact, takes a "both/and" approach in which he encourages both individual action and organizing for collective action, while keeping realistically in the forefront of consciousness that individual actions alone are not sufficient. However, in his next book, *Climate Shock* (2015), Wagner and his coauthor, Harvard economics professor Martin Weitzman, make sure to devote three pages to entertaining the possibility that applying what psychologists term "self-perception theory" to personal environmental behaviors could result in extending individual actions into collective actions. That is, in theory, "Ask people to 'go green' in some small way like bringing a canvas bag to the store, and they may feel a greater moral obligation to do something larger about the environment . . . see yourself as greener, vote greener."[29]

Wagner and Weitzman co-label this the "Copenhagen Theory of Change," referring to the Danish city where over 50 percent of Copenhageners commute to work by bicycle and where an impressive "84% of the city's residents regularly ride bicycles."[30] The theory is that Copenhagen successfully created an ecological culture and an ethos of environmental virtue among its residents in which they saw themselves as "greener," and that more-virtuous image was self-reinforcing, making it one of the greenest cities in Europe. However, Wagner qualifies the theory of "self-perception" and its impact (if any) as not well studied in relation to environmental behaviors, and he and his coauthor frame "self-perception" theory in explicitly conditional terms: "*If* individual, inherently moral, acts of environmental stewardship—like recycling—lead to better policies, sign us up."[31] But it is clear from their treatment of the subject, and from the thrust of the rest of their book, that this is a big and dubious "if."

Reviews of Wagner and Weitzman's book, many of which singled out the Copenhagen theory of change, sparked various streams of public moral debate and a host of online "backtalk." New York biking activist, founder of the bike advocacy group Transportation Alternatives, and transportation economist Charles Komanoff, for instance, railed against the portion of the *New York Times* review that dealt with self-perception theory in the context of Copenhagen's cycling environment.[32] The review had excised Wagner and Weitzman's caveats and conditional statements regarding the "Copenhagen theory of change," claiming instead that after the oil crisis of the 1970s, Danish idealists had made "a personal commitment to ride bicycles rather than drive, out of moral principle, even if that was inconvenient to them" and that merely "the sight of so many others riding bikes motivated the city's inhabitants and appears to have improved the moral atmosphere enough" so that cycling became the main form of transportation for a majority of Copenhagen residents. Ergo, the argument went, it was small individual actions and not laws, regulations, or the implementation of infrastructure that effected massive change on a collective scale in Copenhagen—a city now dubbed a "bicycle paradise."[33]

Komanoff counters the claim that creating a "moral climate" precipitated the changes in Copenhagen, citing and posting instead a 2010 study of Copenhagen cyclists surveyed, in which "environmental/climate concerns" ranked *dead* last at 9 percent in self-identified reasons for why Copenhageners cycle. The chief reason cyclists cited (55 percent) was that cycling was faster, 33 percent said it was more convenient, 32 percent said it was healthful, 29 percent said it was cheaper, and so forth. Building safe bike lane infrastructure, passing laws that give cyclists the right of way, imposing severe consequences on reckless drivers, eliminating parking spaces in the city to make it both expensive and inconvenient to drive, reducing road space for automobiles, and passing laws to reduce permissible driving speeds for motorists, Komanoff writes, all conspired to make Copenhagen a green "supermodel."[34] Copenhageners may feel virtuously green for biking to work, but they are driven *not* by ecopiety but by economic realities (market incentives making it inexpensive to bike and market disincentives making it extremely expensive to drive) and design realities, such as publicly funded infrastructure that makes it both faster and more convenient to bike.

The readers who posted the 131 online comments to the *Times* review of Wagner and Weitzman's book similarly debate and question the Copenhagen model—challenging what Copenhagen's biking majority might or might not suggest about the efficacy of self-perception in addressing climate change. In the very first reader comment, "Rob from Westchester" contests the moral premise of the self-perception model of behavioral change: "It is not poor morals that makes me drive. It is awful pedestrian and bicycle infrastructure, crappy transit, and land use decisions that leave us in Sprawlsville. All my daily destinations are too far and take too long to do anything but drive. My car and the roads I drive on are well-subsidized, convenient, and just plain easier. Until driving becomes expensive, inconvenient, and a pain to the wallet, even a person with the best morals is going to drive." The trend of responses (83 out of 131) echoes "Rob's," with "It's the infrastructure, Stupid"–themed comments: "Sounds like the Danes started their bike culture for economic reasons, not ethical ones"; and "Copenhagen didn't just convince its people to buy a bike, the city made it possible through pedestrian and bike centered urban planning"; and "Amen, amen, amen. It's all about the infrastructure." The frustration expressed in these comments echoes Monterio's design manifesto that "[t]he world is a mess because a certain set of people designed it to be a mess. Now we need a different set of people to design our way out of it."[35] That kind of design requires collective investment, such as the kind implemented successfully in Copenhagen.

Several postings favorably cite Naomi Klein's *This Changes Everything: Capitalism versus the Climate* (2014), which argues for a systemic political, economic, social, cultural, and moral makeover. "Richard from Massachusetts" takes issue with comparing Copenhagen's mild climate to the realities of weather faced in the northeastern United States: "It is going to be hard to convert many commuters to bicycle use in the cities of the northern tier of states in the US and all of Canada. The climate is too cold, wet and snowy much of the year to make bicycle transportation a viable alternative." Just five reader comments side with the view that moral principles and idealism as expressed in small concrete steps are actually effective at addressing climate change, and the rest of the comments relate to other issues or consist of people responding to and refuting five climate-denier postings. "Bill from Toronto" contends that "we, as individuals, need to focus on changing our own way of living to serve as examples and encouragement to others," adding, "As for expecting governments to lead the way? Not until those who demand change have formed a critical mass." "Allen from California" enthuses, "Right on! Stop trying to change other people and simply change yourself," affirming that it is "the individual ideal of morality" that will do a good deal to solve the climate problem. "Bill from Ithaca" also sides with the small, individual changes approach, but it is hard to tell whether he is being sarcastic or not when he exclaims, "No cap and trade or carbon taxes—Great!" As the back-and-forth continues, multiple aspects of this issue are considered, wrestled with, and hashed out, but the debate over voluntary small acts performed out of moral principle, versus the efficacy of laws, regulations, infrastructure, and subsidized economic incentives remains central to the online public discussion.[36]

Regarding the "Thousand Points of Light" model popularized in the late 1980s, in which US conservatives championed the voluntary private charity model as superior to and more effective than robust federal government involvement, political philosopher Patricia Smith observes, "[P]rivate charity is a good thing and should be encouraged," but she argues that it is simply not adequate to the task of solving the problem of those "victimized by an unjust system."[37] Smith sounds not dissimilar to Wagner, who advocates a "both/and" approach when he praises and encourages vegetarians and those who walk to work, urging them to continue these acts of environmental virtue, but does so while emphasizing that addressing the global climate crisis is about fundamentally "changing our systems, creating a new business as usual."[38] It is a brutal fact, says Wagner, but "the collective will and drive of billions voids most if not all feel-good efforts of freelance environmental heroism."[39] Carbon sin–tracking software users may be rewarded as "Climate Heroes," self-offsetting their individual failings with eco-pious good deeds and consumer purchases, but the very stories that promise real solutions and change, as told through these applications and their practice, perpetuate the reassuring deception that small, individual efforts are both efficacious and *sufficient* to addressing the scope of the environmental challenges that face us. One can confess and log one's carbon sins, make amends through pious deeds of green consumerism, and, as *Oroeco* claims, "solve the world's biggest problems along the way." Along the way? Along the way to where? As both Smith and Wagner indicate, individual acts of piety are to be commended, but these are not solutions, and, at worst, an emphasis on *individual* pious acts can effectively distract from and obscure the kind of "heroic" structural changes (a comprehensive and systemic full "turning") and collective action needed for the earth "to notice" in any substantive way.[40]

"Forget Shorter Showers"

When *Grist*, a popular source of environmental news and commentary, published an upbeat article lauding the debut of *Oroeco*'s new carbon sin–tracking application and posted the article to *Grist*'s public Facebook page, the online reception was less than enthusiastic from site followers—who, making use of the Internet's dynamics of interactivity, posed some resistive readings, questioning the app's efficacy and that of others like it.[41] Almost immediately, a reader counterposted an essay called "Forget Shorter Showers" by environmental writer Derrick Jensen, previously published in *Orion* magazine. *Orion* is a kind of environmental version of the *Atlantic* in that the magazine uses longer, thoughtful essays targeted toward well-educated readers. In the "Shorter Showers" piece, Jensen eviscerates the notion of small acts of individual consumption as *political* acts: "Would any sane person think dumpster diving would have stopped Hitler, or that composting would have ended slavery or brought the eight-hour workday, or that chopping wood and carrying water would have gotten people out of Tsarist prisons, or that dancing naked around a fire would have helped put in place the Voting Rights Act of 1957 or the Civil Rights Act of 1964? Then why now, with all the world at stake, do so many people retreat into these entirely personal solutions?"[42] An exchange ensued on *Grist*'s Facebook page following the counterposting of the Jensen article, and although this posting attracted just twenty-one comments, none of the responses defended personal individual actions as being efficacious in solving the global environmental crisis. After the posting of the Derrick Jensen article, another reader colorfully posted to the same conversation, "I eat organic, and grow food, and compost, and all that shit. [More than] 90% of fresh water is used by industry and agriculture. My 5 minute shower doesn't mean shit."[43]

The posting of the Jensen article, however, also leads us via the magical world of hyperlinks into a much larger conversation and moral debate that takes place in the online reader responses to Jensen's "Shorter

Showers" essay. The more than three hundred comments are somewhat mixed, many of them lengthy, constituting thoughtful essays in and of themselves, and some were contributed by other prominent environmental thinkers and writers. Most of the comments acknowledge the salience of Jensen's premise but others, while agreeing with Jensen in principle, counter that individual acts of "voluntary mitigation" (of environmental harm) are an important "symbolic start" and serve to raise awareness. Other readers regard the model of public protest and organized political action as being ineffectual and outdated, arguing fatalistically that small personal voluntary actions are "the only thing we can control," and so it is important to do what we can where we can, even if that only takes place in the realm of personal consumption.

Jonathan Rosenthal, the cofounder of the fair-trade company Equal Exchange, posts a thoughtful response to the online discussion, commending Jensen for making important points about the ways in which "capitalism obscures political action by promoting smaller acts" but qualifies this by saying, "I think you have not included the importance of spiritual practice and individuals entering processes of political change through the doorway of individual action." Cultivating a "spiritual practice" of personal moral acts, argues Rosenthal, "makes it an easy way to begin the journey that sometimes leads to political action."[44] Another response follows in which the commentator addresses Jensen and observes, "You are absolutely right about the double-bind we place ourselves in when we let the powers that be convince us to approach the environmental crisis solely as consumers and not as agents of social change." What ensues is a rich and vibrant online exchange of what social psychologist Jonathan Haidt terms "moral talk," the kind of strategic back-and-forth debate that builds alliances, "recruits bystanders," and "sometimes leads people to change their minds."[45] Although it is a gross understatement to point out that there are certainly dramatically less civil examples of "moral talk" to be found online, in both conversations—smaller (as in *Grist*'s Facebook responses) or larger (as in *Orion*'s online article comments)—new media platforms, and especially social media, provide a kind of space for moral debate, discussion, and exchange that perhaps Adorno could not have imagined during his relatively brief period spent observing media culture in the United States. For all the concerns about corporate control of the Internet, and they are indeed legitimate and well justified, digital online discussions do offer forums for engaging and working out the pressing moral issues of the day.[46]

In a variety of media contexts, Derrick Jensen, Bill McKibben, Naomi Klein, Mike Tidwell, Gernot Wagner, and a growing number of other environmental thinkers and movement leaders have all similarly questioned what Jensen has called the "moral purity" model of activism.[47] "I'm not saying we shouldn't live simply," clarifies Jensen. "I live reasonably simply myself, but I don't pretend that not buying much (or not driving much, or not having kids) is a powerful political act, or that it's deeply revolutionary. It's not. Personal change doesn't equal social change."[48] The powerful appeal of the notion of "individual small personal acts" as an efficacious way to address the environmental crisis should not be underestimated, especially when those solutions promote shopping as consumopiety.

In collaboration with a multidisciplinary team, community health researcher Jan Semenza set out to study public perceptions of climate change in the United States and barriers to behavioral change. This 2008 study found that the urban populations the team studied in two major but geographically distant US cities (Houston, Texas, and Portland, Oregon) reflected significant levels of awareness, concern, and a willingness to help address global climate change: "Awareness about climate change is virtually universal (98% in Portland and 92% in Houston) with the vast majority reporting some level of concern (90% in Portland and 82% in Houston)."[49] The major barrier to the research subjects actually taking action, the team found, was people simply not knowing *what* to do about the problem.[50] In other words, Semenza and colleagues' informants who took no action with regard to climate change and its effects lacked the guidance of an identifiable "track" or a "pathway" to action. On the behavioral survey measure, the residents of Houston and Portland diverged. The research team found that Portland residents as a community were 20 percent more likely to engage in climate-change-mitigation behaviors (reducing energy use at home, burning less gasoline, taking public transportation, buying local food, etc.) because the city had put into place key infrastructural design that supports these behaviors and thus had done more to eliminate "economic, structural, and social barriers" that would impede citywide participation in mitigation behaviors in greater numbers.[51] By contrast, Houstonians who wanted to take action on climate change lacked comparable easily identified infrastructure, support, and community interventions through legislative and regulatory measures that would facilitate such action and so were confused about where to start.[52] That is to say, once again, design matters.

Here is where the design of carbon sin-tracking apps that have offset and green shopping features enable those apps to fit neatly into what we might think of as a cultural lock-and-key mechanism. That is, something that most Americans *do* know how to do, and indeed are already enculturated to do as virtuous practice, is to *shop*. In America, to go out and shop is to do one's civic duty; to be a consumer is to do something for one's country. Two weeks after the horrific events of September 11, 2001, as the United States headed into war, then-president George W. Bush urged families to head on "down to Disney World" and to keep spending money.[53] In a later speech, Bush lamented that some Americans "don't want to go shopping for their families" and feel intimidated to do so.[54] A number of political pundits interpreted the administration's message as one that promised America salvation through consumption.[55] If Americans stopped shopping, then the terrorists had "won." Sociologist Andrew Weigert has written about this kind of mentality as the American "ethic of consumption" that expresses itself best through evangelical popular culture. Ironically, says Weigert, it is not "this-worldly asceticism . . . but this-worldy consumerism" that is a sign of salvation. It is the American Protestant televangelist who most skillfully hammers home this message: "consuming, or helping their televangelist to consume, signals salvation, and thus provides that supreme motive that all believers seek."[56] This "most American of messages, salvation through consumption," Weigert explains, goes beyond the televangelist and his audience and permeates American culture. In so doing, it effectively "takes [sociologist Max] Weber's ['spirit of capitalism'] thesis and stands it on its head."[57] President Bush did not, as some have claimed, use the exact words, "Go out and shop!"; however, his phrasing at a time of extreme crisis about a way to salvation via Disney vacations and family shopping trips evoked, especially for those who share Bush's evangelical faith, a resounding call to an ethic of consumerism. The culturally resonant message of *consumption* as solution, and even as pathway to salvation, thus provides a powerful key that slots easily right into the lock of moral discussions concerning what actions Americans can and should take to address global climate change.[58]

Lifestyles of the Rich and Green

Beyond "carbon apps," reality television shows in the earlier part of the twenty-first century provide a rich archive that reveals the ways in which mediated popular culture has promoted and reinforced stories of individualistic ecopiety as practiced through consumopiety by tapping into the reality genre's voyeuristic appeal, sense of intimacy, representations of authenticity, and outlets for audience interaction.[59] HGTV's (Home and Garden Television Network) *Living with Ed* (2007–2009), for instance, hosted by environmentalist/actor Ed Begley Jr. and his (as portrayed) less-than-eco-pious wife, Rachelle, chronicles Begley's perpetual striving to live a more environmentally sustainable life.[60] Begley has been active in environmental issues since the 1970s but most visibly so when making headlines in the company of other Hollywood stars for pulling up to the Academy Awards' red carpet, driven not in a gas-guzzling limousine but in a Prius or other hybrid.[61]

The first season of Begley's HGTV reality television show chronicles the environmentally conscious actor's daily acts of ecopiety. He dutifully washes the solar panels on his roof and installs a windmill, erects a property fence made of recycled plastic milk bottles, puts high-efficiency drip hoses in his drought-resistant garden, and "rehabs" his (modest by Hollywood standards) kitchen using eco-friendly building products. In the second season, however, Begley branches out, showcasing several enormous Hollywood estates and their owners, whom he celebrates for practicing green virtue through "environmentally conscious living." In an environmental update to the 1980s hit TV series hosted by Robin Leach, *Lifestyles of the Rich and Famous* (1984–1995), Begley, like Leach, provides entrée into Hollywood's inner sanctums but does so specifically to exhibit virtuous green practices in home ownership. Celebrities featured include Larry Hagman (most famous for playing Texas oil tycoon J. R. Ewing on *Dallas*) and his opulent eighteen-thousand-square-foot mansion "Heaven," musician Jackson Browne's "off-the-grid" California ranch, former supermodel Cheryl Tiegs's solar-shingled Balinese-style

Bel Air mansion, and comedian Jay Leno's "environmentally conscious" garage for his famous car collection. The "lifestyles of the rich and green"–themed episodes laud each celebrity as a model of ecopiety for such virtuous acts as adding solar panels to roofs and switching the mansion swimming pool from chlorine to an ozone-technology cleaning system. Leno is lauded for his installation of a hot water heater "cozy" in his antique car garage for more efficient vehicle steam cleanings.[62]

Similar to the content of carbon sin–tracking apps, *Living with Ed*'s tour of celebrity green homes squarely focuses on daily individual lifestyle choices and home "lifestyle purchases" but conspicuously lacks discussions of broad-scale policy changes and/or the collective power of governments to address environmental crises in substantial ways. Instead, the focus of each episode is on individual "star" ecovirtuosi who green their mansions as voluntary, devout environmental practice. Begley shows audiences how Cheryl Tiegs has installed a low-flow shower faucet in her mansion and we, the viewers, receive the voyeuristic pleasure of entering Tiegs's private shower and other intimate domestic spaces as she performs ecopiety for the camera. What impact might it have had if the program instead showed Tiegs and her influential and well-heeled celebrity friends collectively lobbying to make changes in California's agricultural water-use policy, attending local or state planning meetings, and using their prestige as celebrities to meet with California senators and representatives? Instead, the acts of ecopiety featured are small, personal, and depoliticized.[63]

In the show's tour of the now-late Larry Hagman's luxurious estate, Heaven, we learn that Heaven's landscaping sports an ecologically conscious and fire-resistant succulent instead of conventional grass, a field of solar panels to generate electricity, and a recycled-water indoor swimming cave. In the late 1990s, I had the opportunity to visit Mr. Hagman's extraordinary estate to attend a fundraiser (since Hagman's death, the house has been purchased by a Scientology nonprofit organization), and Heaven, on its perch high above California's Central Coast, does indeed exude a sense of both the celestial and the sublime.[64] Mr. Hagman was devoted to solar energy and made charitable donations to support solar panel installation for low-income residents. Even so, no cognizance is taken in the course of our voyeuristic pleasure in touring Heaven with Begley of the wildly profligate consumption of the earth's resources involved in putting an eighteen-thousand-square-foot mansion, encircled by a manmade fresh-water fountain that acts like an evaporating moat, all atop a mountain in arid Ojai, California.

The intentions of Begley and the program's producers to show audiences moral models of celebrity ecopiety for emulation purposes are undoubtedly noble ones, but the focus on the "greening" lives of the rich and famous reinforces the narrative that an eco-pious consumopiety is central to addressing environmental crisis. Ironically, in the show's portrayed economy of virtue, in which expertly green celebrity ecovirtuosi are the prime moral actors, the environmental impact of an eighteen-thousand-square-foot mansion is implicitly "offset" by the installation of expensive and ostensibly ecologically conscious features in that mansion, such as an indoor swimming cave. Jay Leno's hot water heater cozy and various high-tech eco-features in his garage likewise apparently absolve him of the environmental impact of his more than 150 cars and motorcycles and his elite consumer lifestyle as an "auto-enthusiast." Celebrities who switch to ozone technology to clean their mansion swimming pools somehow get cast as perspicacious paragons of eco-wisdom, civic green virtue, and admirable environmental altruism.[65]

The *Living with Ed* interactive web site allowed viewers to "Ask Ed" questions and seek advice about proper "eco-conscious living," particularly the proper products they can consume in order to achieve the lifestyles of the celebrities they have seen on the show. Viewers could also link to Begley's own line of natural cleaning products, his organic clothing line, and his endorsed gadgets through the *Living with Ed* green consumer store, enabling those who watch to "do as celebrities do." These features have now been transferred over to a dedicated site at BegleyLiving.com, where Begley's Best all-natural products continue to be sold. Interestingly, this new and updated site promisingly contains a menu category under the heading "Public Policy," but actually clicking on this heading only leads to instructions on how to perform an environmental audit on one's personal home. With the exception of a posted short promotional spot Begley recorded for *(it) magazine* on how to green one's community, the site is absent of content that has to do with political involvement, legislative proposals, collective organizing, or lobbying for public policy measures.[66] Mr. Begley is lovably earnest and funny on the show, and he is clearly passionately devoted to the environment as his central cause, but it is telling that the show is largely (and safely) apolitical, focusing instead on "green tips" and what new clever environmental gadgets and/or home systems the viewers can purchase to be similarly virtuous.[67]

Ecopiety is inextricably knit up with consumopiety in the stories told each week on *Living with Ed*. In the eco-celebrity episodes especially, the green civic virtue made possible by Hollywood celebrities' extraordinarily deep pockets obscures and implicitly justifies what for most Americans are mind-boggling levels of profligate consumer spending. It would be striking if Mr. Begley had chosen, instead of his mountaintop trip to Heaven, to focus on the many impoverished families living in meager circumstances down below in South Central Los Angeles, where whole communities, by necessity, consume relatively few resources and have a considerably lower impact on the environment than have mansion-purchasing celebrities. These less celestially identified families struggling "on the ground" to survive on meager resources might be legitimately heralded as the real "Climate Heroes," but those stories would not fit American popular culture narratives that idealize celebrities as figures of worship and venerate shopping as good civic practice.[68] Neither would a non-consumer-driven format fulfill what has traditionally been one of the central functions of the television medium—to glorify consumption and deliver audiences to advertisers.[69]

In his now-classic historical essay on the important role played by early network television as a powerful marketing medium for significant socioeconomic transformations in US society during the 1950s, American studies scholar George Lipsitz argues that "in the midst of extraordinary social change, television became the most important discursive medium in American culture . . . charged with special responsibilities for making new economic and social relations credible and legitimate to audiences" by negotiating the anxieties and conflicts stemming from the "clash between consumer culture in the 1950s" and a holdover of distrust toward excessive consumerism that had been forged during the Great Depression and World War II.[70] Employing the tools of narrativity, especially narratives that focused on working-class families' aspirations for upward mobility, television programming became indispensable to molding Americans' *consumer identities* through the marketing of not simply new products but new practices such as "installment buying" and "living beyond one's means."[71] Endemic to this approach was a validation of modern capitalist society and an appeal to "change the world in the present through purchase of the appropriate commodities," a legacy that with a few exceptions persists today.[72] Reality TV's stories of appropriate green "consumer choices" school the audience not only in what to buy but also in the virtues of capitalist consumption and its asserted compatibility with environmental sustainability and planetary well-being.

In many ways, Mr. Begley's reality TV showcasing of expert celebrity ecovirtuosi and his proclivity for modeling elite green behavior to inspire similar behavior in viewers are arguably incisive and strategic. In their research on the evolution of prestige and its links to the mechanisms of cultural transmission, anthropologists Joseph Henrich and Francisco Gil-White contend that "natural selection favored social learners who could evaluate potential models and copy the most successful among them." Thus, "ranked-biased copying," they theorize, emerged as an evolutionary adaptation, inclining us as humans to pay close attention to the behaviors and actions of others, especially those with high levels of prestige.[73]

Henrich, who studies evolutionary approaches to psychology, decision making, and culture, and Gil-White, who specializes in evolutionary sociocultural anthropology, assert that humans are "picky infocopiers" for good reason and copy those who are successful in order to replicate that success. With each generation, "as people copy highly ranked models, the mean behavior of the population will move quickly—relative to genetic evolution and ordinary 'guided variation' (Boyd and Richerson, 1985)—toward the most adaptive behavioral repertoire currently represented."[74] Not only that, but humans are inclined to copy not just behavioral traits but the "ideas, values, and opinions of prestigious individuals," exhibiting a kind of "general copying bias." That is, even though a person of high prestige may be successful in one particular area (say, famously playing a rich Texan oil tycoon on TV like the late Larry Hagman—my example, not theirs), this tends to make the prestigious individual "generally influential," extending his or her perceived prestige in one domain to other domains with which they may have little or no expertise.[75]

If Henrich and Gil-White's observations are correct, having high-prestige persons, such as Hollywood celebrities, model ecopiety may indeed trigger in audiences what the anthropologists argue is a powerful mechanism, not only of cultural transmission but of more rapid infocopying and adaptation. Nonevolutionary psychologists and nonsociobiologists have a different name for this: product-marketing research specialists and advertising experts call these highly sought-out models for infocopying "influentials." Celebrity endorsements and their attachments to branding are powerful tools for selling products precisely because the prestige of the celebrity anoints a product with an aura of desirability and sanctioned efficacy.[76] But "influentials" need not necessarily be affluent or movie stars. "Influentials" are also those who hold prestige and respect in their various neighborhoods, communities, and workplaces. One in ten Americans, say global marketing research experts Ed Keller and Jon Berry, "tells the other nine how to vote, where to eat, and what to buy."[77] Celebrities, more than neighbors, however, possess perceived special divine-like qualities in the minds of fans that can induce intense fan devotion and imitation, so much so that those who are star-struck identify closely with celebrities and even imagine close friendships or relationships with them.[78]

Begley's approach then is arguably shrewd—put a celebrity in a Prius and infocopiers (who have the means to do so, or perhaps even those who do not) will run out and purchase less polluting and more energy-efficient cars. Consider the attention more modest cars received in the fall of 2015 when Pope Francis, a media celebrity in his own right, was shown in countless news stories and Internet postings to have chosen a small, fuel-efficient Fiat over a more lavish "Popemobile" for his US visit. If I see that beautiful model Cheryl Tiegs's tresses get washed under a virtuous low-flow shower head, I might go out and buy one, too. Then again, I might also regard Hollywood celebrities with suspicion and, as the "coal rollers" do, resent them as elitist "nature nuffies," choosing instead oppositional behavior openly hostile to the environment in response to their prescribed virtue.

It is unclear whether evolutionary sociocultural psychological models of infocopying behavior take deep-seated class resentments into account. Regardless, what *is* being modeled in the stories of "Lifestyles of the Rich and Green" featured in Begley's reality television show is still a prescriptive ideal based upon an apolitical, individualistic, privatized model of consumer purchases as civic-minded solutions. But that particular model communicates in a culturally familiar and comfortable register for many American television consumers, especially those DIY-inclined home renovators and decorators most likely to be attracted to the HGTV network.[79] In the mediasphere more generally, however, stories are comparatively scarce of high-prestige individuals actively challenging the political and economic structural impediments to enacting a comprehensive shift in US environmental and economic policies that might make a substantive difference globally to the environmental crisis.

Green Modeling

The "reality" television show *America's Next Top Model* (*ANTM*) first made its debut on UPN (now the CW network) in 2003, and as of 2019, the popular show headed into its twenty-fifth season or cycle, producing more than 319 episodes, in which contestants compete for an elite modeling contract. Successful models progress through successive challenges to return for another week, while "losers" are systematically eliminated and sent home. In the special episode "The Models Go Green" (season 9, episode 2), *America's Next Top Model* tells a story to its viewers of an ecopiety that is fashionable and all about personal consumption choices. "Going green," perhaps not surprisingly, has nothing to do with things like legislation, public policy, or political activism, but it does signal personal good taste, much like the right accessory matched with a little black dress. *ANTM*'s "Go Green" episode reflects a broader "eco-chic" trend and a series of popular style guides penned by self-described eco-fashionistas. Starre Vartan's *Eco-Chick: How to Be Fabulously Green* offers a guide to eco-fashion and chemical-free make-up, and even debates whether latex condoms, which are nonbiodegradable and end up in the stomachs of sea creatures, are okay for an "eco-chick" to use since, in theory, they help to mitigate population growth. Tamsin Blanchard's *Green Is the New Black: How to Change the World with Style*—which *Vogue* magazine advises is "a must-read!"—features movie stars and their eco-friendly styles, while offering sustainability shopping tips and a green fashion shopping guide. In her chapter "Can Celebrities Save the World?" (spoiler alert: they can), Blanchard advises her readers that selecting haute couture is a surprisingly green option. She explains, "[T]he higher up you go, the more sustainable the garment.

A Valentino evening gown [generally costing anywhere from twelve to thirty-one thousand US dollars] will be made using only the best quality, hand-woven fabrics, stitched together by the most highly skilled—and highly paid—seamstresses. Chances are, even the beading, which will probably have been done in India, will be by skilled craftspeople who are well paid and well looked after."[80] Blanchard offers no citations, so it is hard to judge what the paternalistically phrased "well looked after" means for these South Asian garment workers hand-sewing tiny beads onto dresses. Another chapter, "I Shop, Therefore I Am Ethical," offers Blanchard's guide to "shopping without guilt." In it, Blanchard bemoans the "fashionista's dilemma" of loving nothing more than the rush of coming home laden down with heavy bags after a "hard day's shop," but then later feeling guilty about it. She reassures her readers, "We love clothes. We love to shop. Some of us would say we live to shop. And if you're smart, this can be a good thing. The trick is to direct your spending power."[81]

ANTM host Tyra Banks sounds a similarly reassuring note throughout the "Go Green" episode. The models' mode of transportation is even emblazoned with the same phrase that is used for Blanchard's book title: "Green Is the New Black." However, when Banks informs the models at the start of the episode that they will be chauffeured to their photo shoots, not in the luxurious gas-guzzling limousines that previous contestants enjoyed or the tony black SUVs associated with the glamorous life of high fashion, but in a jungle-themed "Green Is the New Black" biodiesel bus, the models have a tough time scratching up even a scant bit of feeble excitement for this revelation. The models only become enthusiastic when they arrive at an enormous "environmentally friendly" Los Angeles mansion with a large swimming pool and hot tub. Green mindfulness cards positioned around the mansion provide the models with environmentally conscious lifestyle tips, such as the suggestion that limiting each of their showers to ten minutes will save twenty gallons of water. At one point, presumably moved by a wave of environmental virtue, six of the scantily clad models suggestively perform ecopiety by all piling into a bathtub together "to conserve water."

The "going green" theme of the episode seems to peak at the scenes of the eco-friendly bus and the group bathing, as the rest of the episode features a "shopping competition" at a clothing store, followed by the models being photographed for a campaign against cigarette smoking. In fact, as framed, the episode's anemically pale green theme marries well with the fashion, consumption, and capitalist themes that form the basis for the show. As the episode progresses, narratives of ecopiety and consumopiety are co-constituting and firmly entwined. One of host Tyra Banks's "challenge cards," addressed to the contestants, explicitly links going green to the models' making money in the fashion industry. Banks instructs the contestants, "An inexperienced model is called 'green,' an experienced model makes 'green,' and a conscious model goes 'green.'" The clever juxtaposition of these different meanings of "green"—inexperienced, money, and environmentally "conscious"—reflects the braiding together of these three strands, especially the connections between labor and the practice of a fashionista ecopiety that will strategically yield more financial success.

Once ensconced in their green manse, complete with floor-to-ceiling tropical waterfall video screens, the models model an ecopiety that is fundamentally one of privatized responsibility, entitlement, chic fashion choices, and self-congratulatory marginal changes.[82] Again, these are the sorts of marginal changes that Tom Crompton, Monin and Miller, Harding and Rapson, and economist Gernot Wagner all warn can lead to pitfalls of single-action bias, in which people feel they have done their bit for the environment because they have acted in one (often minor) area. A psychological study conducted by P. Wesley Schultz and colleagues even identifies and cautions against a consumption-virtue "boomerang effect," in which households that were informed that they were currently consuming less than the average amount of energy consumed by other households in their area, upon receiving this revelation, proceeded to *increase* their consumption.[83] *America's Next Top Model* contestants might spark infocopiers, but what is being modeled is a kind of superficial, faddish approach that makes "going green" the latest style to be put on for social approval and cachet value, only to be taken off again when green is no longer in fashion.

A closer examination of blogs, online comments, and fan discussions following "The Models Go Green" episode, however, tells a different story from an infocopying model. Rather than being swept away by the episode's environmental messages, viewers who watched the show and then participated in post-episode online commentaries seem to delight in pointing out the ridiculousness of the show's premise. Bloggers made fun of the show's overly simplistic modes of "going green" and its superficial messages. One reviewer humorously adapted the episode title to a more watered-down "The Models Go Green . . . ish."[84] Another used "Valley Girl" speak to mock the "dumbed-down" version of environmental concern portrayed in episode: "Tyra [the host] has apparently decided that this season is all about modeling with a message, and the first message is that, like, bad stuff is happening to the earth that is making it more worser and stuff."[85] Fan discussions and viewer comments that follow the "Go Green" episode online reviews go further, not just questioning the sincerity of the environmentally conscious theme but digging into what viewers theorize are the profit motives behind the show's supposed morally motivated "environmental" production decisions. "Irisheyes18" on TV.com's episode comments calls host Tyra Banks a "hypocrite" and comments, "Going Green is a joke on that show. Big deal, bio fuel bus. How much plastic and other chemicals were used to build that house and other sets? How many vehicles are used to film this show, used by every person on the show?"[86] "Starghella," in the same posted discussion, resents the element of "seriousness" interjected into the environmental episode and quips, "I don't watch America's Next Top Model to hear anybody's political statements about being green."[87] In another comment, "N3ll3n" applauds the "Go Green" theme in general but remains skeptical of the motivations behind and the efficacy of addressing environmental issues within the program's genre: "I'm not entirely convinced that a show about models is the correct forum, or that this show's target group is the right target group—but the idea [going green] is sound."[88]

Citing both environmental air pollution concerns as well as health reasons, Tyra Banks announces in the "Go Green" episode that none of the models will be allowed to smoke that season. The online comments sections on both A.V. Club.com and TV.com are not "buying" the ecopiety that Ms. Banks is selling. Instead, discussion participants articulate resistive readings, as they theorize that banning smoking is merely a brilliant strategic move on behalf of the producers to interject conflict and catfights into the program, thereby increasing the season's ratings. After all, what could be more volatile than a house packed with starving fashion models who are *not* allowed to smoke?[89] "Kerouac9," commenting on A.V.Club.com, gleefully observes, "All these girls are going to put on 8–12 lbs. over the course of the show, which is like 10 percent of their total body weight. It's gonna be great," adding sarcastically, "Not to mention the fact that the quitters aren't going to be at all edgy or cranky."[90] In other words, the "elimination" departure scenes will be filled with major meltdowns and contestant drama.

These moments in the A.V.Club discussion that question the authenticity and moral motives for the supposed "green" framing of the show, pointing instead to profit-driven motives for purportedly socially conscious production choices, and clearly employing skepticism when viewing the "environmentally virtuous" content of the show, all make one wonder what the Frankfurt School "A.V. Club" (of which Adorno and Horkheimer were founding members) would have thought of such exchanges. *America's Next Top Model* may indeed be modeling a neoliberal version of voluntary, individualistic, and personal ecopiety that both reflects and reinscribes conformism to an exalted system of capitalist consumerism, but as we know from Hall, Fiske, and other media theorists, the active "readings" of any text are complicated and variegated. Blogger, fan, and viewer comments discussing *ANTM*'s "Models Go Green" episode suggest that these viewers at least critically take cognizance of the show as a strategically constructed cultural product within a profit-driven industry. *ANTM*'s demonstration of environmental virtue in this special episode takes on a superficial fashionably pale hue of green, but it is one that online commentators appear to see through as they engage in "street theorizing" and critiques of the program.[91]

Of course there may also be legions of docile, uncritical viewers who hang on Tyra Banks's every word and never question this special "Go Green" episode, its framing, premise, or motivations, as they watch, but research on "active audiences" and their interpretive capacities would suggest otherwise.[92] Make no mistake, say media sociologists Croteau and Hoynes, "media messages matter" and are "central to our lives." The term "audience" is problematic, though, and increasingly so, as "a large body of research demonstrates that media audiences are active interpreters of information"—more "active readers" than passive recipients of media messages.[93] And yet, Croteau and Hoynes also caution a middle way in the "active" versus "passive" audience debate, proffering that, while audiences are indeed active meaning makers, "[T]hey are not fully autonomous; a sociology of the media needs to be sensitive to both interpretive agency and the constraints of social structure."[94]

Other "reality" television programs from this same period exhibit a bit more substance in their stories of "greening" than does *America's Next Top Model.* ABC's *Extreme Makeover: Home Edition* did a "going green" episode (2009) in which it built a green model home and performed an eco-rehab on a school, explaining geothermal heating and cooling systems along the way and providing advice on how to get new energy technologies through local construction permit review offices. In 2005, *E-Force* was the Outdoor Channel's second-top-rated show and followed tough Florida Fish and Wildlife Conservation officers as they enforced wildlife regulations on state residents and interacted with a menagerie of exotic and dangerous tropical creatures. A spin-off of *Living with Ed,* TV1's *Mario's Green House* (2009) centered on actor/director Mario Van Peebles's efforts to build a green home for his five children and extended family in the Los Angeles area. Like Begley's series, however, the story of Van Peebles's "green rehab" is intimate, domestic, and voyeuristic as we see the family's bedrooms and personal living space, becoming flies on the wall as the family deals with daily struggles to make green home-design consumer choices.

The "reality" television series *Greensburg* (2008–2010) adopted the popular format of the "makeover show" to provide a window into the conflicts and in-fighting between local financial interests and ecopiety in practice, as the show documented the rebuilding process of a small Kansas town that had been flattened by an EF5 ("Enhanced Fujita Scale 5"–i.e., BIG) tornado in 2007.[95] Episodes frequently depicted a stand-off between those who wanted to get out of their FEMA-provided disaster trailers as soon as possible, rebuilding structures quickly and cheaply no matter what the process, and those who wanted to take the time to plan a sustainable community, making Greensburg a pious *model* of environmental "best practices" for the nation. The tagline for the show's trailer suggested themes of salvation, rebirth, and renewal, inviting viewers to "watch this town transform from devastation to a green rebirth."[96]

One of the more intriguing experiments in green reality television is a show that never quite made it. Aimed at making a substantive, politicized media intervention into the moral superficialities of other environmentally themed reality shows, and pitched as "interactive eco-adventure reality TV," *The Nautical Tribe* was conceived as a "reality TV series strategically designed to create a cultural tipping point toward a more sustainable and collaborative world."[97] The program was to be hosted by Fabien Cousteau, the grandson of famed oceanographer Jacques Cousteau. Sixteen college students were recruited to sail aboard the *Nautical Tribe* boat, where they would be issued environmental and humanitarian challenge projects and then graded by eminent university professors (in the areas of marine ecology, biological architecture, international affairs and development, anthropology, etc.) in conjunction with viewer audience text voting to see who would be deemed the most "Green Philanthropic Adventurer" at the end of the season. No one was to be "voted off" or eliminated from the boat. Instead, two eighty-thousand-dollar scholarships would be awarded at the end, one to an individual and one to be shared among the members of the best team. Instead of scantily clad, skinny fashion models ready to perform a scripted bathtub scene of girls in wet t-shirts, candidates who applied to be on the proposed first season of *The Nautical Tribe* were STEM-smart young women and men studying marine biology or development studies and wishing for careers in conservation science or in working with NGOs.

Drawing from Mahatma Gandhi's famous directive, the show's treatment and web-based marketing asked contestants to "be the change they wish to see in the world."[98] Be the change, however, did not mean "buy the right stuff." Unlike the competitive "shopping challenges" of *America's Next Top Model*, teams competing on *The Nautical Tribe* would respond to such challenges as rescuing marine mammals, restoring mangrove ecosystems and coral reefs, setting up a center for sustainable development in Haiti, and designing anti-gang programs in Jamaica. Whereas *America's Next Top Model*'s producers appear to cultivate and reward "catfights," pitting individual models against one another, *The Nautical Tribe* producers, judging professors, and audience voters would reward constructive collaborative team efforts, collective action, and productive work on environmental solutions.

Perhaps not surprisingly, funding *The Nautical Tribe* and greenlighting its production were a challenge. Producer Jim Rizor has now reduced the project's six-year web presence, taking down its beautifully designed and curated project marketing site, complete with series trailers. Now

The Nautical Tribe exists only as a social media presence on Twitter and Facebook, with about four thousand followers on the regular Facebook page and about a thousand or so members of the public group, many of whom have followed the project's ups and downs over the years and now share information with each other about how to participate in ocean-related conservationist efforts. The group's description now says, "The Nautical Tribe is an effort to ROCK THE BOAT toward a cultural tipping point in united stewardship."[99]

The very image of "the tribe" (albeit problematic in its romanticizing of "neotribalism," a dynamic pervasive in reality TV programming) conjures a very different model from that of individual consumer choices and personal pieties.[100] Producer Jim Rizor's initial 2011 trailer for *The Nautical Tribe* posits a series of questions to the show's audience that intervene in much of the "business as usual" modus operandi of reality TV, asking, "What if the world were your country?"; "What if what binds us together is more important than what separates us?"; "What if education [were] an adventure?"; "Where we valued experiences more than possessions?"; "What if there [were] a game that was won by the person who gave the most?"[101] The trailer suggests the ideal of a whole-earth "tribe," a collectively shared and cared-for planetary commons, served by a collaborative piety and devotion in play to create real systemic change.

The premise is a radical departure from those that share the reality TV genre. Not only were no consumer products to be sold on the show but also, more fundamentally, the show itself, in its treatment, did not sell a narrative of salvific green capitalist consumerism as the solution to global problems. In short, it sought flagrantly to violate the reality TV formula Barry King says fosters "the learning of dispositions, habits and interests that reproduce a larger cultural formation."[102] Seminal reality TV shows, such as *Survivor*, *Big Brother*, and, most notably, *The Apprentice*, in effect create a socially Darwinian world of vulture capitalism, in which gladiator-like competition undergirds all else, conflict is encouraged, competitors must be eliminated, individual self-interest rules the day, and the player with the most resources at the end wins.[103] The *Hunger Games*–like framing of the manufactured "reality" in reality TV programming reinforces a "sacred money and markets" narrative and a libertarian politics of the rugged individualist "survivor" (left to his/her own devices) at the expense of the common good.[104] It is striking, then, that a television series that *was never made* and was merely an idea communicated through web-based marketing and prospective trailers somehow, albeit in a different social-media-based form, has in some sense "survived" over the years with a small but devoted following of those committed to its vision.

Television, contends media theorist Neil Postman, is "our culture's primary mode for knowing about itself."[105] In the age of the Internet and digital new media, one might or might not argue for modifying Postman's statement, or at least qualifying it with regard to the now-changed delivery mechanisms for television.[106] Many forms of mediated popular culture offer us sources and modes of cultural self-gnosis. We have seen here "sightings" of ecopiety in the spectacle of green reality TV, including perhaps its proposed and failed interventions, carbon-sin software application designs, and even "eco-chic" popular style guides. All of these cultural productions in some way keenly reveal the ways in which laissez-faire economic philosophies and "the world is yours to survive" political ideologies in the United States are reproduced and reinforced through mediated popular culture, even when reality program producers, software designers, and fashion writers aspire to "progressive" aims such as "saving the earth," "going green," or "solving the environmental crisis." Environmental stories that champion extra-political, privatized, consumer solutions as the most effective solutions for addressing planetary problems dominate the mediasphere. The intensely networked, interactive, participatory, hypermediated realm of new media, however, increasingly affords spaces to trouble, contest, and talk back to these essentializing narratives, engaging in resistive acts of restorying. As we see in the next chapter, stories of the supernatural, told through and across multiple media platforms and engaged in by participatory fans, put a curious twist on moral engagement, mediation, and the complex entanglements of ecopiety and consumopiety in the popular restorying of earth.

5

Vegetarian Vampires

Blood, Oil, Eros, and Monstrous Consumption

The opening scene of Lynn Messina's *Little Vampire Women* (2010) begins with Louisa May Alcott's four young March sisters—Meg, Beth, Jo, and Amy—all gathered together, lamenting that "Christmas won't be Christmas without any corpses!" The narrator explains that due to the March family's impoverished state, the girls rarely had the "luxury of a living, breathing animal to feast on." What's more, "A human had never been on the menu, even when the family was wealthy and lived in a large, well-appointed house, for the Marches were humanitarians who believed the consumption of humans unworthy of the modern vampire. Humans were an inferior species in many ways, but they deserved to be pitied, not consumed."[1]

In this satirical twist on Louisa May Alcott's classic, *Little Women* (1868), Messina gives us many such absurd demonstrations of vampiric moral conscience and social responsibility to make us laugh. But there is something fitting about Alcott's (human) nineteenth-century vegetarian and socially concerned March family being recast and rewritten—*restoried*—in contemporary vampire terms. Today's "vegetarian" vampires would likely find common cause with the virtuous abstinence practiced by the March sisters in Alcott's classic novel, and indeed historically by Alcott and her own family. Louisa's transcendentalist father, Amos Bronson Alcott (1799–1888), cofounded New England's first utopian vegan commune, Fruitlands.[2] The Fruitlands community predated the term "vegan" by more than a hundred years and anticipated, as historian Richard Francis points out, the "twin preoccupations of our own time, ecology and environmentalism."[3] Tied to the Fruitlands experiment was Bronson Alcott's advocacy of temperance, vegetarianism, and celibacy, and his general repudiation of the sins of excessive consumption. Alcott equated overindulgence with moral turpitude, contending at one point that "robbers and murderers usually overindulge [consume large meals] before going out to commit their crimes."[4] Excesses in consumption in one generation could be passed down and cause harm to future generations—a notion we might discuss today in terms of a kind of "genetic damage." Nineteenth-century Fruitlanders, writes Francis, "were deeply concerned with the danger of environmental degradation and pollution, the harm it represents for people and for the world itself. This awareness was linked to their sense that all phenomena were interlinked, that life can be seen as one force [what Emerson spoke of in transcendentalist philosophy as the 'Over-soul'] running through a multitude of differing manifestations."[5]

In Messina's *Little Vampire Women*, Father March (also cast as a vampire), far away from home and absorbed in his charitable work with Civil War soldiers, writes to his daughters, urging them to work daily "to overcome their vampire natures."[6] So, too, a contemporary breed of *socially conscious* vampires, as portrayed through mediated popular culture, struggles with a voracious appetite to consume. Today's vampires battle unbridled bloodlust and wrestle with an "addiction" that imperils species survival, mars ecological health, and precipitates the extraction and depletion of vital resources. Nineteenth-century transcendentalist counterculture and the Fruitlands experiment remind us that these moral concerns about the violence of extractivist (forcibly draining) economics are not new, nor are the connections between abstention from meat eating and critiques of capitalism.[7] Portrayals of the contemporary green/vegetarian vampire significantly update these moral concerns and reformulate them as relevant to today's media audiences. Green vampires embody environmental moral critiques of *extractivist capitalism* and its resource-sucking planetary consequences. And yet, ironically, the very presence of socially conscious vampires is inextricably tied to practical, market-driven/profit-driven motivations and interests.[8]

Audience encounters with these vampires, expressed through online participatory fan cultures and their commentary, tellingly highlight the contradictions between moral *interventions* into the excesses of capitalistic overconsumption and the vampire's role as media merchandising cash cow. Vampire-themed mediated cultural works provide compelling restorying tools for engaging popular moral imaginations about the nature and future of human/earth relations, yet they fuel this moral engagement through the very extractivist consumerism they purport to trouble. This chapter illuminates myriad ways in which vampires, as historically floating signifiers mediated through the field of cultural production, have come to play a meaningful role in restorying human/earth relations in the Anthropocene.

The "Story Problem"

Much as aforementioned economist David Korten has advocated a global narrative shift as critical to transitioning a dominant "extractive money economy" to an "emerging generative living economy," Canadian journalist/social activist Naomi Klein similarly identifies the current environmental crisis as *the* "story problem" of our time. In the video documentary version of her book *This Changes Everything: Capitalism versus Climate Change* (2015), Klein poses the question, "What if human nature isn't the problem? What if even greenhouse gases aren't the problem? What if the real problem is a *story*, one we've been telling ourselves for 400 years?"[9] A variety of vampire texts in mediated popular culture, in turn, take up this "story problem." Their vampire narratives present interruptions and interventions into dominant consumer narratives, warning of the vampire within us and urging socially responsible moral restraint from our consumer addictions. Contrapuntally, some of these vampire-themed cultural works encourage this restraint while lavishly ensuring the addict's "supply" of consumer goods. As the content of vampire narratives dares audiences to face and self-reflectively temper their monstrous consumer desires, the very media franchises and industry tied to these cultural works concomitantly profit wildly from cheaply manufactured fan products made possible by exploited labor and the voracious extraction of the earth's resources. The sheer variety and volume are mind-boggling.

In the vein of "do as I say, not as I do," vegetarian vampires are not alone. Film scholar Ellen Moore's indictment of Universal Pictures' 2012 version of the Dr. Seuss classic *The Lorax* points to just such an example of moral messaging and consumer marketing at odds. While the content of *The Lorax* film adaptation may have stayed true to Seuss's warnings about overconsumption and its disastrous impacts on life on earth, Universal Studios' embedded advertisements, a brisk online merchandising shop, and Universal Studios' Orlando "Seuss Landing" were all busy hawking countless Lorax plush toys and other mass-manufactured, landfill-bound, but supposedly environmentally themed movie tie-in paraphernalia. Moore makes parallel critiques of such eco-friendly children's films as *Ice-Age: The Meltdown* (2006) and *WALL-E* (2008), which track along similar marketing patterns. Mercenary merchandising, deployed by what Moore characterizes as a consolidated "American media oligarchy," eviscerates the film's narrative content and supposed environmental messaging.[10] To the extent that any of these films are making environmental narrative interventions into the machine of hyperconsumerism, we are left to question whether these interventions have been, effectively and disturbingly, "swallowed" by controlling corporate behemoths. The best creative intentions may indeed be to produce a cultural work that does its bit to, as Korten would say, "change the story." But that storied intervention is delivered through a complex network of market interests that often resist, counter, exploit, and simply absorb its message.

If anything, vampire narratives of moral restraint, rather than inspiring actual restraint, appear to stoke what both film scholar Mary Ann Doane and advertising historian Roland Marchand identify as the powerful manufactured driver at the heart of mass culture: "the desire to desire."[11] The premise of Judith Williamson's groundbreaking work, *Consuming Passions* (1980), was that in a consumerist society, people's desire for meaningful *social change* is strategically redirected instead into a compelling desire for consumer products. These purchases provide the illusion of personal control, autonomous power, and self-efficacy, while safely diffusing collective social, political, and economic criticisms and transforming them into neutered and self-contained personal issues. "The whole drive of our society," writes Williamson, "is to translate social into individual forms: movements are represented by 'leaders' . . . economic problems are pictured as personal problems ('too lazy to get a job'), public values are held to be private values ('let the family take over the Welfare State')."[12] In a similar vein, the problems of global consumption and its effects on the planet frequently get recast in contemporary vampire narratives as the personal moral and ethical struggle for ecopiety by a virtuous vampire who strives to temper his or her extractivist nature.

Vampires R Us

Historically, the mythic power of the vampire has repeatedly provided a powerful way to discuss a variety of socially conflicted and politically charged subjects. Each vampire, says literary critic Nina Auerbach in *Our Vampires, Our Selves* (1995), "feeds on his age distinctively because he embodies that age."[13] At various cultural moments, the vampire in literature, film, radio, television, video, and now on the smaller screen of the digital mobile device has provided a flexible trope for talking about such subjects as shifting sexual and gender norms, xenophobia, immigration, science and technology, evolution, war, feminism, queer identity, exchange of bodily fluids, syphilis, HIV/AIDS, extractive (or "vampire") capitalism, unbridled consumerism, addiction, and anti-semitism, among other salient issues of the day. From the Byronesque Lord Ruthven in "The Vampyre" (1819) to Bram Stoker's iconic *Dracula* (1897); from vampire Barnabas Collins of the 1960s popular Gothic soap opera *Dark Shadows* (1966–1971) to the sympathetic seductive monsters of Anne Rice's best-selling vampire novels (1970s–1980s); from Wesley Snipes's signature role as the conflicted 1990s cinematic vampire superhero "Blade" to Joss Whedon's turn-of-the-century TV series *Buffy* and *Angel*; and from *Twilight*'s forever-teenaged Edward Cullen to HBO's *True Blood*'s mainstreaming Vampire Bill, there is no dearth of corroborating evidence for Auerbach's contention that, in any given age, *we create the vampires we need*. She observes, "What vampires are in any given generation is a part of what [we are] and what [our] times have become."[14] Following Auerbach's line of reasoning, in an age of environmental crisis, we need ecologically concerned "vegetarian vampires," and it appears they have arrived right on cue. Shading the horror genre with a distinct hue of green, vegetarian vampires—monsters struggling with their own predatory and extractive nature—are the story of who humans have largely become in the Anthropocene. The vampire continues to be an ambivalent signifier—if not one that floats, then one that at least hovers—relational and shape-shifting depending on different historio-cultural moments.[15] Even the virtuous, ecopiety-practicing green vampire is still a relatively recently reformed and responsible vampire—freshly out of rehab with his or her powerful extractivist addiction lurking underneath.

"The Western vampire is a creature of capitalism," writes media scholar Milly Williamson, arguing that the vampire enters "the Western imagination concurrently with the emergence of the culture of the bourgeoisie, giving expression to its fears, denials, and contradictions." Thus, contends Williamson, the vampire was effectively "birthed in capitalism." She reminds us that Karl Marx indeed compared the very nature of capitalism to the vampire.[16] As supernatural creature, the vampire is a shapeshifter, and so we should not be surprised, as Williamson chronicles, that the vampire "enters the permanently transient culture of capitalism, adapting and evolving in order to keep pace with the cultural moment."[17] And yet, the very meat the *vegetarian* vampire eschews is also a historical signifier of American culture and capitalism.

In Meat We Trust, a history of meat consumption in the United States, demonstrates how meat and its consumption came to represent the "American standard of living." The robust presence of meat in the American diet has served as a kind of testimony to the abundant yields of capitalism and its successes.[18] Times of meat shortages in the United States were often viewed in desperate terms and considered a threat to capitalism and democracy.[19] The award-winning advertising campaign in the 1990s, "Beef. It's What's for Dinner," effectively tapped into these iconic cultural connections between meat consumption and American identity by showing down-home scenes of family beef dinners set to composer Aaron Copland's quintessential Americana anthem of frontier nostalgia, "Hoe-Down," from his 1942 *Rodeo* ballet score.[20] Chicago-based advertising firm Leo Burnett's artful and award-winning television and radio spots were so effective that the campaign produced one of the most recognizable taglines in media advertising history, with roughly 88 percent recognition among Americans.[21] The vampire may indeed be a "creature of capitalism," but when identified as vegetarian, a meat-abstaining vampire also implicitly challenges something iconic and embedded in American culture. The vampire as symbol can be purposefully wielded in the field of cultural production as a subversive tool to expose capitalism's most monstrous extractivist traits. Even so, that same vampire still appears compelled to do its master's bidding. That is, as we encounter the vegetarian vampire's narrative displays of ecopiety, we discover a vampire who is simultaneously a model of moral restraint and a seductive enabler. This is a paradoxical but alluring dance.

Energy Vampires R Not Us

"Vampires and zombies are just everywhere. Don't turn around—there's probably one behind you right now," Richard Greene and Silem Mohammad warn readers in their introduction to *Vampires, Zombies, and Philosophy*.[22] And it does indeed appear that vampires are *everywhere*—popular teen novels, films, prime-time soap operas, cable series, punk bands, jewelry, clothing, cosmetics lines, condoms, personal hygiene products, body modifications, gaming, advertising, social media groups, nightclubs, and even a company that markets a special line of Vampire red wine.[23] A mere cursory glance across the field of cultural production shows us that the undead are alive and well and seemingly omnipresent among us. Depending on who is wielding these vampires and for what purpose, they are also clearly not all virtuous and green.

"Energy vampires," so designated by government agencies and energy companies alike, have—unbeknownst to us—"invaded" our electronics and appliances. An external hostile force, they have "attacked" our homes and are secretly draining the power current. "Watch out!" warn messages that appear on the US government's Energy.gov web site and alert consumers to adopt greater vigilance about vampire pervasiveness. A subsequent explanation informs consumers that "stand-by" electronics (aka, "energy vampires") that draw energy even when turned "off" are currently sucking energy from unsuspecting American households and thus draining millions of dollars from Americans' pockets.[24] The US Department of Energy began posting warnings about so-called energy vampires on their web site between 2010 and 2015, during the Obama presidential administration. When the transfer was made to the Trump presidency, controversial changes to wording were implemented on a number of governmental agency web sites, including reducing or eliminating language that referenced climate change, but "energy vampires," for the most part, survived the administrative transition. In 2018, the Department of Energy's web page devoted to "Reducing Energy Use and Costs" greeted visitors with a bold, forty-point font headline urging Americans, "Control Vampire Loads!"[25] Previous features celebrating National Energy Action Month, aimed at helping homeowners reduce their energy consumption, asked web site visitors, "Are Vampires Sucking You Dry?" To illustrate this article, the department provided a creepy still frame from F. W. Murnau's classic German Expressionist horror film *Nosferatu* (1922), in which we see vampire Count Orlok's candle-lit shadow in a stairwell as he advances with fangs and claws bared upon his victim.

The Energy.gov caption to this image reads, "[Actor] Max Schreck's *Nosferatu*, presumably climbing the stairs to plug in some unused appliances."[26] A government-sponsored energy web site might seem like an odd place to encounter cultural references to German Expressionist cinema, but here we see vampires depicted in their more traditionally horrific sense—monstrous and not (at least not *yet*) socially conscious or environmentally enlightened. More importantly, these monsters lurk *among us* in our private domestic spaces, sucking us dry. The site cautions, "[A]n actual terror may already be in your home, leeching off your energy supply. Unlike the creatures commonly associated with Halloween, these 'energy vampires' are all too real and all around you, running up your energy bill even when you're not actively using them."[27] To illustrate this point, the Department of Energy has used a signature cartoon graphic that is now borrowed and utilized by a number of educational environmental web sites and pamphlets. The graphic shows a fang-bearing vampire concealing sinister cell phone chargers, laptops, cable modems, and game consoles under his classic vampiric black cloak.[28]

Energy Star, a US Environmental Protection Agency (EPA) voluntary program to promote energy conservation, also warns consumers of unseen and unacknowledged terror within their homes, urging consumers to look around their houses and workplaces for hidden "energy vampires" lurking and preying upon them. Throughout "energy vampire" educational materials, "lurking" is the most often used word to characterize these sinister external forces stealing energy from us—implicitly a situation that is no fault of our own.[29] The EPA makes a big push to get this "vampire" message out, particularly around Halloween.[30] Set to spooky music, a 2016 audio podcast on the EPA web site told Americans, "BEWARE: Energy Vampires Could Be All around You." Various helpful strategies for "hunting down" and "slaying" energy vampires within one's own home accompany this dark revelation.[31]

In the 1950s, popular culture renditions of vampires became symbolic vehicles for expressing fears of communism and the threat of "the Red Scare," as well as anxieties about McCarthyism and its own brand of in-

Figure 5.1. Count Orlock from *Nosferatu*. *Nosferatu*'s sinister Count Orlock heads upstairs to plug in "vampire energy" appliances. Credit: F. W. Murnau (1922).

timidation and terror. "Commies" could be anywhere–"reds under the bed"—cleverly camouflaged as your nice suburban neighbor, or even as a family member in your very own home.[32] Infiltration, invasion, syphoning secrets, and a subversive weakening of America's energy and vitality—themes also found in classic horror films such as *Invasion of the Body Snatchers* (1956, remake 1978)—figured prominently in stories of vampiric draining and depleting monsters. In our own day and age, as in the past, intruding others (most often immigrants) are still cast in national rhetoric as a threat and portrayed as stealing, depleting, draining, and exhausting our national energy and resources.[33] But significantly, in the case of the US government energy sites during both the Obama and the Trump administrations, the culpability for that draining "monstrous" force is externalized to the consumer (an invasive force not the fault of the homeowner), while the responsibility for eradicating the monster is relegated to a personal domestic chore akin to rodent extermination. These "monstrous" consumers, says the national energy conservation program, infiltrate our private spaces and hide among us. Such infestations are apparently widespread, according to the figures offered as to just how many energy vampires lurk in US homes.[34]

And yet, citizens are told we are *individually responsible* for hunting down these vampires and eradicating them. No mention is made in the government's energy-vampire literature of the possibility of government *regulatory* action that could swiftly, and across the board, restrict the manufacture of "phantom-load"/"stand-by power"/"vampire" electronics in the first place. Indeed, many older generations of Americans grew up in an age when it took a few moments for their TV sets to "warm up" and, arguably, these generations were not onerously burdened by this brief wait time. Regulatory measures on electronics manufacturing could effectively prevent the production and sale of appliances (plasma TVs, play stations, cell phone chargers, etc.) that are designed in such a way as to suck energy from the power grid even when the appliance is ostensibly "off." Instead, government-hosted energy conservation web sites and conservation programs place the onus on the individual consumer to be educated and aware of the problem, and then personally vigilant and environmentally virtuous in response—demonstrating *ecopiety* by "slaying" what are technically avoidable "monsters" distinctly external to us. In the visual, verbal, and aural rhetoric of the EPA, the Department of Energy, and energy utility company web sites, Americans are assuredly *not* the energy-sucking vampire but mere unsuspecting and innocent victims prey to the vampire.

The self-identified Real Vampire Community (RVC), a subculture of those who consider themselves to be biologically real vampires, has a cognate to the Department of Energy's depiction of a monstrous extractive energetic threat among us. There is a recognized, pronounced, and practical difference in the RVC between "blood vampires" and "energy vampires." So-named "sanguinarian" vampires are those who—out of necessity—feed off human blood. In contrast, "psy vampires" (or "psychic vampires") instead drain their hosts of vital energy.[35] In occult researcher and pagan community leader Corvis Nocturnum's popular guide to the vampire subculture, one of Nocturnum's vampire informants explains that psy vampires "feed off of and manipulate energy and aura. Whether they take it from human beings, living organisms or even other psy-vamps, the effect on the victim is one of weakness and draining. The [energy or psy] vampire takes the energy into himself, thereby gaining vitality and strength." Tellingly, Nocturnum goes on to say that psy vampires "can also feed off of other energy sources (electricity, storms, even sunlight)."[36] From the RVC's perspective, the EPA and Energy.gov's caricature of the "energy vampire" is not fanciful, then, but more representative of the power of vampires than ordinary mortals might think.

Nocturnum, who, as a kind of spokesperson, points out that "real vampires" span the range of lifestyles from dark vampire nightclub goers to suburban soccer moms, takes pains to explain that within the contemporary Real Vampire Community there is most definitely a "code of ethics," and "energy vampires" have been known to violate this code.[37] Ideally, the relationship of a vampire to a host is a consensual one, arranged by contract, and feeding methods should ethically emphasize both disease prevention and disease transmission.[38] One community leader tells Nocturnum that the relationship to the "blood donor" should reflect the "vampire community values [of] community, respect, integrity, awareness, and safety."[39] In fact, Nocturnum frowns upon those in the community who "have no ethics and prey on the unwary; deeming it their right to use anyone like a 'living' battery and discard when drained."[40] This articulation of a socially responsible contemporary vampire moral code of reciprocity among those who feed off human blood or energy sources (or both) reads much like a moral treatise on environmentally *sustainable* forestry or fishing best practices—don't take too much, too fast, give back, and allow the ecosystem (in this case, the human host) time to regenerate.[41] In other words, if you choose to bite the hand that sustains you, make sure you do not drain dry your host and kill it. This moral code for the modern vampire promotes a kind of sustainability model of consumption that is arguably a far cry from the "old-time" ethical benchmark of WWDD (What Would Dracula Do?). Despite this shift in moral consciousness toward responsible consumption within the "real vampire" subculture, vampires continue to appear in popular storied forms as a way to characterize the nature of serious environmental threats. The narrative entanglements of vampires and the extraction and consumption of natural resources evoke a powerful moral discourse about multifarious monsters in our midst.

Climate Vampire

In writing about the devastation caused by the 2012 "Frankenstorm" Hurricane Sandy, York College chemistry professor Keith Peterman, who also serves on the American Chemical Society's Committee on Environmental Improvement, coined the term "Climate Vampire" to refer to the "monstrous face" of the "impending consequences of our fossil-fuel-addicted society." In the United States, criticizes Peterman, "we have largely ignored the cries of global society and of our youth, 'Please don't suck the life out of Mother Earth!'"[42] Using a similar image, science writer and mechanical engineer Justin Hovland has argued that it is time for "our fossil-fuels vampire" to meet the true death (its final extinction) via one of the traditionally effective weapons against vampires—the power of the sun. If solar energy, the natural enemy of vampires, says Hovland, were used to produce renewable, storable hydrogen to run our homes and cars, we would vanquish our energy-sucking coal, oil, and natural gas enemies once and for all.[43]

Hovland and Peterman are not alone in their strategic use of vampiric idiom to talk about energy and environment. In his best-selling book *Greedy Bastards* (2012), former global managing editor for corporate finance at Bloomberg, L.P. and MSNBC host Dylan Ratigan indicts the fossil-fuel troika (coal, oil, natural gas) as "vampire industries." And civil rights advocate Deborah Dupré employs the image of the vampire throughout her controversial book, *Vampire of Macondo*, on the disastrous environmental health consequences and communities suffering as a result of the April 2010 explosion of the British Petroleum (BP)–operated Macondo oil and gas prospect located off Louisiana's coastline in the Gulf of Mexico. Dupré's book, which refers to BP's infamous platform as the "Vampire of Macondo," frames the severe health consequences of the Gulf environmental disaster in terms of human rights violations. In doing so, she draws particular attention to the disaster's impact on the region's children, amidst repeated company assurances that water, seafood, and chemical dispersants were all safe.[44] To publicize her book, Dupré drove a solar-powered Tesla, the "anti-vampire" automobile, across the country, interviewing victims of environmental pollution along the way. She posted reports of these stories on her YouTube channel and interacted with her audience via tweets and vlogs as her vampire-vanquishing tour progressed. With a reputation for behavior such as home invasion, energy stealing, climate crisis causation, and exploding oil rigs, how then does a green, socially conscious, environmentally concerned vampire find purchase in American popular cultural imagination?

#NotAllVampires

The vampire may still be vilified as predator, perpetrator, and quintessential monstrous consumer in mediated popular culture portrayals, but an eco-pious vampire has also emerged. Like the popular social media Twitter hashtag, "#NotAllMen"—short for "Not All Men Are Like That"—contemporary vampire narratives protest that *not all* vampires are blood-sucking monsters.[45] This turn in vampire nature, the narratives suggest, reflects hope for our own species. In fall 2013, NBC and the BBC rolled out their coproduced *Dracula* series, introducing a remixed, greener version of the character Dracula. This Dracula would, strangely enough, have much in common both with Deborah Dupré's eco-friendly Tesla and its namesake, electrical engineer/inventor Nikola Tesla himself. Tesla (1856–1943) might have indeed found cause to *befriend* the NBC/BBC version of Dracula, who embodies many of Tesla's own dreams and ambitions.

Set in 1880s London, this *Dracula* centers on the clean energy struggles of vampire/scientist/inventor/American industrialist Alexander Grayson, who is a reanimated version of Bram Stoker's vampire, Count Dracula. Grayson is played by dark and handsome British actor Jonathan Rhys Meyers, who is styled in such a way as to bear an uncanny resemblance to nineteenth-century photographs of Nikola Tesla. Grayson plans to take down an evil cabal of British petroleum barons (a not-so-subtle reference to today's BP) that is known as "The Order of the Dragon."[46] To unseat these barons from their positions of power, Grayson plans to introduce clean, wireless, renewable geomagnetic alternative energy to Victorian England and ultimately the world. In the series pilot, Grayson reenacts Tesla's nineteenth-century glowing wireless lightbulb demonstration (also depicted in the film *The Prestige*) to an audience of high society guests. The demonstration of renewable energy strikes fear and ire into the British petroleum barons, whose discussions about the future of oil in the nineteenth century might read just as easily as overheard dialogue from contemporary petroleum corporation boardrooms. Sounding very much like a Gulf War profiteering scenario, the Order of the Dragon has England's military commanders in its pockets and conspires to make war in Europe in order to drive up oil prices and thus increase the petroleum barons' fortunes.

Not only are we as viewers asked to reconsider who the "real" vampire is—Grayson or the bloodsucking oil men—but clean alternative energy is made very *sexy* through Meyers's portrayal of the dashing Tesla Dracula. One reviewer refers to Grayson's "Prince-like sexuality" that makes one expect him at any moment to burst into a "slinky *Purple Rain*."[47] Following the first few episodes of *Dracula*, TV.com's staff posted a number of humorous Internet memes poking fun at the show, including one that shows a moment of erotic tension between Dracula and his love obsession, Mina Murray (Jessica De Gouw). Superimposed over a shot of Grayson gallantly kissing Mina's hand, a caption says, "Don't you find geomagnetic technology so . . . attractive?" In response, Mina has "*swoon*" superimposed over her head. (Recall how the popular erotica novel *Fifty Shades of Grey* similarly features sexy BDSM CEO Christian Grey discussing eco-sustainable recycled graywater systems with his brother and lecturing his submissive lover on the global impacts of unsustainable farming and food waste.) Although teasing *Dracula* for, at times, an overly serious approach to a somewhat absurd premise, the meme identifies a very real message communicated in the program's performances and aesthetics—that is, the promotion of eco-friendly renewable energy as not simply virtuous but downright hot!

This messaging feat—managing to make the practice of ecopiety exciting and *sexy*—has been less than successful more generally when attempted in the public media campaigns executed by real-life environmental advocates. Think, for instance, of dedicated but wonky Al Gore and his dry PowerPoint presentations on climate change. In many ways, it should be no surprise then that green vampires lend ecopiety a certain erotic appeal. Rather than being repelled by "the bloodsucking, foreign monster" in Tod Browning's 1931 film *Dracula*, writes American cultural historian W. Scott Poole, Americans instead "welcomed Bela Lugosi's version of the vampire as a new and exciting sex symbol. Today the vampire and its related mythology serve as one of America's primary erotic symbol systems."[48] One specialist in Victorian popular literature succinctly states, "Vampires are pretty much always about sex."[49] Contemporary vampire stories, in all their erotic allure, thus commingle *eros* and *environment* in popular imagination to conjure hot green vampires who make moral engagement with environmental issues both sexy and playful.

Still, there is a fair amount of critical commentary voiced in the viewers' comments sections that respond to series features posted to the NBC/BBC–hosted *Dracula* web site, to television reviews of the series, and to popular culture blogs that engage the series' environmental themes. In response to Tim Surrette's TV.com review of *Dracula*, "I Vant to Suck . . . Away Your Energy Costs!" "peterspoor 33" posts a message that makes fun of the premise that Dracula's body must be raised from the dead in order to save the world from global climate crisis, dubbing this scenario a case of "break glass in [case of] Order of the Dragon environmental emergency." Another commenter, "Rai 101," recounts "rolling on the floor with laughter" upon hearing the following line come out of Dracula's mouth: "But from the moment we demonstrate the viability of geomagnetic technology . . ."

Other viewers posting comments, however, were quick to defend the show and its environmental content. Postings by "CO711" show more tolerance for the show's messy overlay of remixed elements, rhetorically asking, "Can I buy into Dracula being a 'champion for eco-friendliness?' Why the heck not, my favorite supernatural tone shows are *Sleepy Hollow*, *Supernatural*, [and] *True Blood*. So, I can overlook something that might not come together smoothly and make you go huh." A comment from "Muzrub" finds the "green stance" in *Dracula* an appropriately "modern twist for a modern audience." Defending the Dracula/Tesla connection, "Drop-dead-gorgeous" (DDG) points out that Tesla, like Dracula, was also a "tragic figure" and had been driven out of business by his technology rival, Thomas Edison. DDG observes, "[H]e [Tesla] is a historical character and his demonstration of wireless energy transmission took place on different occasions in exactly the same timeframe as [NBC's] *Dracula*," adding that "we're on supernatural territory anyway," so an eco-friendly vampire is not so far-fetched. Other posted comments in the discussion further develop DDG's points, noting that in Tesla's day, wireless energy must have seemed *supernatural* and that Tesla himself was thus seemingly possessed of supernatural powers. Referring to the 1990s PBS children's cartoon, *Captain Planet and the Planeteers* (produced by Ted Turner), "MunchletteBelle" sees NBC/BBC's eco-friendly figure of Dracula as more Captain Planet than Tesla. Whether laughing or swooning or both, participants in this engaged discussion—in Michael Saler's terms, a "public sphere of imagination"—shift easily back and forth among considerations of science, environment, literature, and the workings of popular culture, debating the power of the vampire image and its use as a captivating delivery system for creative social and political commentary.[50]

The kind of lay cultural theorizing and moral discussion demonstrated in the *Dracula* online comments exemplifies what social psychologists Gün Semin and Kenneth Gergen have termed "everyday understanding"—that is, the theorizers' own interpretive understandings of themselves and their social worlds.[51] "Everyday understanding" runs through a number of fan cultures, and the rich allegorical reservoirs provided in vampire media make such interpretive social and political discussion streams in relation to the environment seem to flow quite naturally. What is often referred to as "new media" actually look a lot more like the kind of unpolished cultural processing engaged in "ancient media," such as the kind of running commentary scribbled on Roman walls or found in their brothels or taverns.[52] The hegemonic, largely unidirectionally delivered mass media of the twentieth century, argues media theorist John Durham Peters, is something of an "exception" in communications history, whereby everyday processing and active participation in media have been more the norm.[53] In this fan forum, we witness viewers challenging, defending, interpreting, and—at points—ridiculing the narrative presented in this network series; but above all, they are actively engaging the moral issues raised.

Oil-sucking, energy-draining, climate chaos–causing vampires who prey upon and drain their living hosts to the point of mortal peril arguably reflect very real contemporary anxieties about powerful corporate and industrial forces that are not always readily *visible* but can feel overwhelming in their underlying omnipresence. In other words, "Don't turn around—there are vampires everywhere!" This may be, in part, why the socially responsible green vampire has emerged as a heroic figure. In his history of monsters in America, Scott Poole contends that "[c]ultures frequently employ an iconography of death to deal with moments of historical horror and rapid social change."[54] He cites examples ranging from the visual culture and aesthetics of fourteenth-century Black Death–traumatized Europe to the eighteenth-century guillotine-obsessed culture that accompanied the bloodbath of the French Revolution. Perhaps in a time when our current rate of rapid species loss is being termed the "Sixth Great Extinction," news of increasing climate chaos and the death of forests, coral reefs, and other ecosystems reaches us on a daily basis, and accounts of a planet warming so alarmingly quickly that scores of animals and humans are already on the move, seeking refuge on higher ground and in northern climes, this is the death and horror that conjure today's green vampire. Literary and cultural historian Nina Auerbach's work suggests that the monster is simultaneously the "other" *and* us. If so, much of the *green* vampire's appeal may lie in being the monster transformed—the energy-sucking, life-destroying, consumerist, blood-lusting vampire reformed, if not *redeemed*. Implicit in contemporary vampire media is a theme that is also prominent in Stoker's *Dracula*—that of evolution. Just like us, the green vampire struggles to temper a voracious appetite, an epic desire to consume, trying to figure out how to feed from a life-supporting host in a sustainable way that does not drain it dry.

Temperance Vampires and Erotic Moral Restraint

In March 2015, Alan Rusbridger, editor in chief of the British daily newspaper the *Guardian*, launched the paper's "Keep It in the Ground" campaign. This now-global campaign draws attention to "do the math" arguments being made by climate scientists, environmental economists, and a variety of environmental groups that calculate, at a minimum, that 75 percent of the world's remaining identified fossil fuels need to stay in the ground if the planet is to have a chance of maintaining the two-degree warming limit considered to be in "the safety zone" for a global temperature rise.[55] In the roll-out of the *Guardian* campaign, Rusbridger challenged two of the world's largest philanthropic organizations—the Wellcome Trust and the Bill and Melinda Gates Foundation—to divest from more than two hundred fossil fuel companies within just five years, selling roughly 1.4 billion dollars in oil, coal, and gas stocks.[56]

Features and opinion pieces that followed in the *Guardian*, contributed by climate change activist leaders like Bill McKibben, reinforced that the key environmental priority must now be to force fossil-fuel-extraction companies to "keep it in the ground."[57] Like Nancy Reagan's iconic antidrug "Just Say No" campaign of the 1980s, or the evangelical "True Love Waits" sexual abstinence campaign that gained such popularity in the 1990s, messages to "keep it in the ground" and messages to "keep it in your pants" bear striking resemblances in their emphasis on moral restraint—i.e., "*not* doing" as moral virtue. The connection between idioms used to advocate sexual abstinence and those used to press for fossil fuel abstinence has not been lost on a variety of Internet meme makers who have taken up the new climate change slogan "Keep it in the Ground!" by cleverly superimposing this slogan onto decidedly phallic images of smokestacks ejaculating carbon emissions.[58]

The language of moral restraint is often associated with English political economist and demographer, clergyman Thomas Malthus (1766–1834). Malthus's book *An Essay on the Principle of Population* (1798) asserted that population growth, when left unchecked, inevitably outstrips the limits of nature, resulting in catastrophic consequences such as pestilence, disease, starvation, and war. To avoid what is now referred to as "Malthusian catastrophe" or "Malthusian mortality," Malthus prescribed moral restraint, which consisted of postponement of marriage and strict celibacy until such means were acquired to satisfactorily support progeny. Individual self-control, self-denial, temperance, and chastity were moral duties that Malthus directed especially toward the lower classes, disproportionately placing on the shoulders of the poor the burden of "preventative checks" on population that would ultimately lead to greater economic stability and higher standards of living.[59] His theories were most famously, and nefariously, used to justify withholding aid to Ireland during the great potato famine, as the British government adopted a noninterference policy—based on the idea that starvation in Ireland was a natural correction to unchecked population growth.

More recently, attempts have been made to rehabilitate elements of Mathus's theories, making them more attuned to dynamics of power, economic inequality, technological innovation, racism, classism, and social justice, while retroactively trying to remake Malthus as a kind of proto-environmental economist.[60] Malthus's principles have been recast, in part, by environmental sociologist William Catton in terms of a given ecology's maximum load or "carrying capacity" to sustain life. Once carrying capacity has been exceeded and resources outstripped, Catton has argued, this "overshoot" would result in a steep population "J curve"—a dramatic population drop, or "die off," and ecological collapse.[61] For Malthus, moral restraint largely referred to childbearing, but in the eco-pious rhetoric of vegetarian vampires, "moral restraint" is viscerally tied to acts of consumption.

The negative moral act of not doing—abstinence, temperance, restraint—resonates with conservative moral values that prize individual pious acts of "just saying no." Understandably, Malthusian models of moral restraint constitute what are known in the world of advertising as a "hard sell." What is more, if such restraints are not distributed fairly through public policy across the socioeconomic strata, elites are often rewarded for dilettantism and symbolic displays, while the day-to-day sacrifices made by those at the lower end of the socioeconomic spectrum are rendered invisible. As we have seen, socioeconomic elites receive praise in mediated popular culture as eco-pious environmental virtuosi for refraining from acts such as chlorinating their swimming pools or for foregoing comfort in their mansions by installing low-flow showerheads and toilets. Meanwhile, working-class people, living from paycheck to paycheck, and already consuming exponentially less than their celebrated elite green counterparts, face environmental admonitions to refrain from eating meat, to turn their thermostats down, to drive less or to buy the right car, and to have fewer children. As in the Victorian era, the burden of moral restraint too often is disproportionately misplaced and misapplied.

Environmental media messaging trying to marshal enthusiasm more broadly for "keeping it in the ground" (eco-pious fossil-fuel abstinence) might do well to borrow creative marketing strategies from American evangelical culture's innovative and aggressive use of *sex* to sell abstinence. In *Making Chastity Sexy*, communication scholar Christine Gardner's study of the strategic rhetoric employed in contemporary evangelical abstinence media campaigns, Gardner demonstrates the clever ways in which these campaigns have recast *not* "doing it" in terms of a choice that is active, empowering, and even erotic. That is, *not* having sex, the *active* restraint of choosing to "keep it in your pants," becomes a tantalizing way to intensify the delayed gratification of sex once enjoyed freely in the confines of a godly Christian marriage. Gardner finds that contemporary messages of Christian abstinence directed particularly toward teenagers focus on the tantalizing prospect of mind-blowing hot sex in marriage as the ultimate pay-off for the godly but erotic discipline of withholding.[62] There is, ironically, a quasi-tantric principle suggested in the sexy marketing of evangelical chastity, in which refraining from sexual activity only *intensifies* the lure and promise of "truly epic" married sex.[63]

But how might climate change campaigns, in turn, make the moral restraint of "keeping it in the ground"—the chaste ecopiety of *not* drilling—just as sexy and alluring? One word: vampires. Erotic vampire narratives as told in popular television series, mass-marketed teen novels, and Hollywood films are already engaging in this kind of environmental moral messaging about the "hot" allure of practicing fossil fuel abstinence, consumption temperance, and environmental virtue. In the course of these narratives, the conventional associations of environmental virtue with a piety of self-denial gets restoried as an *ecstatic* piety of the senses. This transmogrification effectively mirrors the dual-sided nature of piety, which can be associated as much with virtuous renunciation (fasting, vigil, hermetic isolation, celibacy, voluntary poverty) as with virtuous ecstasy (Sufi spinning/ecstatic dancing, Pentecostal speaking in tongues, Hindu erotic devotionalism, etc.).[64]

Green Vampires and Erotic Restraint

Casting alluring, sexy vampires as the bearers of environmental values and practitioners of ecopiety significantly recasts the seemingly dour obligations and collective moral restraint associated with environmentally virtuous behavior in terms of erotic self-interest. This kind of self-interest is tied to what social psychologists and behavioral change researchers P. Wesley Schultz and Lynnette Zelezny argue is the strong value Americans place on "self-enhancing life goals" and "self-enhancing values." Self-enhancing values held by "a sizable percentage of the US population," write the researchers, "have largely been considered incongruous with the values that lead to environmental concern and to environmental behavior."[65] The researchers argue that strategic use of self-interest in reframing "well-crafted" environmental messages to *harmonize* with America's "self-enhancing values" effectively motivates greater environmental behavior among groups that have been historically so unmotivated. This includes actions such as energy conservation and proper hazardous waste disposal as well as becoming active in supporting environmental ballot initiatives. To those who balk, contending that such an approach is "pandering," and that it would be much better to focus efforts on changing American values to those that embrace deeper concern for collective well-being (i.e., "self-transcendent values" or "socio-altruistic values"), Schultz and Zelezny say they agree. However, the transformation of American values is a long-term project with a "long-term solution," and time is short. Rather than being concerned about motivating people for the "right" reasons, the immediacy and gravity of environmental problems necessitate, depending on audience, "framing the appeal in a way that is consistent with self-enhancing values."[66] The most effective approach, the researchers contend, is to target audiences with a "diversity of messages that will appeal to people with a different range of value orientation"—from socio-altruistic values to self-enhancing values.[67] Bluestem Communications (originally formed under the title of "The Biodiversity Project" by concerned scientists in the 1990s) is an environmental public opinion awareness nonprofit that specializes in working with values communication in environmental messaging across diverse and varied demographics. The organization's public environmental messaging campaigns have successfully demonstrated the efficacy of Schultz and Zelezny's self-enhancing messaging strategy as the shortest distance to motivating environmental civic engagement in typically resistant or uninterested demographics.[68] The popular enthrallment with vampires in the field of cultural production, in effect, deftly deploys the power of this self-enhancing values messaging strategy, coupling environmental moral restraint with alluring desire and erotic pleasure.

Twilight: Vegetarian Vampires in Volvos

The tensions between epic desire and restraint that make for such delicious suspense in Bram Stoker's Victorian Gothic *Dracula* infuse contemporary stories of vampires who struggle to temper their desires by abstaining from drinking human blood. For a number of reasons, including love, social responsibility, or simply just bad PR, these vampires must seek out alternative resources. This leads to what is often characterized as a more ethical vegetarian lifestyle, even though it ironically often necessitates subsisting primarily on animal blood. That is, "vegetarian vampires" have successfully disciplined their desire for human blood, but in the process, nonhuman animals are rendered figurative tofu and their value as living, breathing, feeling creatures largely denied—a notion that is contested in some vampire media narratives and is further challenged by fans themselves. What kind of work is the green "vegetarian vampire" in popular culture doing as both mirror and engine of contemporary environmental ethical concerns and sensibilities? Even as stories of vegetarian vampires evoke the environmental problem of a voracious consumer culture, and its unsustainable consequences, they do so while tapping into audiences' own erotic desires to be all-consumed. Or, as historian Scott Poole points out, "Popular fascination with the vampire provides the best example of the strange human tendency to want the thing hiding under our beds to be in bed with us."[69]

In the film version of *Twilight*, based upon Mormon Stephenie Meyer's vampire fiction trilogy, Edward, the vampire object of teenage human Bella's affections, describes his family of "good" vampires as being "vegetarians," since they abstain from consuming humans and instead hunt animals for sustenance. Edward's commitment to consuming alternate resources, and his decision to resist or temper his desire for human blood, are echoed by Bella's own practice as a vegetarian, which is repeatedly contrasted (in the film version of the story) to that of her meat-consuming father. Much as Fruitlands' cofounder Bronson Alcott once contended that "outward abstinence is a sign of inward fullness," Edward struggles between his love and protective feelings for Bella and his strong desire to make her his next meal.[70] As viewers and readers, we are schooled in the depths of longing and the eros of restraint, a lesson that has implications for America's consumer lifestyle and its effects on planetary well-being.

At one point in *Twilight*, in order to save Bella, Edward must suck vampire poison out of her body. Once the poison is extracted, he must fight his nature so as not to suck her dry and leave her dead. Going against every vampire instinct, Edward manages to stop himself from feeding on her any further, tempers his hunger, and tears himself away in the middle of intense pleasurable engorgement. He successfully restrains himself in the interest of the larger payoff—not only Bella's sustained and sustaining love but eventually the intensified gratifications of a delayed erotic intimacy. It is here that we see some of the self-interested pay-off, or "self-enhancing values," for pious restraint and temperance.

Bella clearly and repeatedly communicates her desire for a consummated sexual relationship with Edward, but it is Edward who, in a gender-role reversal of the cock tease, while professing his intense longing for her, insists on abstinence until marriage. Although restricting sex to the confines of marriage is certainly in keeping with the Mormon values of the series' author, there is also a clever way in which Edward's enforcing abstinence, his *not* allowing Bella what she wants—phallic penetration of both fangs and penis, the intimate exchange of bodily fluids (both blood and semen)—ties into power and control dynamics of BDSM erotic play. That is, Bella wants "it," but Edward holds her back, physically at times, restraining her in his strong arms that can and *do* crush a car.[71] Meyer's narrative conveys a common theme promoted in Mormon sexual purity literature–that waiting for it, that *not* having it, that being held back from it—makes it all that much more hot and intense. The BDSM world is very closely linked with vampire popular culture and, in this world, it is the tying up, the restraint, and the begging for it while *not* getting it that makes the "it" finally delivered an intensified reward.

When Edward finally "turns" Bella and makes her a vampire, she becomes, as it were, the Mormon convert who is then granted immortality via being sealed to her husband in celestial marriage. Once within the confines of marriage and sealed to her new vampire immortality, Bella has fangs of her own, and she is no longer the weak and vulnerable one. She immediately becomes the phallic aggressor, the penetrator, and can give as good as she gets. All restraint is off, as neither partner now has to hold back sexually, and yet both still remain pious, temperance-practicing "vegetarians" in their food sourcing, refraining from consuming humans. The married and thus now (from a Mormon perspective) morally sanctioned sex of this couple is explosive, and Bella wonders to herself how they would ever get themselves to stop ("doing it"), since she and Edward do not require sleep. Were they not blessed with a child to be responsible for (another key Mormon message in the series), Bella muses that they might never be able to rein in their all-consuming passion and get anything else done. This promised intense pleasurable reward for the moral restraint of desire is typical in a number of teen purity/chastity cultural productions that market abstinence, using the savvy alluring messaging of self-interested pleasure. Might sexy environmental messaging of anticonsumption and eco-pious moral restraint promise a similar payoff?

Leah Lamb thinks so. Lamb is a former founding producer and host of Al Gore's Current TV channel (2005–2013), a cable station devoted to environmental issues. After binge reading the *Twilight* books on vacation, Lamb summed up her take-away from the series in her blog: "The earth is our host and we are one helluva mega coven of vampires."[72] She proceeds to enumerate all she learned about environmental sustainability from reading *Twilight* and notes how "it's a hell of lot more fun" to read teen romance novels than dry books on subjects such as dolphin slaughter, drowning polar bears, fish kills, virgin forest clear cuts, sharks going extinct, and oceans filled with plastic debris. Despite its delivery via modes of mediated popular culture, Lamb finds that *Twilight* communicates a powerful and pronounced message about overcoming our "true nature" and refraining from killing the host we love. "I think I love this story," confesses Lamb, "because in so many ways (not to be melodramatic or anything), I live this story. Every day, I fight my impulses in the name of saving the planet as we know it. The story of the times we live in is that the nature of our culture and society is to consume past the point of sustainability." She enumerates her daily struggles with ecopiety in her efforts to buy the right stuff: how she sleeps on an eco-friendly mattress, wears mostly consignment store clothing, uses nontoxic paint in her house, and drinks sustainably harvested fair-trade coffee. She then lists her eco-transgressions and the details of her heavy carbon footprint: owning consumer products from China, drinking tap water that comes from the Hetch Hetchy dammed reservoir, enjoying Hawaiian orchids in her home, and eating organic bananas. "Nearly everything I touch feels like it has a destructive element to it," she regrets; "either it was made via unfair labor, used the land through unsustainable practices, has a mammoth carbon footprint, or is dangerous to throw away. It makes me feel like . . . a monster. A vampire monster to be exact, sucking the life and soul out of the planet in order to sustain my own life."[73]

But here is where Lamb, who describes herself as "a sucker for vampires," finds hope. She draws a number of lessons from the vampire media she consumes, including taking the long view of the planet, since, as she reasons, vampires have lived through many generations and centuries and have the kind of longitudinal perspective that humans lack. It also makes it easier to learn from the mistakes of the past—something she advocates humans do more of. But the main lesson she draws is this: if vampires can make "conscious decisions to find new alternatives to sucking blood and killing the things they desire, then so can we." Vegetarian vampires, in particular, model the struggle to stop killing the things we love. Lamb emphasizes that vampires can help humans develop a new skill of "saying no in the face of plenty."[74] If stories of sexy teen heartthrob vampires can convey that message in an alluring way, then Lamb is not picky or moralistic about the delivery system. If it takes slinky vampires to seduce the public into environmental moral restraint, so be it.[75]

Lamb is but one of a number of environmentalists who have invoked vegetarian vampires as models of moral restraint, while forging "partnerships" between vampires and organizations that promote sustainability and conservation. The Nature Conservancy's "*Twilight* Conservation" marketing campaign, for instance, attracted financial contributions to protect the Washington coast's old-growth forests by "branding" these forests as critical and sensitive "vampire habitat."[76] In conjunction with the 2012 film release of *Breaking Dawn: Part Two*, the Nature Conservancy strategically launched a series of press releases and online media advertisements touting its campaign to protect and restore coastal rainforests. The campaign raised "awareness" with lines like, "Did you know that The Nature Conservancy protects hundreds of thousands of acres of vampire habitat in this region? It's true. Have you seriously forgotten the treetops upon which Bella and Edward first flirted and cavorted? We're hard at work to make more of that deep, dark forest where vampires can prowl. . . . And it's not just vampires. The coast needs werewolves too."[77]

The Nature Conservancy cleverly punctuated this media campaign with breathtaking photographs of giant old-growth trees in mysterious boreal forests, where we can picture Bella and Edward on their first "date." This kind of creative use of mediated popular culture, strategically drawing from and repurposing iconic franchises, is precisely what Henry Jenkins and his research team from the University of Southern California argue is sparking creative opportunities to motivate civic engagement.[78] Conducting cross-cultural research, the team formulates a global "Atlas of the Civic Imagination," studying how civic-minded storytelling becomes critical and fertile inspiration for civic engagement and social change.[79]

Beyond "vampire habitat," *Twilight* fans have been concerned about the conservation of the actual town of Forks, Washington, the setting for *Twilight*, as a pilgrimage site for film/book tourism. Fans contribute to the preservation of key scenes in Forks depicted in the book and films, sponsoring conservation projects and pooling resources for road trips to Forks, where they revel in the chance to retrace Bella and Edward's steps. Once a fairly depressed logging town, Forks has now reinvented itself in the image of *Twilight*, and the boom in tourism has had real-world environmental and economic implications.[80] Less dependent on the logging economy, the town now hosts activities like "vampire habitat walks" and nature experiences both in the rainforest and on the local beach at LaPush, both of which tie into the novel and adopt the language used by the Nature Conservancy's vampire conservation media campaign.[81] Bella Italia, the restaurant featured in the film version of *Twilight*, is a tourist favorite where Bella's favorite vegetarian dish—mushroom ravioli—is de rigueur dining for visiting "Twi-hards."[82] Being swamped by film and eco-tourism poses its own environmental challenges to the region, but conserving instead of cutting the surrounding forest—preserving the scene of young romantic vampire love for Twi-fan tourists to experience—has provided a boost to Forks' economy.

People for the Ethical Treatment of Animals (PETA) is another organization that has aligned itself strategically with the eros and romance of *Twilight*. Actress Christian Serratos, who plays one of Bella's friends in *Twilight*, famously posed nude in a *Twilight*-like forest scene for a PETA advertisement that is part of a series combatting the human wearing of fur. In the ad, Serratos leans coquettishly against a tree, with her bare backside and breast in profile, staring doe-eyed into the camera, looking like a woodland creature herself—captured and consumed by the camera's gaze. The blood-spattered tagline has Serratos declaring, "I'd rather go naked than wear fur."[83]

The actress looks back at us, seductively inviting us into the eros of abstinence, making the "not doing" of not wearing fur hot and alluring. The advertisement's visual rhetoric locates Serratos in the primal woods—naked, sexy, and available—and yet "just saying *no*," her restraint is simultaneously pious and sexy. In an interview posted on PETA's web site, the actress foreswears "slaughtering, eating, and wearing carcasses" and reveals that she, like *Twilight*'s vampire Cullen family and human character Bella, is vegetarian. She recounts, "When I stopped eating meat, I noticed that it was easier for me to focus, and I was really proud of myself for being green also."[84] *Greenness*, here a virtue and moral accomplishment to be proud of, does not read in this advertisement as an act of dour asceticism, grim denial, or buttoned-up virtue. *Not* consuming is instead deliciously erotic.

The rhetoric of virtuous vegetarian vampires in *Twilight* does not go uncontested by online vegetarian "Twi-fans." "Equating eating animals with vegetarianism is just plain wrong," writes Laura Wright, a postcolonial literature professor, both a *Twilight* fan and a blogger on representations of vegans in popular culture at the Vegan Body Project web site. Wright points out that, "in a moment that has received much attention in the veg press and blogosphere," Edward Cullen explains to Bella that "[w]e [the Cullen vampire family, to which he belongs] call ourselves vegetarians because we don't drink human blood. But it's kind of like a person surviving only on tofu; you're never really satisfied." A number of critics have howled at the hypocrisy of rendering animals as figurative tofu. Wright indeed cries foul at casting vampire "vegetarianism" as the "very antithesis of vegetarianism—eating animals." She goes on to reason, "[J]ust because you're not a cannibal doesn't mean that you're a vegetarian."[85]

Tensions over the moral virtues of vegetarianism also get debated in the realm of *Twilight*'s online fan fiction at FanFiction.net, in which authors play with potential moral conflicts by creating new/speculative narrative arcs for Bella and Edward. In a fan fiction piece titled "Green Bella," author "SilverCloud234" poses a scenario in which vegetarian Bella "loves the environment and animals," but Edward has neglected to break it to her that, although he identifies himself as a vegetarian, this really means he eats animals. An awkward moment ensues. Bella has just come from an animal rights activist protest, in which she has blocked an oil tanker with her body. Edward sheepishly confesses he has something to tell her and reveals, "I don't eat people. I eat animals." Bella flies into a rage and argues that she would much rather that he eat *humans* instead of animals and then pleads with him to make her a vampire so she can "kill all the humans and save the earth." Later, Edward complains to Jacob, his werewolf romantic rival, about Bella wanting him to become a real vegetarian (the vegetable kind). Jacob commiserates, saying, "You too! She was trying to get the pack to go green. Do I look like I can live without meat?" In frustration and annoyance at Bella's sanctimonious outrage, Edward ends up pouncing on Bella and sucking all her blood out once and for all. FanFiction.net participatory expressions of "talking back," or, in this case, parodying narratives of virtuous vegetarianism, righteous restraint, and the austere piety of "going green," all speak to the contested reception of these notions as moral absolutes. Clearly some fans, both vegetarian and not, are not buying it and are thus producing restoried and subversive versions of the core text to share with others.[86]

Product tie-ins with the *Twilight* series engender even more audience skepticism, as well-dressed, affluent Edward drives a shiny silver Volvo C30 sports coupe and later insists on buying Bella a Mercedes Benz. Capitalizing on the opportunity for its reputably reliable and safe cars to gain a bit of younger sex appeal, auto manufacturer the Volvo Group made a product placement deal with the *Twilight* franchise and ran commercials in conjunction with the release of the film *New Moon*, which featured scenes of Edward driving his sporty Volvo and authoritatively placing Bella within it. Ad copy for the campaign read, "There's more to life than a Volvo. There's the power to keep safe what you hold most dear. That's why you drive one."[87] Fans could go to WhatDrivesEdward.com and enter to win tickets to the *Twilight* premiere and their very own version of Edward's Volvo.

Stephenie Meyer may have cast heroine Bella Swan as an antimaterialist and thus, in many respects, an outsider to her teen consumer cohort, but Bella certainly trades up in consumer lifestyle when Edward becomes her boyfriend, and again, when she marries up into the Cullen family. The couple even honeymoons on the Cullens' private island. We are not quite sure just how wealthy the Cullens are, but in 2010, *Forbes* magazine was sufficiently intrigued and admiring of family patriarch Carlisle Cullen that the magazine rated him their "#1 Wealthiest Fictional Character," estimating his net worth at $34.1 billion. At one point, Bella goes to see a zombie movie with her high school friend, who remarks in a satirical "meta" moment as they leave the theater, "I don't know why you'd want to sit through all those zombies eating people. . . . It's gross. Is it supposed to be a metaphor for *consumerism*? Because, don't be so pleased with your self-referential cleverness, you know? Some girls like to shop . . . not *all* girls apparently [referencing Bella]." Ironically, of course, this nod to the critique of voracious capitalist consumption communicated via both zombie- and vampire-themed storied media is executed through a genre and media franchise that fuels the multi-million-dollar marketing success of endless consumer tie-in products. Nordstrom's department store has carried *Twilight* series clothing and make-up for that chic undead look.[88] Despite the series' message of sexual abstinence until marriage, *Twilight* condoms became popular with fans, as did items such as *Twilight* soap, underpants, toilet paper, tampon cases, diaper covers, cookbooks, and more exotic items such as sparkly Edward freezable dildos (for that authentic cold vampire touch).[89]

Fan sites provided links to full-sized Edward wall decals for girls' bedrooms so that, like Bella, they could feel watched at night by him.[90] These items do not even touch on the usual media tie-in merchandise hawked to fans: t-shirts, soundtracks, DVDs, hats, keychains, pens, notebooks, action figures, lunchboxes, candy, costumes, sheets, pillows, and pajamas. Although not directly tied into *Twilight*, such specialty items as red vampire contact lenses and even "salt and vampire-flavored" Pringles potato chips have enthralled devotees.[91] This consumerist feeding frenzy, in addition to the over one hundred million books sold in the series, prompted university librarian and writer J. Turner Masland, veteran blogger at *Dewey's Not Dead*, to declare author Stephenie Meyer to be "the worst ecological catastrophe of the 21st century."[92] Boggled by the sheer volume of books sold, Masland invokes the famous Dr. Seuss environmental advocate, remarking, "The Lorax is weeping. . . . [A]ll those trees add up."[93]

Fan tourism to Forks, Washington, may provide the town with alternative sources of economic sustenance to logging, and vampire media may popularize the clever notion of conserving "vampire habitat" in the Olympic peninsula, but the resources to churn out seemingly endless *Twilight* merchandising come from somewhere. As the work of de Certeau, Fiske, Hall, Jenkins, and numerous other cultural theorists reminds us, media reception is always tricky and contested, so perhaps it is no surprise that messages of moral restraint in consumption end up driving sales of vampire condoms and sparkly frozen dildos.[94] What is apparent is that the narratives embedded in the cultural production and the messaging used to market and sell those products are frequently contradictory, and an anticonsumption message of restraint can be used effectively to stroke and even intensify the consumer desire to desire.

Addictions and Interventions

Green vampires as tropes in popular cultural narratives explicitly reference or evoke the language of vampirism employed in environmental discourse to signal the depleting, and thus unsustainable, nature of capitalist consumption as being that which sucks the very life from the planet. Addiction is a main theme of the CW Television Network's teen-targeted *Vampire Diaries* (2009–2017), a television series in which vampire hero Stefan is yet another contemporary "vegetarian vampire." Throughout the series, Stefan struggles with his addiction to human blood, exhibiting many of the symptoms of a heroin addict. Despite its etymological connections with vegetation, "vegetarian," once again, signals ethical temperance *not* from animal blood but from human blood. Like Edward Cullen, Stefan maintains a steady diet of woodland creatures in order to survive. This trade-off, in which nonhuman animals are more morally acceptable to consume than human blood, is hotly contested by Stefan's human-blood-consuming brother, Damon, in the ongoing series' debates about consumption ethics. In a season 2 episode called "Brave New World," Stefan chastises Damon for stealing human blood from a blood bank. Damon turns the tables on Stefan and asks, "Aren't you worried that one day all the forest animals are going to band together and bite back?"

The question of whether Damon's consumption of freely given/harvested human blood, albeit stolen from blood banks, might actually be more ethical and sustainable than Stefan's rampant depletion of the forest ecosystem continues to surface in the narrative. The series also draws symbolic connections between blood addiction and unsustainable consumption of natural resources. Art mimics life mimics art as, in the series' second season, the cast member who plays the vampire Damon teamed up in real life with the renewable energy equipment leasing company Go Green Mobile Power to replace *The Vampire Diaries* set's polluting gas generators with portable renewable "green-energy" generators, citing the moral imperative for humans to wean ourselves from addiction to fossil fuels.[95]

Vampire Damon is not the only one who contests the legitimacy of applying the label "vegetarian" to human-blood-abstaining vampires. In the digital realm, series fans actively engage questions of animal ethics and environmental sustainability. They talk back to Stefan's supposed vegetarianism by posting memes depicting him preying upon poor, helpless bunnies. Following an episode in which Stefan is trying to teach Caroline, a newly made vampire, how to be "vegetarian" by hunting for bunnies, a "Stefan, don't eat the bunny" theme emerged in fan forums. In the episode, Caroline questions whether bunny consumption truly is ethically preferable to human consumption, and whether this might not be the kind of act a real monster, like a "serial killer," would commit. Caroline's conflicted exchange with Stefan follows:

> CAROLINE FORBES: So what do I do when I see the rabbit?
>
> STEFAN SALVATORE: Chase it, catch it, feed on it.
>
> CAROLINE FORBES: Isn't killing cute defenseless animals the first step in becoming a serial killer?
>
> STEFAN SALVATORE: Well, you sort of skipped "serial killer" and went straight for "vampire." Hey Caroline, if you're not serious about this then I think you should tell me.
>
> CAROLINE FORBES: No, I am. Look, I swear that I am! Okay, but it's just I haven't been in the sun for days, and everyone's at the swimming hole having fun and Matt is there and he finally told me that he loved me, but I've been blowing him off, and now YOU want me to eat bunnies and I'm kinda freaking out, okay?[96]

A series of fan memes ensued. One memorable one, posted in the VampireDiaries.net forum, depicts a terrorized bunny screaming beside a photo of Stefan in monstrous form declaring, "I Eat Bunnies!" Another image reposted by fans from Tumblr.com shows Stefan looking down at an adorable bunny, who looks into the camera with sad eyes and pleads, "Please don't eat me."

Fans of *The Vampire Diaries* have also engaged in lively online debates about "the powers of nature," what constitutes "nature," and what might be the bounds of its moral agency. This exchange was sparked by various storylines in the series that involve "the powers of nature" as the source of power for witches, who draw power from nature to cast their spells. In *The Vampire Diaries* universe, witches work in harmony with the powers of nature and strive to maintain nature's balance, while vampires circumvent or interfere with the cycle of life, which is antithetical to nature's laws. Sometimes the powers of nature are referred to by the series' characters as simply "the Universe," a practice also common in theosophical tradition.[97]

On *The Vampire Diaries* wiki, a collaboratively fan-edited web site that publishes multiple descriptive entries on myriad aspects of the television series, participants in the wiki's "Earth" entry engage in an animated discussion about the nature of nature and what creates balance. One contributor questions why vampires would be antithetical to the balance of nature and not humans, since humans pollute the earth. A subsequent exchange deals with whether nature "has a conscience" or not. Another contributor to the discussion reasons that if nature "has a plan," then it must have a conscience. A reply to this statement agrees, saying, "[I]t [nature] does have a conscience like in real life, I mean what controls the weather?" A cantankerous reply to this posting says that the wiki page should be deleted, arguing, "There is no evidence that supports nature has consciousness. This is pure speculation." A participant in this exchange adds, "I think the concept of Nature is foolish," while another proclaims, "Nature itself is actually neither good nor evil because it is both."[98] One fan speculates what it would be like if nature were a person, played on the series by an actor, "sort of Nature incarnate in human form." A reply to this pooh-poohs this idea, countering, "[N]ature is earth, so how can it be a person . . . you cannot say nature is a person, it's not." A second reply offers, "Nature is not a deity, or a spirit, but pure energy!" The conversation continues, as many different concepts and understandings of nature get debated: Does nature have a plan or does it not have a plan? Are the forces of nature good, evil, or just trying to restore balance? Are humans disruptive to nature's balance, like vampires, or simply part of nature and that ongoing balance? When nature is threatened, does it defend itself? What kinds of powers does it use?

In VampireDiaries.net forums, ongoing fan debates about the series' recurring theme of the "balance of nature" actually led to the posting of a viewer poll that asked, "Do you believe in the balance of nature line?" Participants in this forum again hotly debated the concept of the "balance of nature," and some fans expressed anger at "Nature" [usually capitalized by fans] when, in the storyline, Nature was responsible for killing off their favorite characters in order to restore balance.[99] For a television series that is often dismissed as teenager brain candy, *The Vampire Diaries* sparks discussion on the wiki that certainly moves beyond the stereotype of fans gossiping about whom the series actors are dating. Online forums provide communal spaces where viewers can compare notes, as they theorize series concepts and mull over the moral, ethical, and existential dimensions of the show. Sexy actors penetrating each other with fangs may attract a kind of concomitant pleasure and enthrallment with these topics, especially among a younger demographic to whom environmentalists like Al Gore and Bill McKibben, with their well-ordered slideshow lectures, are a much tougher sell. *The Vampire Diaries*' and *Twilight*'s teenage fandoms are not unique in this regard, as adult vampire drama has also provided a kind of narrative social lubricant for social-environmental critique and shifting social energetics.

"*True Blood* Is Making Me Want to Go Vegan"

Producer Alan Ball depicts in the Home Box Office (HBO) series *True Blood* (2008–2014) a phantasmagorical world of supernaturals: vampires, faeries, werewolves, maenads, shape-shifters, voodoo queens, shamans, and witches. To create *True Blood*, Ball, who previously won acclaim for his disturbingly beautiful and violent portrayal of American suburban life in the film *American Beauty*, adapted best-selling author Charlene Harris's popular novel series, *The Southern Vampire Mysteries*. In the series, heroine and telepathic waitress Sookie Stackhouse solves crimes amid romantic vampire liaisons in her small Louisiana bayou town of Bon Temps. In *True Blood*, vampires have "come out of the coffin" and begun to "mainstream" now that the Japanese have invented a synthetic (albeit untasty) blood substitute called "Tru Blood" that meets vampires' nutritional needs. As with the greater availability of foods such as Tofurky and seitan, which technically satisfy humans' nutritional needs and thus render human consumption of animal protein unnecessary, vampires no longer need to consume humans to survive. More socially conscious Tru Blood–drinking "vegetarian" vampires practice temperance, pay taxes, and agitate for civil rights to marry whom they please.

The source of conflict in much of the first season is between bloodthirsty merciless vampires (who continue to kill and consume humans) and series hero Vampire Bill, a respectable (khakis-and-button-down-wearing) vampire who cultivates self-restraint as he tries to figure out the ethical limits of desire. Bill struggles to restrict his diet, feeding only, at first, consensually from his girlfriend, Sookie. Even this limited temperance becomes unbearable, since it turns out that Sookie is a human/faerie hybrid, thus endowing her with irresistible faerie blood—the vampire version of crack cocaine. In Bill's struggle to sublimate the desire to kill and consume (and his struggle with backsliding), a moral imperative plays out—the pressing need in the Anthropocene to adapt to and rapidly adopt alternative resources in the face of changing social and environmental conditions. Bill models not only the imperfect struggle toward pious restraint but the embrace of environmental civic responsibility, as when he admonishes the young vampire under his custodial charge not to hunt humans to bring home to drink. He instructs Jessica on making the moral choice to sublimate her desire for instant gratification and engorgement—consuming bottles of synthetic blood substitute instead of humans. With frustration, Bill then picks up an iconic blue recycling bin and sternly adds, "We also *recycle* in this house! Tru Blood and other glass products go in the blue container, and paper products in the white container."[100]

In response to Bill's performance of such a recognizable mundane act of ecopiety, environmental bloggers, Pinterest.com pinners, and "Truebie" fan chat rooms, among others, went wild with exclamations such as, "I Love Environmentally Correct Bill!"; "Even Vampires Recycle! . . . What's next? Is Blade going to install solar panels?"; and "Bill is so eco-friendly!"[101] *True Blood*'s official Facebook page and Twitter account have repeatedly wished Truebies a Happy Earth Day by reposting this

Figure 5.2. *True Blood's* Vampire Bill, "We Recycle in This House." *True Blood's* Vampire Bill instructs his ward Jessica on eco-pious recycling etiquette. Credit: Home Box Office.

clip with the caption, "Bill Compton recommends you honor Earth Day today, Truebies."[102] As with HBO's *True Blood* viral marketing campaign, borders and boundaries blend between the on-screen world and the off-screen world, as the official series web site has posted features on how environmentally active the actors who play the vampires are in real life, and Stephen Moyer's publicist has placed stories that tout Moyer and his costar/wife Anna Paquin as "recycling obsessives."[103]

Beyond moral directives on recycling, consumptive desire troubles the bayou waters of Bon Temps, as the town's citizens develop intense addictions to vampire blood or "V," a drug that functions like a combination of Ecstasy and Viagra, leaving the user with a voracious appetite for consuming more. Desire and its insatiable partner addiction pervade the small Louisiana town in the realm of sex, drugs, blood, and death. Both humans and vampires in Bon Temps are consumed with the desire to consume, a not-so-subtle social critique of America's own vampire-like, extractivist, consumerist addictions that pervade the series. The visual rhetoric of the series' vampire bar, Fangtasia, further dramatizes this critique, as a framed illustration behind the bartender depicts former president George W. Bush as a vampire, fangs bared, feasting on the jugular vein of the Statue of Liberty. This iconic image, painted by comic book artist Alex Ross, was taken from an October 2004 *Village Voice* cover that accompanied an article, "It's Mourning in America," by *Nixonland* author Rick Perlstein.[104] "Vampire George" ensconced on the set of *True Blood* did not go unnoticed. The episode aired on a Sunday night, and by Monday social media had already shared this image widely, accompanied by pointed political critique and commentary.[105]

"Can a corporate behemoth like HBO be the source of important media interventions?" asks Shayne Pepper, a media scholar of "public service entertainment."[106] Pepper argues that mainstream media outlets are too often dismissed as "incapable or, at least, unwilling to take on the task of producing and distributing programming that makes decisive critiques of social inequality or demands collective (if not radical) action from its audiences." Insulated from many of the dicey concerns of individual advertising sponsors, and, unlike PBS, free from federal funding and its constraints, HBO has a subscription-based model that enables it to take political risks, pushing the envelope on controversial social issues from HIV/AIDS to global poverty. Although many of these risks are also clearly taken for profit-driven reasons, not simply altruistic ones, Pepper traces HBO's history of media innovation, social critique, and media intervention.[107] This history spans funding not only edgy documentary films but also dramatic films and miniseries infused with socially engaging content. It is in this context that Alan Ball's environmental and social commentary in *True Blood* and, as we will see, his *Six Feet Under* HBO series are made more legible as both media interventions and moral interventions.

True Blood character Russell Edgington, a former "vampire king" of Mississippi, articulates directly and synthetically one of the series' most powerful interventionist moments. This comes in season 3, in an episode called "Everything Is Broken," via an Edgington monologue, in which the former king enumerates the many ways in which humans' inability to temper consumption makes them, effectively, no different from vampires. Edgington commandeers a television news anchor desk and addresses the US human population live on TV: "We [vampires] are narcissists, we care only about getting what we want no matter what the cost, *just like you*. Global warming, perpetual war, toxic waste, child labor, torture, genocide. That's a small price to pay for your SUVs and your flat-screen TVs, your blood diamonds, your designer jeans, your absurd garish McMansions—futile symbols of permanence to quell your quivering, spineless souls. . . . You are not our equals. We will eat you. After we eat your children."[108]

A powerful, erotic, ecstatic, and deadly compulsion to consume saturates *True Blood*'s narrative arc. In the fifth season of the series, we learn that this vampiric compulsion also constitutes no less than "a *religion*." Russell Edgington is part of a larger movement of vampires called the "Sanguinistas," religious fundamentalists who oppose vampires mainstreaming into human culture, read literalistically a sacred text called the *Vampire Bible*, and regard vampires' consumption of human blood as a divinely endowed gift and sacred sacrament. In a twist on fundamentalist Christian interpretations of biblical "dominion," the *Vampire Bible*, which predates the Jewish and Christian Bibles and which is referred to as the "Original Testament," states that humans are "no more than food," and that humans were created for the specific purpose of vampire consumption. According to the Sanguinista fundamentalists, vampires thus have been divinely granted dominion over humans and can use them as they wish, farming them like veal, if they choose.

More than simply storylines and dialogue, *True Blood*'s blood-and-guts aesthetics of drained corpses, half-consumed bodies, and exploding vampires meeting their "true death" prompted some viewers to write about how the show was making them rethink their meat consumption. This is perhaps not surprising, since many Americans eat while watching television, and much of the flesh consumed as prey on the show is human. Journalist Andrea Chalupa's "*True Blood* Is Making Me Want to Go Vegan" recounted in the *Huffington Post* her dilemma of having friends over for dinner to watch *True Blood*: "Just as we were about to bite into our chicken pot pies, there's a blood bath on the screen." She confesses, "Eating meat and watching *True Blood* doesn't always sit well." Her solution was to create a weekly ritual of "vegan Sundays," in which she would serve her friends vegan versions of traditional southern dishes, such as "Veggie Biscuit Pot Pie." Willing converts empathize and share similar reactions within the article's comments section, posting their own meat-based dinner recipes and asking for suggestions for *True*

Blood–watching modified meatless conversions. Established vegetarians who comment take the opportunity to share links to documentary exposés of the cruelty and unsustainability of the meat industry and provide ways for fans to connect to activist organizations.[109]

Vegan and vegetarian bloggers and discussion sites such as VeganMainstream.com and VeggieBoards.com began "lay theorizing" various vegan ethical elements in the series as well as "outing" who the vegan or vegetarian actors were.[110] They paid particular attention to episode 2 of season 4, "You Smell like Dinner," a reference to Sookie's appetizing scent to vampires, who struggle with maintaining the moral restraint to date her, not dine on her.[111] The online discussion and attention paid by *True Blood* fans to the ethics of meat consumption, some indeed changing their eating habits while watching the program, would seem to indicate among fans at least some limited movement from moral engagement with series content to real-world action.

In examining HBO's media interventions, media and politics scholar Shayne Pepper points to the ways in which the subscription-based cable channel successfully brings critical social issues to the fore of national consciousness. Pepper issues a caveat, however, that "mainstream media interventions are more likely to call viewers to 'fight the system' by working within it rather than tearing it down." While these interventions might not "invite radical action to tear down systems of inequality, they at least strive to engage the viewer and make him or her actually care enough to take part in the shaping of those systems—not always an easy task in today's world."[112] What is more, while much of the action taken by fans rethinking meat consumption might be personal and individual—changing their own Sunday night dinner—online participation with other fans has led to encounters with resources and links to opportunities for more collective-based action.

Reaching Peak Blood

In Michael and Peter Spierig's film *Daybreakers*, the connections among vociferous vampire-like consumption, the depletion of earth's natural resources, and animal rights drives a horror sci-fi plot in which the bulk of humanity has willingly chosen to convert to being vampires in their quest for immortality and perfection. As humans become scarcer and scarcer, however, the now-vampire-dominated earth experiences a global blood shortage akin to a massive oil shortage. As the vampires reach "peak blood" and military conflict ensues, the vampires hunt and trap humans (like animals) to be factory farmed by multinational pharmaceutical blood companies. Humans are heavily sedated and warehoused in cramped factory conditions—the vampire equivalent of veal—where they are machine "milked" for their blood. As the few remaining humans are hunted out of existence, vampire hematologist Edward Dalton (played by Ethan Hawke) is missioned by his pharmaceutical company to find a synthetic blood substitute for human blood before world war erupts over the battle for scarce resources. Instead of a viable substitute, Dalton finds a "cure" for vampirism that will convert resource-hungry consumers and return them to their humanity, if only he can administer the cure and rehumanize the world before it is too late. In the midst of his quest, Dalton himself suffers from severe blood deprivation but continues to abstain from human blood on principle because, like the "Little Vampire Women," he feels compassion for suffering humans. *Daybreakers* presents a dark portrayal of our "monstrous" oil culture and capitalism's "consuming passions," while suggesting a humanizing "conversion" from resource-depleting vampirism to more sustainable environmental ethics and practices.

In a twist on *Daybreakers'* premise, following the film, Resilience.org, a real-world digital resource site for various social ecological activist groups, issued a parody press release, "Vampire Coalition Unveils 'Save the Humans' Program." In it, the "International Vampire Alliance for Human Survival" announces that overnight they have captured the human 2014 Climate Summit delegates and turned them into vampires. Why? To give them the long-term view, which will finally make them enact biting climate legislation that would conserve human habitat and ultimately save humans. The press release explains, "[V]ampires have been lobbying for significant and legally binding carbon emissions cuts, along with water, topsoil and biodiversity conservation measures. The immortals are personally invested in the future of the planet, as they can live thousands of years. . . . [W]e care about humans—they're our primary food source. And if you guys are all crowded up around Siberia and Canada, fighting for space and getting drowned in tsunamis and dying of malaria and famines in fifty years or so, well, let's just say that things are going to get ugly."[113] Tongue is of course firmly in cheek in this press release, but the rhetorical power of the vampire in mediated popular culture texts gets us humans to think about what it means to be prey, to be another more-powerful species' food source, to be "farmed" for someone else's consumption, and even to have another species try to conserve our "habitat," even if it means saving us from ourselves so that we continue to serve as food. In vampire narratives, we imaginatively get to "try on" being both the vampire and the prey.

Reading across these cultural works, a vampire environmental ethic emerges that equates the deep monstrous desire to consume and deplete the earth's resources to the vampire's voracious hunger to consume and drain the life out of the body that sustains him/her. In a world of shifting norms, the new green vampire seems to be searching for a kinder, gentler, less bloodthirsty existence, one that has more peaceable and eco-friendly implications. This is new ground for vampires, and it is a struggle to find their way to adapt to new conditions, just as it is for humans to figure out how to meet their needs without sucking the earth dry. In both the human and the vampire case, the path toward greater sustainability involves learning temperance and restraint.

What actually sells this otherwise unappealing message of restraint? Green vampires savvily take a page out of conservative Christian abstinence literature, in which "no" is somehow hotter than "yes." For the vegetarian vampire, in the Sartrean sense, the "trouble with desire" is also "earth trouble," and the great challenge to temper consuming passions has a broader and more longitudinally promised payoff in the creation of a more peaceable earth community for all beings.[114] To be sure, this kind of thematic restorying is not what is traditionally understood to constitute media intervention, but it does, in an expanded definition, as media scholar Nick Couldry puts it, "challenge *existing* inflections of the general will."[115] Popular vampire narratives are interventionary in drawing pronounced attention to and owning our vampiric nature in the Anthropocene, while posing various scenarios of adaptation and/or possible "cure." Restorying the vampire for an age of environmental crisis provides a sexy, fun, appealing floating signifier for restorying the earth and humans' moral relationship to the more-than-human world. However, the unbridled bloodlust of big media marketing and sales machines for mammoth merchandising profits undercuts the interventionary aspects of vampire media narratives and their potential restorying impact on human consumption. Vegetarian vampires and their media makers may preach moral restraint, but they do so while, ironically and unscrupulously, bleeding the earth dry by manufacturing and marketing mountains of cheap, disposable, fossil-fuel-based landfill refuse. If freed from the shadow of the insatiable merchandising machine, green vampires have the capacity to captivate our attention, shift social energetics, and seduce us delightfully and playfully in the direction of social change.

6

Composting a Life

Green Burial Marketing and Storied Corpses as "Media"

In *Composing a Life*, cultural anthropologist Mary Catherine Bateson analyzes the life stories of five female artists, focusing on how these women have adapted to radical life changes—dramatic and often spontaneously shifting environments, work, and family circumstances.[1] The final chapter of her book is titled "Enriching the Earth," and in it Bateson explores how her five subjects strive to do their work, make their art, and live their lives in ways that contribute to the well-being of the planet. It is in the work of "composing a life," contends Bateson, that "[e]ach of us constructs a life that is her own central metaphor for thinking about the world."[2] So, too, "composing a death," or at least the logistical arrangements surrounding the materialities and meanings of that death, function as a lens through which we might examine cultural mediations of death in the Anthropocene and their links to tropes of ecopiety.

Like Bateson's subjects, those who exercise green burial options strive to compose green deaths that enrich the earth—literally, through fertilizing the ground as compost. Such options usually include refraining from injecting chemical embalming fluids into the body and forgoing the practice of encasing the grave inside a concrete, metal, or plastic vault. It can also entail using a readily biodegradable shroud or casket, fostering land conservation and restoration by supporting green cemetery preserves, and even, in some cases, using cremation ashes to grow new trees or coral reefs. There is a dynamic of reciprocity involved in "composting a life," as those who envision and plan green burials are themselves enriched by the process. Narratives of those who reflect on making their own green burial arrangements speak of sensations of "warmth" and the deep satisfaction they feel in envisioning "returning to the earth."

In a culture that psychologists and historians alike have argued is steeped in death denial, a phenomenon that, as we will see, marketing researchers find can actually drive consumption, the green burial movement is nudging Americans into thinking about something we are famously known to go to great lengths to avoid. In 2004, after the American Association of Retired Persons' *AARP Magazine* ran an article on green burial choices, 70.4 percent of the magazine's surveyed elderly readership said that they would choose green burial over other options. In 2005, the first narrative depiction of a green burial appeared on television in the widely viewed and acclaimed Home Box Office (HBO) prime-time drama series *Six Feet Under*. By 2018, a much wider national sampling than that conducted by the AARP showed that 54 percent of Americans reported that they were considering a green burial for their final disposition; and 72 percent of American cemeteries reported that they were receiving increased customer demand for green burials and eco-funerals.[3]

How did we get to the point, in a relatively short period of time, from conventional burial being the contemporary norm to more than half of Americans being open to and considering "composting" themselves? As it has done historically more generally with regard to funerary customs, mediated popular culture has played a critical role in the shifting social energetics around burial and notions of more eco-friendly arrangements for human remains. Storied representations of green burial in television, film, and streamed video, as well as stories formulated and communicated via the realm of consumer marketing, are actively reshaping notions of what constitutes a "good" death and disposition.

This chapter marks a pivot in the book, as it examines with deep ambivalence the marketing of green burial and its messages of ecopiety as performed through consumopiety. It questions whether the environmentally conscious ideals of more eco-friendly funerals have merely been swallowed by the logics of consumerism within a profit-driven marketplace of funerary goods and services. The answer to this is complicated. Yes, a bumper crop has popped up of green burial products and modes of professional assistance in dealing with the dead. In effect, the green burial movement has launched a new "green death" industry and realm for consumerism. However, the movement in both its more idealistic and its consumerist forms resists easy characterization. Like the other "sightings" in this book, its movements are simultaneously multidirectional and demonstrably contrapuntal. Among green burial advocates and evangelists, a strong streak of anticonsumerism and Yankee frugality persists that fiercely resists the notion that one must buy the right funerary accoutrements in order to do good for the earth, reduce one's carbon imprint, and execute an "eco-friendly" final act. Not surprisingly, consumer marketing for green burial and its related paraphernalia emphasizes saving the earth, one single, individualized, heroic, voluntary act a time, as does the marketing of green products more generally. Even so, in the consumer activism promoting green burial, there is marked movement from individual piety to broader public policy. Advocates state by state have lobbied to remove restrictions—many originally lobbied for by the funerary industry—that prevent or discourage DIY corpse care and alternative types of interment. These changes in law have made it easier for green cemeteries to get established and for conventional cemeteries to reinvent themselves as hybrids.

What of economist Gernot Wagner's question, "But will the planet notice?" According to the auditing of the Green Burial Council (an environmental certification organization for green cemeteries, funeral homes, and product manufacturers), each year, American consumers purchase and then proceed to bury 20 million feet of hardwoods, 4.3 million gallons of toxic embalming fluid, 1.6 miles of reinforced concretes for vaults, 17,000 tons of copper and bronze for casket ornamentation, and 64,500 tons of steel.[4] In other words, in terms of scale, if half of all Americans chose green burial, or even a third, it would be more than a mere drop in the bucket. In buying a hybrid vehicle or purchasing organic food, the consumer often must spend more money to choose what has been marketed as a "greener" option. For the most part, such eco-pious practice requires consumers to possess the real disposable income needed to make these choices and to go above and beyond regular spending. Within the realm of green funeral products, though, the allocation both financially and materially for a green funeral can be considerably less than the median cost of eighty-five hundred dollars for a conventional funeral.[5] Green burial is thus not a consumer phenomenon dependent on the small percentage of ecologically minded virtuous affluent, though marketers most definitely aggressively target that demographic. Even so, the "less is more" ethic of green burial makes it both accessible and appealing to those of more meager means who often face going into debt in order to bury a loved one. What is more, there is a qualitatively different consumer experience between the shame of having to scrimp to bury your loved one on the cheap and the dignity of choosing a burial that virtuously gives back to the earth. In terms of shifting burial consumer practices on a scale that the planet will actually notice, price point is key for the nearly 80 percent of Americans who have no savings and report living from paycheck to paycheck.[6] The old joke "I can't afford to die" is no laughing matter for most Americans.

In open land conservation efforts, buried green corpses have also begun to be strategically utilized as "talismans" to ward off big-box, strip-mall, and condo developers. Once human remains are lawfully interred on a property, the legalities and family permissions required to dig up and move said bodies, robbing them of their eternal rest, are hairy to say the least and can thwart even the most determined of developers, or at the very least slow them down.[7] In this and other ways, more than simply a personal act of piety, the impact of committing to green burial can extend and has extended into the realm of public policy and community planning. Billy Campbell, the founder of Ramsey Creek Preserve, the first official green cemetery woodlands created in the United States, advises that "[b]y setting aside a woods for natural burial, we preserve it from development. At the same time, I think we put death in its rightful place as part of the cycle of life. Our burials honor the idea of dust to dust."[8] Florida-based Glendale Memorial Nature Preserve's motto is "From Eden we come . . . to Eden we return." The preserve's web site promotes the power of those interred on its land via the practice of green burial to stave off the cannibalization of open space and to thwart the unchecked expansion of real estate development: "Every day across North America alone, 6000 plus acres of open space are leveled in an ever widening and endless swath of highways and airports, condos, malls, and factories. While the costs of purchasing, restoring, monitoring, and preserving land keeps steadily climbing, millions of acres of natural beauty are lost. Glendale Memorial Nature Preserve, however, is an enclave of escape from that trend."[9]

In visiting this nature preserve in northwest Florida, I drove past an enormous nearby Walmart shopping center and witnessed a sea of other retail developments and accompanying massive parking lots. I came to learn that these more recent developments were all located on paved-over fertile agricultural land that had mostly been family-owned farms and ranches, like the ranch at Glendale. The proprietors at Glendale have fought tooth and nail to hang on to the land, whereas others in their place could not. Green burial has provided the mechanism for doing that. In a win-win, Glendale has also provided a place where low-income families in the Florida panhandle can and do bury their loved ones without going into debt, while the preserve simultaneously protects open space and vital wildlife habitat from developers.

Lastly, there is a larger question taken up in this chapter, one that is candidly much harder to get at. Although both ecopiety and frugality clearly appear to be drivers for those who turn to green burial, there is also something less tangible and more mystical articulated by those who embrace the prospect of "merging with nature," "going back to the earth," or "living on" as another life form in the surrounding ecosystem. With this more amorphous dimension in mind, this chapter posits, albeit without clear conclusion, whether making burial more eco-friendly, and in so doing cultivating pleasing visions of death as "natural" rather than aberrant, might actually counter some of the factors that drive Americans to overconsume in the first place. By teaching us, as climate change philosopher Roy Scranton articulates it, "How to die in the Anthropocene," can green burial concurrently edge us toward learning how to *live* more sustainably in the Anthropocene?

New/Old Practice

Of course there is nothing *new* about green burial, which, as a number of green burial proponents point out, was previously known simply as "burial." The practical need to transport large numbers of dead Civil War soldiers back home for burial over long distances, combined with the heat and unbearable stench of rotting corpses, generated a demand for embalmers. Borrowing a technique invented by the French, these men stabilized the rate of corpse decay through a series of invasive procedures that included arterial injection of a preservative cocktail that contained arsenic, mercuric chlorides, turpentine, alcohol, and creosote, among other ingredients.[10] Widespread popular media coverage showing photographs of Abraham Lincoln's posthumous funerary tour, and the public display of what appeared to be his miraculously preserved corpse, helped to market chemical embalming to American consumers, which eventually became nearly ubiquitous in mortuary practice.[11] Prior to this period, burials in the United States were for the most part what we would today term "green," and embalming is still not widely used in countries outside the United States. Green burial is thus what we might call a new/old practice.

The very word "human" is connected to "*humus*" (earth, soil) and our term "humanities" comes from the Latin "*humando*" (burying). From an anthropological standpoint, "To be human means above all to bury."[12] In death and in funerary preparation, the body itself becomes a *story*. A corpse is also a "symbol system," a system anthropologist Mary Douglas asserts to be "a model which can stand for any bounded system. Its boundaries can represent any boundaries which are threatened or precarious."[13] With these cultural, symbolic, material, and communicative frameworks in mind, we can consider how bodies and their accompanying funerary accoutrements in the mode of contemporary green burial become purposeful media that tell a different story than do conventional contemporary burials about humans' relationships to earth and its cycles. Conventional burial products segregate human remains from the rest of nature, strategically deploying the language in marketing and ad copy of "protecting" and "defending" the body from the forces of rot and decay. By contrast, the marketing of green burial products promotes their role in facilitating a wholesome release of the body more fully into the reclaiming forces of nature. Conventional embalming services are predicated on the consumer accepting funerary industry narratives that communicate something inherently wrong, contaminating, and dirty about a dead, decaying body. To address this pollution requires the body to be drained of fluids and "cleaned," as it were, through disinfection with chemicals. The body is then injected with artificial preservatives in order to appear alive so as to correct visually for its abnormal state of being dead.

The chemically unadulterated green corpse similarly tells a story but one that resists the conventional narrative. Green burial customs tout noninterference with natural processes of decay, affirm that an intact corpse is healthy, nutritious food for the earth, and reassure consumers that it is actually okay for a corpse to look dead because it is. As media interventions, green corpses do cultural work to disrupt and challenge pervasive death-denial narratives.[14] Rather than being symbolically set apart from nature, a green corpse ideally also undergoes burial in such ways as to reflect the person's essential identity as mere humble part and particle of the great biotic whole. Like materiality itself, the "stuff" of green burial constitutes "a fundamental platform, and medium, of sociality."[15] In something of a reversal of what marketers refer to as "the consumer journey," when entering the world of green burial, the consumer is asked to become willingly, readily, and literally that which is consumed.[16]

Bodies, Graves, and Caskets as Media

Death, writes media theorist John Durham Peters, "is a great revealer of infrastructures."[17] All recording media, he contends, are, in some way, an attempt at immortality, an attempt to defy death and to generate something that lives on beyond the grave. In his expansive theorizing of media, which includes a subject range from clouds and dolphins to fire and sea vessels, Peters analyzes graves as media and indeed bodies themselves as media. Simultaneously natural and cultural, media, as understood by Peters in reciprocal terms, are environments, while environments (water, fire, sky, earth, etc.) are media.[18] Of all media, Peters argues in *The Marvelous Clouds*, "The body, a mix of sea, fire, earth, and sky, is our most fundamental infrastructural medium."[19] In this vein, both the eco-piously prepared bodies of green burial and their accompanying mortuary accessories have much to tell us in both their inscribed messaging and their consumer marketing.

There are a number of catalysts for the growth of greener ways to "die artfully."[20] A primary one has been that the environmental consequences of conventional burial are increasingly untenable to those morally committed (in life) to caring responsibly for—and lessening their impact on—the earth.[21] In an age of widespread use of formaldehyde and other toxic embalming fluids, steel-lined caskets, and concrete-lined vaults or graveboxes, the realization of a more environmentally conscious dust-to-dust death has become a considerable challenge. The grave, Peters contends, is "one of the most basic of all human meaning-storage devices."[22] As "media," green-burial corpses and their funerary product accompaniments, or lack thereof, all become signs of commitment to ecopiety. They effectively restory humans' relationship to earth and to their own mortality, while heretically challenging the ultimacy of consumerism as a social value. In fact, Peters reminds us that the ancient Greek word for "sign" ("*sēma*," the root of "*semantic*" and "*semiotics*") means "tomb," as tombs served as visible markers or signs to the living, communicating the presence of the unseen dead.[23]

The choice of a greener death can also be about practicing ecopiety by harmonizing the body, as medium, with the larger ecosystem and its natural cycles of decay and rebirth. In short, for those who espouse the virtues of green burial, getting back to nature in death may be just as important as getting back to nature in life.[24] In contrast, hermetically sealing one's remains away from the reclaiming and recycling forces of water, soil, and worms, as executed in most conventional burials, constitutes yet another manifestation of modern humans' problematic alienation from the rest of the earth community. There is a narrative interplay in green burial marketing materials between consumers making an empowered, self-styled, individualistic statement ("I did it my way") and signifying moral concern that extends beyond the mere individual to an integrated whole earth community. Many green burial planners, especially those who choose woodland burial locations, opt to have a tree, a spray of flowers, a rough "found" stone, or simply no grave marker at all. Historically, funerals and models of burial have demonstrated the social and financial status of the dead while they lived. Those of high status were the subject of elaborate funeral rituals, often ensconced in tombs that provided visual markers of earthly wealth, power, and success. In contrast, the remains of commoners were more likely to end up in unmarked and/or general graves. For those who choose green burial to end up electively in what is essentially an unmarked commoner's grave edges toward a kind of countercultural shift in attitudes toward death, status, and funerary display. The visual anonymity of the unmarked grave in the forest burial preserve rhetorically bespeaks a kind of humility in the marker's absence and renders the entire earth as *fosse commun*—a communal ossuary that ultimately receives us all.

In the marketing of conventional burial industry products, casket product names often communicate a martial quality of defensiveness. Sturdy models such as "The Centurion" or "The Citadel," made of tropical hardwoods or metals, promise *fortress*-like containment that will guard and vigilantly defend the body from the forces of death and decay.[25]

Companies that manufacture and market "grave vaults" or "grave liners" that encase caskets in the ground similarly advertise armored properties of their products that will "do battle" with the onslaught of nature's decomposing forces by providing an impenetrable protective barrier. Companies that manufacture and market both stainless steel and copper grave vaults will boast in their ad copy that their products provide "eternal protection."[26] To drive home this point, the ad copy for Clark Grave Vault, a popular vault company, offers consumers the helpful historical information that "[a]rcheological digs often recover items made of pure copper in superb condition after 8,000 years."[27] The implication is that one's loved one or one's own body will enjoy similar longevity of "protection" from the aggressive elements of decay. The company's tagline is, "Only memories last longer." Their ad copy reassures the consumer that "[e]very Clark vault is guaranteed to protect the casket from water and other elements through the use of an air seal principle."[28] The air-sealed design of the vault, much like similarly advertised air-tight caskets, like air-tight Tupperware for that matter, constitutes its major selling point—its fortification to repel natural elements and preserve the contents.[29]

Ironically, this kind of air-tight engineering has become sufficiently vigilant against the elements so as to become a liability. As anyone who has ignored food leftovers in a sealed plastic container in the refrigerator too long can attest, gases from decaying organic matter build up and cause the container to bloat under pressure. Industry publications now warn of the very real dangers of "exploding casket syndrome," in which the gases released by rotting corpses build up in air-tight containers to such a degree as to blow the lids right off caskets and the doors off mausoleum compartments, thus posing a "shrapnel" danger to visitors.[30] (Google images and YouTube are filled with visual chronicles of the gruesome gooey results.) Conventional sealed caskets must now incorporate what are termed "burping" technologies in order to allow for the gradual release of the natural gases produced by decomposition.

Rather than battling and resisting the elements of decay like their conventional burial counterparts, green burial product marketers position their "natural" caskets, or other readily biodegradable containment options, such as linen shrouds, in terms of their working *with* the reclaiming elements of nature. Marketing materials emphasize how these products support and even hasten a kind of sacred recycling process that transmogrifies the body into life-giving nourishment (human humus) for the earth community. Green caskets are made of readily biodegradable materials that range from simple pine to basket-woven willow branches, bamboo, sea grass, papier-mâché, and even "leaf cocoons."[31] Online shopping sites for these products frequently employ the words "natural" and "harmony," while suggestively and sensually speaking about "giving oneself to the earth."[32] The ad copy at Memorial.com online funerary shopping describes its banana leaf casket option in this way: "The Slate Banana Green Casket is hand woven from coiled banana leaves. The casket features rattan and seagrass accents and is lined with a soft cotton fabric. This biodegradable casket is made from 100% natural materials, no glues or metal fasteners, making it perfect for cremation or a green burial. Green burials are becoming more popular, they are seen as a way to give a person back to the earth in a natural and dignified way."[33] This "giving" of the body to the earth is the antithesis of the conventional burial consumer messaging of protection, defense, fortification, and armor, deployed against a hostile and relentlessly aggressive enemy "attacking" the deceased. However, Memorial.com retails the banana leaf casket at fifteen hundred to two thousand dollars, despite its use of much cheaper materials and lack of metal fasteners—a comparable price to many of the web site's more economically priced conventional caskets. Charging comparable prices or even charging more to consumers who wish to be eco-pious and give themselves or a loved one to the earth is not unusual, particularly among larger, more conventional retailers that carry an eco-friendly sideline of products. Funerary supply companies have correctly ascertained that greener deaths can spell big profits, especially for morally and ideologically committed consumers who are not price sensitive in their practice of consumopiety.

On the other end of the spectrum, for about a hundred dollars, consumers can purchase online a plain muslin burial shroud, complete with sewn-in body board, and instructions. Etsy.com has become a treasure trove of home-crafted green burial products, and numerous blogs, Pinterest boards, and YouTube videos demonstrate ways family can decorate and personalize shrouds with sewn-on mementos, painted designs and messages, leaves, pressed flowers, aromatic oils, and herbs. A strong DIY participatory culture on the Internet of green burial advocates resists the high-end consumer marketing of designer burial products from mega manufacturers and, as we shall see, challenges its prescribed versions of consumopiety.

The Whole Death Community

A comprehensive cultural guide to death-related topics, *The Whole Death Catalog* (2009), derives its title from its popular countercultural predecessor, *The Whole Earth Catalog* (1968–1998), which for three decades provided readers with tools and resources for self-designing and living alternative ecological lifestyles. In its media reviews, *The Whole Death Catalog* humorously recommends Home Box Office's iconically morbid series *Six Feet Under* as "must-see TV for morticians."[34] Reaching far more than simply morticians, however, *Six Feet Under* "killed" its cable competition with the series' fifth-season 2005 finale, attracting nearly four million viewers.[35] Many more viewers have since watched the Alan Ball–created series via streaming, online video sharing platforms, and post-market DVD products. *Six Feet Under* tells the story of a small, multigenerational, family-run funeral business in California, as the family deals with the meanings, absurdities, and materialities of life and death. While dealing with a new corpse each week, the Fisher family struggles with tensions between tradition and change, legacies kept and broken, big economic challenges and small family business values—all interwoven into perpetual cycles of endings and new beginnings.

As HBO's dark drama series came to an end, *Advertising Age*'s post-media analysis touted the "perfect product placement" Toyota Corporation had "scored" for the Prius in the last six minutes of the series' iconic finale. The finale concludes with an extended montage that depicts the deaths of all of the series' characters, interspersed with shots of twenty-something, fiery-haired daughter Claire taking to the open road in her brand-new sky-blue Prius. Prior to the series finale, Claire had driven the family's hand-me-down 1971 lime green Cadillac "funeral coach" (hearse), complete with silver skull hood ornament. After totaling her gas-guzzling, carbon-belching hearse, Claire embarks on a new life direction and purpose, choosing a Prius to get her there. Saying a final goodbye to her brother Nate, to the family funeral business, and to her life in California, Claire embodies what *Advertising Age* identifies as a marketer's ideal adopter of the Prius. A young, edgy, cool cultural creative, she models the kind of "influencer and trendsetter" whom audiences read as "moving forward and having a better life."[36]

What *Advertising Age* failed to notice, however, in its laudatory coverage of Toyota's Prius placement coup was another important product placement in the same series—a seminal plug in popular culture for the virtues and aesthetics of green burial practices. In "All Alone" (episode 10, season 5), creator Alan Ball succeeded in showing the first narrative depiction of green burial on national television. Baby boomer Nate Fisher, the central character of *Six Feet Under*, suffers a brain hemorrhage, becoming a very public and popular poster corpse for green burial and its target demographic.[37] When making pre-planning decisions about his own after-life-care funerary preferences, Nate stipulates, "No caskets, no toxic chemicals to leach into the soil. I just know that when I die, please wrap me in a shroud, plant me next to a beautiful tree so that nobody can build a mini mall there."[38] Viewers ultimately watched Nate Fisher's family "plant" his body next to a tree in a nature preserve, lowering his handle-equipped green burial shroud into a DIY hand-dug grave. And many of these viewers were indeed baby boomers, the prime market for funeral planning.[39]

In the burial scene, the presentation and promotion of Nate's eco-piously prepared corpse evokes environmental sensibilities through the performance of a final act of green virtue—eschewing toxic embalming chemicals, rendering one's corpse all-natural compost, and caring for the earth by deploying green corpses as tangible defensive wards or guards against sprawl development and mini-malls.[40] Nate makes his body in its death a mortuary media intervention into the capitalist-driven destruction of wildlife habitat and open space. He prophecies to both his brother and his business partner before he dies that what he characterizes as a "natural approach" to burial is what is coming next for the American way of death.[41] If today's burgeoning market for green burial products is any indication, his prophecy was both correct and self-fulfilling, since *Six Feet Under*'s high-profile product placement helped to publicize green burial more widely as a consumer option.

Esmerelda Kent, a former film costume designer who created character Nate Fisher's burial shroud for *Six Feet Under*, has gone on to form her own successful green funeral products company, one of a number of new companies now supplying burial materials to funeral homes and di-

Figure 6.1. *Six Feet Under*, "Nate's Green Burial." Nate Fisher's funeral in the series *Six Feet Under* provided the first depiction of green burial on national television. Credit: Home Box Office.

rectly to eco-conscious consumers. Marketing copy on Kent's Kinkaraco Green Funeral Products web site capitalizes on an "as seen on TV" selling point for its green burial shrouds: "The first prototype 'KINKARACO SHROUD' was purchased by the production of *Six Feet Under* and debuted as the burial shroud for Nate Fisher! I was greatly honored to also be hired on the set as the Shroud wrangler/Green Burial expert for the shooting of SEASON 5/ EPISODE #10-All Alone . . . the first televised GREEN BURIAL was beyond thrilling!"[42] Kent's shrouds bear names such as "Tru-Green Believer"; and the pampered-sounding "Botanika DeathSpa" model comes with an all-natural herbal liner.[43]

A sturdy "Endfinity" board can be ordered to be placed under the shrouded corpse to make handling easier and comes with special straps to assist DIY non-machine-assisted lowering into the grave. These items range in price from about three hundred to seven hundred dollars, but Kent's Mort Couture line of finer all-natural silk fabrics runs closer to a thousand or more, appealing to its more well-heeled customers. This is a small burial products company with nowhere near the volume of the mega retailers. The company's stated goal is to provide "Affordable Green Funeral Products MADE WITH LOVE for people who want their funerals to be as natural as their lives."[44] The company takes its name from the Sanskrit word "*kankara*," which Kent translates as a servant to the "charnel" or cemetery grounds. The web site is dedicated to a Kandampa Buddhist master, and one section of the web site outlines the company's philosophy of practicing "Buddhist economics," which it explains often means free shipping on shrouds urgently needed and espousing the value of putting "people above profit."[45]

Like *Six Feet Under* creator Alan Ball, Kent is a Buddhist convert, and Kinkaraco's marketing and products reflect an embrace of Buddhist notions of "impermanence."[46] The story Kent's products tell about the body and mortality, the marketed consumer journey toward her products, is one that emphasizes cosmic unity and an unperturbed approach to change. Death is not aberration, punishment, or something gone awry; it simply is. Kinkaraco's marketing disrupts funerary narratives that promise "eternal preservation" and replaces them with a narrative that reframes ethical "corpse care" in terms of natural "rapid decomposition." Woven into the company's web copy is that ecopiety (my term, not theirs) is practiced through virtuous green purchase, a message echoed across green-burial product marketing.[47] But there is something more communicated in the marketing of Kent's products beyond mere consumopiety. Thin, silky shrouds communicate a bodily intimacy with the surrounding earth of the grave site. The fragrant, colorful herbs that surround the body resemble a festive holiday feast preparation. Rather than dour or mournful, every aspect of the company's product designs and their vibrant displays are filled with abundant signs of life. Although the company now provides more elaborate and ornate options, Kent's basic designs are still in keeping with the vision of green burial that Alan Ball first publicized via Nate's burial in *Six Feet Under*—one of facing, not denying, the metamorphic forces of decay, and one of messy intimacy with, not separation from, earth's reclaiming processes.

As Seen on TV

Television and radio host Gail Rubin reassures her viewers that "[j]ust like talking about sex won't make you pregnant, talking about funerals *won't* make you dead." Rubin uses this tagline in each episode of her public-television show, radio show, and Internet-based series, *A Good Goodbye*.[48] The host has a tough challenge, tackling a subject no one likes to think or hear about. Strategically, Rubin frames her programming on death in a friendlier way, as "[f]uneral planning for those who *don't* plan to die." She emphasizes throughout the series that logistical planning well ahead of one's death reduces stress and conflict in the family at a time of grief, saves money, and creates a meaningful goodbye. More importantly, it reduces day-to-day death anxiety while one still breathes.

The show's opening and closing graphics depict an idyllic pastoral field with a verdant tree at its center. The tree's roots visibly sink into the soil below the green turf and light up. The roots glow as with life and fecund promise. In the television series' episode "Going Green," Rubin specifically concentrates on a variety of eco-friendly burial options that she also attends to more generally throughout the series. She comments that green options are of particular interest to her as a Jewish woman, in that they espouse many of the same customs required of Jewish burial: no embalming, a fully biodegradable casket, and burial in such a way that the body comes into direct contact with the earth. Indeed, a number of green burial product and service marketers make the connection Rubin makes, crediting Jewish burials as "the original green burials" and characterizing Jewish law as naturally eco-pious.[49] The various green burial products Rubin features on her show, all desirable for their quality of hastening rapid decomposition, speak in design and purpose to an embrace of the body itself as the ultimate "fully recyclable container."

Rubin's special guest for her "Going Green" episode was Darren Crouch, president of Passages International, Inc., one of the largest wholesale suppliers to funeral homes in the country. In the episode, he talks to Rubin about the company's expanding line of green burial products. Crouch characterizes each product in a language reminiscent of natural-foods marketing, and indeed consumer demographics overlap for high-end green burial consumers and what marketers refer to as LOHAS (Lifestyles of Heath and Sustainability) consumers.[50] At one point, Rubin notes the growing problem in the conventional funeral industry with manufacturing caskets capacious enough to accommodate ever-larger, obese American corpses that now outstrip even extra-large casket measurements. Crouch reassures the host that oversized caskets are not usually a problem for the consumers of green burial products, as they tend to be more health-conscious about what they consume and therefore readily fit in standard-sized caskets.

Introducing the audience to a cremation urn that is made of solid Himalayan rock salt (also marketed as a health food at Whole Foods Market), Crouch emphasizes a virtuous feature of this product: "it contains no additives." He notes that a cornstarch urn, another of the company's products, which also rapidly biodegrades for burials at sea, uses the same kind of cornstarch that is in the "forks and spoons one finds at Whole Foods Market." These pronounced links drawn among green corpses, natural burial products, and Whole Foods Market consumers are savvy. "Advertising rarely constructs situations out of nothing," observes Tricia Sheffield, as she locates religious dimensions in the history of advertising; "rather, [advertising] mirrors back to its audience forms of culture that will be easily recognized."[51] I half expect Crouch next to produce a "gluten-free" funerary urn for consumer purchase, but instead he goes on to demonstrate an eco-friendly ash scatterer, which, using another food metaphor, he observes works much "like a large salt shaker." Explaining the popularity of this product, he offers that "people no longer live in the city they grew up in thirty years ago," and so scattering has become more and more popular. He also notes that there has been "a large increase in demand" for green funerary products and services. Despite economic fluctuations, the market has steadily grown each year, calling for an ever-expanding green burial product line and an expansion of related eco-funeral services, such as relocating bodies of the deceased for burial via "carbon-neutral UPS shipping."[52]

To respond to this increase in green consumer demand and to guide customers toward their product options, the company has launched a web site, AGreenerFuneral.org, which outlines myriad choices for funeral personalization and implementation of varying degrees of environmental virtue. Invoking organic food and hybrid automobiles for comparison, Crouch outlines the rationale for the web site: "People know when they want to buy organic produce at the grocery store where to buy that. If they want to buy a hybrid vehicle, they can go to Ford or Toyota and buy a vehicle like that." When a loved one dies, however, he laments, most people don't know how to plan a green burial or where to go to buy greener products. That is where his company comes in. Crouch also delineates a spectrum of virtuousness, or shades of green, and cautions that "green burial" can mean many things to many consumers. The company counsels its retailers to investigate what a client *means* when he or she walks into a funeral home asking for a "green burial." A "light green" option, for instance, might consist of being buried in a conventional graveyard and in a conventional concrete vault but in a biodegradable casket. A "dark green" option might be to eschew the chemical embalming process, choose a biodegradable casket or shroud, skip the vault, and occupy a forest-based cemetery in which there is ongoing wildlife restoration.[53] His comments on this suggest that the retailer should be careful to follow the client's existing hue of green and not to present the client with greener options than the client has indicated. He further schools the viewing audience in various gradations of green cemetery ecopiety—cemeteries with push mowers instead of gasoline mowers, hand-dug graves instead of those excavated with backhoes, and the choice of trees, wildflowers, and natural stones as grave markers instead of quarried, cut, and polished headstones.[54]

Programs like Rubin's *Good Goodbye*, a burgeoning number of online marketing sites for green burial products, and a host of do-it-yourself and how-to YouTube vlogs on green burial logistics, as well as independent documentaries, television series and specials, and journalistic media coverage support sociologist Glennys Howath's contention that "[w]hatever the relationship between society and the individual, the media is significant as a marker of popular cultures and social mores surrounding death and dying."[55] Mediated popular culture, once again, acts as both mirror and engine producing cultural works that provide a space to work out what it means to choose a greener death as a moral choice in an age of environmental crisis.

Gail Rubin's programming effectively mediates between the consumer's strong fearful emotions toward death and steps for practical funerary planning, while in this particular episode, she specifically mediates between green capitalism and prospective consumers.[56] As president of a funeral product supply company, Crouch mediates between products and purchasers, packaging and marketing green burial products in a way that green-conscious consumers will understand, will recognize, and for which they will feel affinity. Purchasing a green casket, he suggests, is much like purchasing a more eco-friendly vehicle, although presumably with a somewhat more final destination. Likewise, buying an all-natural urn, free of additives, is familiar to those who already shop for all-natural or organic foods.

Here we see Crouch and his industry making shrewd application of what Nobel Prize–winning economists Amos Tversky and Daniel Kahneman term "System 1 thinking." That is, when faced with new decisions and asked to evaluate new materials or data, we tend to employ heuristics or mental shortcuts. Rather than conducting new research or evaluating new decisions case by case, and on their own terms, we instead default to existing patterns with which we are already familiar. When presented with an unfamiliar social practice, such as green burial, and new consumer choices, such as green caskets, prior decision patterns of a similar nature shape our new decisions and evaluations.[57] In a narrative sense, this is also why building a new story on top of an older, already familiar story (an existing pattern) tends to be a highly effective marketing strategy—one we see at work in the history and formation of successful new religions. It is not that far a stretch for consumers who already have the "all-natural, additive-free foods file" open and in use in their brains subsequently to accept and develop similar positive associations with concepts of the unembalmed body as all-natural, additive-free food for the earth.

Media marketing that stories green burial taps into an already established and readily available eco-pious consumption narrative that will be familiar by now—like the organic food and farming notion of individuals voting with our forks, green burial consumers can "vote" with their corpses. Again, this is a model of personal piety, not public policy, and promotes the reassuring idea that every little bit helps, and individual consumer choices will be sufficient to healing and restoring the earth. It is also a story, somewhat paradoxically, of the virtues of simplicity, naturalness, and frugality, even as that simplicity and naturalness get increasingly complicated, and the price point on frugality rises on designer green burial items.[58] And yet, beyond the realm of consumer marketing, on-the-ground green burial advocates are not just voting with their forks but voting with their votes. They are showing up to community planning meetings, writing letters, agitating to get local ordinances changed, publishing editorials, getting interviewed on local radio programs, and forming community meet-up groups and "death cafés" in order to organize.[59] Thanks to a feisty group of green burial advocates, Vermont's law mandating that bodies must be buried five feet under the ground was changed in July 2017 to permit bodies to be interred at just three and a half feet under. This may not seem like a big deal, but it is. Burying bodies five to six feet under puts them at a depth too far for the nutrients in the body to reach the surface and nourish the topside plants and insects. A shallower grave means that the body can actually be used as a food source for the local biotic community, so this change in legislation was a big victory for green burial activists.[60]

In publicly conveying his wishes for his own green burial in 2018, Vermont pancreatic cancer hospice patient and green burial proponent Jack Griffin's comments reflect the importance of knowing his body's nutrients will be made bioavailable to the life community: "Eventually, you will go back into the food chain—into a nice big elm tree, or a crow, or an eagle, or a robin in the spring. I just want to be useful."[61] The Home Funeral Alliance, a free membership consumer group that supports and advocates for home funerals and green burials, supplies downloadable sample letters on its web site for citizens to send to legislators. Its "Advocating with Action" page further urges members into civic action: "Participate at your local and state level. Learn how to track legislation in your state, [use] tools for meeting with your state legislators, and how you can take action now. Send a letter to appropriate policymakers."[62] Whereas large-scale corporate retailers market personal ecopiety and the fantasy that individual consumers, by buying the right products, can purchase a greener and better world, green burial advocates on the ground organize, educate, evangelize, and resist these messages.

Ecopiety and the McFuneral Makeover

"You can't consume your way to conservation," admonishes funeral consumer advocates Lisa Carlson and Joshua Slocum, "and you shouldn't have to spend more in order to waste less." The duo advises that "the greenest casket you can buy is the one you *don't* buy."[63] Slocum is the executive director of the Funeral Consumer Alliance, and Lisa Carlson is executive director of Funeral Ethics Organization. Both of these organizations serve as consumer watchdogs, advocate for legal and regulatory reform with regard to funerary policy, monitor ethical abuses in the funerary industry, and support consumers' rights to a funeral that is "meaningful, dignified, and affordable." In their consumer guide, *Final Rights: Reclaiming the American Way of Death*, the authors share insights and tips with consumers about how they can bury their loved ones and plan for their own death in a way that is frugal, simpler, personalized, and reclaims the funerary process of death from industry professionals. In their guide, the tagline for which is "the book the funeral industry doesn't want you to read," they walk the reader state by state through corpse-related permitting processes and outline what is and is not permissible, educating consumers about their burial rights. For instance, most consumers are unaware that there is no law requiring embalming in any of the fifty states, and a funeral director who tells you it is mandated is breaking the law.

Slocum and Carlson also model how consumers can write complaints and agitate against local ordinances such as requirements for leak-proof caskets, or cemetery policies that mandate universal embalming. They feature consumer advocacy successes in a variety of states where green burial advocates have agitated for change and prevailed. "A surprising number of people treat dead bodies as if they were one-person Superfund sites," they protest. Pointing out the absurdity of some of these concerns, they continue: "Until 2010, the Vermont Veterans' Memorial Cemetery actually required all bodies to be embalmed, out of concern for groundwater quality. That's right—the state supposed that burying carcinogenic formaldehyde close to town wells was a safer idea than letting people decay naturally."[64] Their message to their readers is straightforward: know your rights, get active, articulate your wishes, and if there are ordinances prohibiting a simple, green funeral, get civically engaged in changing them.

Both authors support green burial as a consumer option but are also concerned with the creep of what they consider the misleading marketing of green capitalism: "There are all manner of 'green' coffins coming on the market—made of recycled paper, sustainably harvested woven willow, etc.—but some of them are imported from overseas. How 'green' is it to have something flown in for your funeral on a jet? And the cost? The 'Ecopod'—which looks like a cross between a willow seed and a science fiction hibernation chamber—retails for several thousand dollars."[65] This consumer caveat may bring to mind the aforementioned banana-leaf casket marketed by Memorial.com as an eco-friendly alternative suitable for green burial. Some retailers do specify that their banana-leaf caskets are fair-trade, but the products are still shipped to the United States all the way from Indonesia, racking up a considerable carbon footprint. Slocum and Carlson debunk the eco-friendliness of many industry-marketed green burial products and suggest instead a homemade box from scrap lumber, which is both cheaper and more sustainable. As with basic shrouds, the Internet teems with myriad crafty ideas about how to decorate and personalize basic pine boxes and even recycled cardboard containers.

Joe Sehee, executive director of the Green Burial Council, a US group that sets green burial ratings and standards, raises similar concerns about burial products being greenwashed, as green burial marketing increasingly involves what he sees as "tarting up products and services as environmentally sound when they aren't."[66] Sehee cautions about some so-called green burials now being more expensive than a conventional burial and in actuality proving not to be all that green. Slocum and Carlson, too, bristle at vault makers whose advertising copy markets their lined concrete burial vaults and conventional burial as the "more environmentally friendly" choice by protecting the surrounding environment from casket seepage.[67] Slocum and Carlson also cast their consumer skepticism upon so-called "green" disposition products such as Promession, a Swedish company's patented process in which a corpse is reduced to a fine powder by dipping it in liquid nitrogen and then shaking it in a machine. They comment, "Emissions-free, certainly, but the process does seem like a Rube Goldberg approach to burial," asking, "why go to so much trouble when nature will do the same job underground (if more slowly)?"[68] Why indeed?

Of course such complicated processes can then be rationalized as adding value and marketed as providing a needed service to consumers. In the marketing machinations that hawk alternative green products and services, the swallowing forces of late-capitalist consumerist peristalsis are in full force. Marketing consumopiety has become a tactic for large, corporate retailers to neuter, depoliticize, and co-opt what had begun as a countercultural movement. Bodies function as media, and corpses, graves, and funerary material culture, literally or implicitly, are inscribed with final statements of values, meaning, and sociality, including of course market logics. As countercultural as green burial practices and products purport to be, their aesthetics and mediation still in many cases reflect those logics.

In the realm of self-produced citizen-streamed video, and in other areas of the Internet, such as Pinterest boards, Etsy.com vendors, and social media pages, shared information about low-cost green-burial options, DIY funerals, and "how-to" instructions thrive. Wendy Miller's YouTube–posted burial video, "A Very Natural D.I.Y. (Dig It Yourself)," a popular resource to consult and link to for those contemplating green burial for a loved one, makes just such a resistant statement. In the video, Miller honors her mother's wish that she be buried as simply and thriftily as possible. Miller collects her mother's body in a cheap rented camper, transports it to the burial site, and then prepares it for burial. She documents for the viewer a how-to process as she buries her mother in a large field, where burials are permissible. The handling of the body and the digging of the grave are demonstrably hard work, and Miller struggles and sweats through much of the video. She ultimately shows her satisfaction in "doing for her mother" by having spent as little as possible to fulfill her wishes, while still "planting" her somewhere peaceful and beautiful.[69]

Online accounts of DIY or at-home funerals are almost always eco-funerals. "Everything You Know about Funerals Is Wrong" documents one family's care for a deceased mother. Children decorate a cardboard casket with markers, drawing flowers and infinity signs on the casket, while an adult has written in bold blue marker on the side panel, "The Earth Does Not Belong to Us. We Belong to the Earth."[70] Documentary films like *A Will for the Woods* have also told the stories of families' intimate experiences with green burial and reflect a pronounced resistance toward consumerism on the part of the featured subjects. The documentary details the excruciating but hauntingly beautiful process of musician and psychiatrist Clark Wang planning his own green funeral. Repeatedly Wang and his wife speak of consuming few resources and leaving a legacy of forest conservation, as Wang boldly faces his death with humor and grace. The ultra-nonconsumerist funeral his wife conducts for him in their modest living room shows Wang lying in a plain box on top of the coffee table, surrounded by friends who laugh, cry, play music, and honor his life. The very composition of the scene is a powerful representation and endorsement of what authors Joe Dominguez and Vicki Robin articulate in their now-classic anticonsumerist manifesto, *Your Money or Your Life*, as the meaningfulness experienced in a "low-consumption/high-fulfillment lifestyle."[71]

In Lisa Carlson's guidebook *Caring for the Dead*, to provide a model, the author shares with readers her own letter of funerary directive to her family—one that voices a determined anticonsumerism. Her pronounced resistance to the pressures and promises of consumerism is coupled with a mystical delight in the prospect of her postmortem communion with the earth.[72] She directs,

> I want to make clear that, when I die, I want the absolutely least expensive funeral you can possibly arrange. I don't want some funeral director telling you that you ought to have "the best" for your mother. If you want to spend money to show how much you love me—for goodness sake do it while I'm still alive. . . . "Plant" me under an apple tree, or better yet—a flower garden. . . . That's where my spirit would be most happy. It feels strangely warm to feel myself becoming one with the earth, to picture my elements feeding new life. That's the way I want to go—that's the way I want to come back again—as nourishment for a beckoning flower. . . . I want to be free to go . . . in a plain pine box, one that's not too perfect . . . just like me.[73]

More than simply the duty and obligation often associated with piety, there is real expressed delight in this passage as she relates the "strange warmth" she feels when envisioning a kind of mystical unity in which she will one day become "one with the earth." Carlson's book goes on to share transformative experiences of those who, rather than outsourcing to impersonal professionals the intimate acts of caring for the dead, look death straight in the face and tend directly to their loved ones' bodies by washing, dressing, combing hair, arranging the body for burial, and physically participating in lowering their loved one into the ground. One of Carlson's sources characterizes this kind of care as deeply spiritually and emotionally rewarding—"the most awful wonderful thing."[74]

In *The Undertaking*, poet/undertaker Thomas Lynch similarly contrasts the deeper emotional connection fostered by green burial to the empty generic consumerism of "McFunerals, McFamilies, McMarriage, and McValues."[75] In recounting his experience of personally caring for his father's dead body, Lynch recounts, "[I]t had for me the same importance as being present for the birth of my sons, my daughter."[76] In a burial blog entitled "Green Burial: My Body, My Self," hosted by the Conscious Elders Network (a nonprofit of volunteers who donate their service in retirement to social, economic, and environmental justice activism), the author paints a vision for his readers of green burial that goes beyond the piety of responsible earth stewardship and instead evokes sensual mystical union: "Your body, your one and loving vessel in this life, is treated with grace and care and respect by people who know and love you. And they lay you gently in the grave that has been prepared for you, with the ceremonies and in the manner you have chosen. And your body will relax and sigh, feeling the loving embrace of her mother as she becomes one once again with the great soul of the world."[77] Notice the elements of relief and satisfaction in this account, as the imagined corpse, though dead, is still sensing and feeling the pleasures of a green death. In another blog, "My Big Fat Green Funeral," a Florida man pleads, "So help me, God, do not bury me in a casket from Costco" and makes known his wishes to have his body released and returned to the ocean to nourish the fishes and marine ecosystem, an option that will be further explored below.[78]

Throughout visions and accounts of green burial both online and compiled in consumer guides, it is evident that something "more" is being expressed in green burial experiences that transcends mere virtuous practice of piety. That "more" is alluring and somewhat unquantifiable and yet presents a significant draw for new adopters to approach thoughts of death with less fear and anxiety and more pleasure and peace.

Eternally Green

Not simply consigned to land, green burial options also extend to burial at sea, where organizations such as Eternal Reefs enable consumers to turn their cremated remains into coral reef ecosystem restoration "reef balls." The marketing of these ocean memorial services and their media testimonies go far beyond mere "do-gooding" of pious practice and instead speak movingly of a sense of mystical "merging" with or even of "becoming" an ecosystem. The deceased body arrives at Eternal Reefs already reduced to ash, preferably from a high-efficiency, eco-friendly crematorium, equipped with scrubbing technology to minimize air pollution. Cremation, although it saves on space, is not considered as green as full-body green burial. It takes roughly twenty-eight gallons, or two SUV tanks, of greenhouse-gas-producing fossil fuel to reduce a human body, including the bones, to ash. Pollution emitted from crematoria can also release lead, mercury, and other toxic substances, making them airborne, so green burial advocates encourage carefully vetting the technology and emissions system when choosing a crematorium.[79]

For a cost ranging from one thousand to seven thousand dollars, the cremains are then mixed with cement, sand, and sometimes ground up seashells and cast into artificial coral reef balls that the company sinks in the ocean to provide vital habitat for tropical species in areas where coral reefs are dying due to a number of factors, including bleaching coral from warming ocean temperatures. It is important to make the distinction here that unlike for-profit companies that purvey green funerary products and services, Eternal Reefs is a registered 501c3, it does not profit from reef ball purchases, and the costs of reef ball burials at sea are tax deductible as charitable contributions. In short, the organization funds its ocean habitat restoration work via the sale of its ocean memorial products and experiences.

Each reef ball can be personalized with a plaque of the deceased person's name, and personal message or slogan, as well as life mementos (military medals, sports team emblems, items of jewelry, and even a baby sock or shoe), which can be cemented into the top of the ball as identifying objects. In some places, family members can follow a dive trail to go "visit" their loved one on the ocean floor.[80] Through marketing pamphlets and videos, nonprofit organizations like Eternal Reefs feature before and after shots that demonstrate to consumers how, in just a few months, they can "become a coral reef." What starts out as a sterile-looking concrete ball within a few months transmogrifies into a living ecosystem with swaying plants and tropical fish.[81] After shots show the loved one now hosting blue tangs and sea urchins in their midst, and Eternal Reefs' literature and promotional videos demonstrate how, within three months, your loved one has become a thriving ecosystem.[82] This presents a vibrant image of postmortem eternal life. As with green burial on land, one of the selling points for becoming a reef is the dissolution of the human/animal divide that often painfully separates pet owners from their pets in death. When you become a coral reef, your pets' ashes can be mixed together with yours (at no extra cost) and you can all become a reef together. Or, spouses/partners can wait for each other, have their ashes comingled, and then become an eternal reef as one unit. Families can also participate and hand-stir the ashes of the loved one into the concrete mixture, thus also appealing to DIY and participatory-culture inclinations. In one of the marketing images, a family surrounds a reef ball of a family member who had been a veteran. Inscribed into the concrete with chalk next to his name are the words, "I Did It My Way."

As environmental media, "vessels of storage, transmission, or processing," mortuary reef restoration balls, and their accompanying visual and verbal rhetoric, communicate a sense of ecopiety on the part of the consumer, but this sense is combined with the alluring suggestion of immortality as a tropical reef.[83] Inscribed into the reef ball is a strong ethic of reciprocity and connection with the whole life community. Eternal Reef's marketing tag line is "Giving back to Mother Nature." Its web site tells stories of how those who choose to become reefs become "living legacies preserving marine life for future generations."[84] A number of the testimonials from family members quoted by Eternal Reefs' web site speak of an ethic of "paying it forward." A repeated theme that comes up in reef ball marketing videos among family members who speak on camera is their loved ones' strong desire to "be useful" even after death, or their wish to "become part of something useful."[85] Reef ball companies stress that the balls are used for restoring coral reefs, for coral transplanting and propagation, oyster bed development for mangrove restoration, shoreline erosion control, and fishery management.[86] Eternal Reefs' reef-design partner, Reef Ball Foundation, has placed more than half a million restoration reef balls in more than fifty-nine countries. In the marketing of these reef balls, an eco-pious call to conservation is mingled with rhetoric that suggests the promise of eternal presence within the new living system of the reef. One testimonial cited on the Eternal Reefs web site proclaims, "Remaining useful after death for the millennia seems to be such a gift to offer others."[87]

Might this urge to merge with the oceanic ecosystem—the desire to gift oneself for use in the service of other species—constitute a critical step in what Roy Scranton characterizes as a human "hive dance" that directs us toward learning how to die in the Anthropocene? Ideals of eternity and immortality pervade the reef ball marketing materials, but the visual rhetoric of the time-lapse (before and after) transmogrification of humans into living ecosystems bespeaks a very different existential understanding of personhood, immortality, and the sanctity of the human body from the media of conventional burial. Circulated images of sunken human/coral living reef balls work as media interventions, subverting conventional narratives of human embattlement and defensiveness toward death. Instead, human remains are cast as collaborating in death with the surrounding ecosystem. Where copper grave vaults, mahogany hardwood caskets, sealer technologies, and injected chemical preservatives promise stasis and status quo—an intact, unspoiled body being defended against the corrupting forces of nature—the eco-reef ball tells a story of relaxing into and inviting transmogrification as an exciting new step of being useful in a new vocation. Charred bodily cremains merged with concrete, sand, and seashells form an unrecognizable (yet personalized) host for a different kind of preservation and restoration—one that serves threatened species and their vulnerable habitat.

"This is not a funeral," exclaims a testimonial on the Eternal Reefs web site, "it's a new beginning." A family member in one of the company's promotional videos similarly observes that this is not the end of her husband; becoming a living reef is simply what he is doing next, as with a second career or retirement job.[88] These words are echoed in the organization's web marketing as they tell their story of becoming a reef restoration charitable organization: "Military veterans, environmentalists, fishermen, sailors, divers and people who have been active all their lives or whose lives have been cut short, are comforted by the thought of being surrounded by all that life and action going on around them. It's really more like—look at what they're doing now."[89] The words of Carlton Palmer, the organization's first customer to become an "eternal reef," also conjure an envisioned immortality through human/reef ecosystem merger and active afterlife service. In preplanning his journey to the bottom of the ocean, Palmer declares, "I'd rather be on one of those

Figure 6.2. "Before": A personalized reef restoration ball has been created from the loved one's cremains and is ready to "plant" in the ocean. Credit: Eternal Reefs, Inc., Sarasota FL, www.eternalreefs.com.

Figure 6.3. "After": The loved one takes on a "new existence" as flourishing reef ecosystem. Credit: Eternal Reefs, Inc., Sarasota FL, www.eternalreefs.com.

reefs with all that life and action going on around me than lying in a cemetery with a bunch of old, dead people."[90] The reef ball marketing tagline implies not simply merging with "creation" (the natural world) but also actively participating *co-creatively*, in death, in an ongoing process of biotic creation. Again, a sense of "more" pervades the images of afterlife in these media materials, which also convey a relaxed sense of embrace rather than denial of death.

Becoming a Tree

For landlubbers who are not drawn to the coral reef option, "becoming a tree" is a popular choice for the afterlife that combines ecopiety with a delight in the prospect of an envisioned mystical merging. The slogan of Greensprings Natural Cemetery Preserve in Newfield, New York, is, "Save a forest, plant yourself."[91] Greensprings' online marketing positions its preserve as part of a moral model of reciprocity between humans and the more-than-human natural world, identifying green burial as "part of the circle of life" and "a natural return to the earth." Web site copy emphasizes to prospective clients that "[y]our choice for natural burial is a choice for natural renewal and growth—a way to give back to the Earth that sustains us all."[92] But the decision to become a tree, like the decision to become a coral reef, is one that goes beyond the mere practice of ecopiety.

As eco-friendly grave markers go, a number of options are used in green burial, ranging from a "found" stone (unquarried, uncut) to a memorial tree to a shrub or a spray of wildflowers. Of these, by far, the *tree*, and its powerful symbolism, stands out in green burial as a key "actant" in the human/vegetable/mineral transmogrification process.[93] The gravesite tree marker—much like the ecologically prepared corpse underneath it that feeds the tree with organic nutrients from fat, bone, and blood—is both vessel and medium. German media theorist Friedrich Kittler identified media as "not passive vessels for content, but ontological shifters," and the media involved in posthumous human/tree transmogrifications play a role in just such an ontological shift.[94] In the Natural Burial Company consumer guide, "Be a Tree: The Natural Burial Guide for Turning Yourself into a Forest," founder C. A. Beal offers insight into what the nature of that ontology might be. He argues that in contrast to modes of disposition such as incineration of human remains into ash (cremation), freeze-drying (Promession), or toxic chemical dissolution (resomation), composting the organic matter of the body as food for living systems constitutes the most "useful" and thus most moral means of recycling of our human containers into another living entity's "dinner." "Becoming a tree," writes Beal, "for no other reason than to offset our own CO_2 emissions during our lifetime, might just be the ticket we need."[95] Just what sort of ticket is this, though? And where is it a ticket to?

For Beal, becoming a tree enables humans to "continue to play a role in the web of life." Choosing to "reintegrate into a forest," contends Beal, sends a "signal" to the market by changing "purchasing behavior."[96] The growing pool of green burial consumers is "increasingly putting its money where its mouth is, and now we're putting our bodies there, too," declares Beal. He argues that the spread of green burial practices cultivates a shift in consciousness that just might "nudge our culture in another direction." Composting a life means realizing both as consumers and as those to be consumed, that "packaging counts." "Dying to do the right thing," as Beal contends, also makes us more aware of and causes us to take responsibility for the kind of deaths that we as humans cause in the web of life.[97] According to the consumer guide, consciously, willingly, joyfully even, becoming "food for the regenerative Earth system," leaving behind our remains as a "gift of energy" to other life forms, transforms us, not just biophysically but ontologically. Green burials reflect the environmental adage that, as with garbage, "there is no away," and that natural, unimpeded decomposition initiates the corpse's "reintegration into the interdependent biological organism that cycles elements on planet Earth."[98] Beal's guide for "turning yourself into a forest" presents a detailed and compelling vision of the funerary consumer journey—one that carries ontological implications for a shift in human/earth relations.

Designer/entrepreneurs Anna Citelli and Raoul Bretzel share this transformative vision. Together, they have created a unique egg-shaped, seed-like, organic, biodegradable burial pod for humans that they call the "Capsula Mundi." This "world capsule" makes use, once again, of the tree as compelling medial image for passages in between linked worlds or states: life to death, death into rebirth, or subsequently into new life.

In the study of religious iconography, comparative religionist Mircea Eliade famously wrote about the tree as a kind of "axis mundi," a world axis that connected heaven and earth, the sacred and the profane, the realm of the gods and the realm of humans, the unseen world and the quotidian.[99] In the Capsula Mundi, the human body is folded up into the fetal position and placed in the egg-shaped capsule, adopting the position that the person once occupied inside the womb. The egg/womb/burial pod is then planted in the ground and a specially selected seed for a tree, or a new sapling, is planted just above the pod. As the outer shell of the corn- and potato-starch-based pod decomposes, so does the body, providing the nutrients or fertilizer to feed and support the tree. One life thus feeds another, as the tree makes use of the nutrient-rich decomposing flesh, now fertilizer, up through its roots, and over time, even absorbs calcium drawn from the corpse's bones. The designers envision future generations not walking through graveyards of the dead but instead walking among a forest of living ancestors—what they refer to as "sacred forests" of living memorials to those who have gone before. As media, Capsula Mundi pods in their visual rhetoric have much to say about how to die in the Anthropocene, and that is to recycle oneself, morphing into a medium that gives back to the life community. The capsula's design evokes the full circle of life from fetus to corpse, from mother's womb to mother earth's womb/tomb—cradle-to-grave composting. Aesthetically, it renders the cycle of life/death/rebirth in extraordinarily beautiful and artful terms.

Citelli and Bretzel, who are based in Italy and crowdfunded the Capsula Mundi project through the Internet site Kickstarter.com, say their product "has the aim of fundamentally changing the cultural approach to death and to comply with a holistic vision: after dying we'll still be part of the cycle of life and we should leave behind a positive legacy for our loved ones and for the future of the Earth. . . . We think that people need to be free from the taboo of death that weighs on the Occidental culture, like an immovable stone, and from the reaction that we've had, it seems that people are open to a paradigm shift in this area."[100] The designers unmistakably emphasize and reinforce this vision of a paradigm shift in cultural views of death in their marketing copy. On the web site "check out" page for Capsula Mundi's online shop, for instance, just above the purchasing options, a centered title reads, "By purchas-

Figure 6.4. Image of the Capsula Mundi. Italian designers Anna Citelli and Raoul Bretzel's stylish egg-shaped pods transform human corpses into trees. Copyright of Capsula Mundi, Rome, Italy, www.capsulamundi.it.

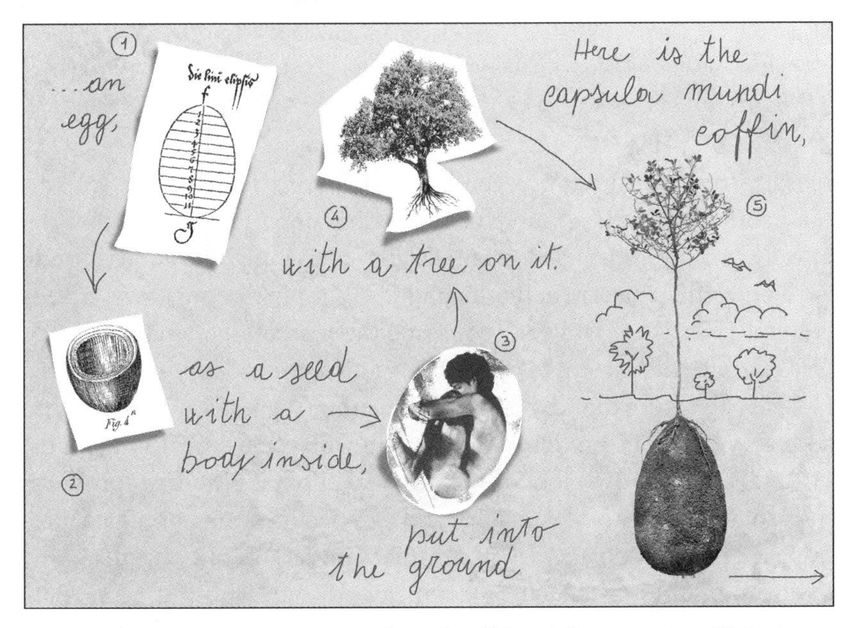

Figure 6.5. Loved ones continue on in the cycle of life as a "living memorial." Credit: Copyright of Capsula Mundi, Rome, Italy, www.capsulamundi.it.

Figure 6.6. "I Love You Grandma." Capsula Mundi envisions, "Cemeteries will acquire a new look: no more cold grey tombstones but living trees creating a forest, a holy forest." Credit: Copyright of Capsula Mundi, Rome, Italy, www.capsulamundi.it.

ing the Capsula Mundi urn you are contributing to an important cultural shift."[101] This is a powerful marketing message, especially for the already socially and environmentally conscious—that the goods we buy are a powerful means of activism, and that our purchasing power breeds cultural resistance. And yet, in reading the marketing literature for the Capsula Mundi and similar beautifully and artfully designed products, it is easy to forget that this personal, voluntary purchase is not tied to stricter regulations on poisoning the soil with toxic embalming fluids, or to rebate incentives for consumers who purchase funerary products made out of readily biodegradable and recycled materials, or to penalties imposed on casket makers who manufacture caskets out of endangered hardwoods, or to tax benefits for cemetery owners who engage in wildlife conservation and restoration. It is an alluring vision and a beautiful product, but it is still that—a product.

As of 2018, Capsula Mundi egg-shaped urns were available to purchase from four to five hundred US dollars, and the full-sized body pods were in the works, presumably, taking scale into consideration, at a much greater cost, probably in the thousands of dollars, not including the shipping costs from Italy. Once again, Slocum and Carlson point out in their burial consumers' guide that the greenest "packaging" for a corpse is the packaging we *don't* buy. A body wrapped in a simple shroud and laid in the ground to feed the worms is as eco-pious as it gets. And yet, it is admittedly more complicated in the case of the Capsula Mundi and other creatively designed disposition alternatives to calculate the undeniable power of art to transform culture and its degree of influence in nudging shifts in social energetics. Whether wrapping a corpse in a simple linen shroud or a specially designed flesh-digesting biodegradable "mushroom suit," compressing ashes into synthetic diamonds or mixing them into artificial coral reefs, the choice of method of disposition and container are critical decisions for green burial consumers. These decisions reflect a variety of affective relationships to death and responses to coping with the realities of mortality.[102] "Even the most commonplace object," says sociologist Ian Woodward, "has the capacity to symbolize the deepest human anxieties and aspirations." In analyzing material culture and the stories it has to tell, Woodward explains, "[P]eople construct a universe of meaning through commodities, they use these objects to make visible and stable cultural categories, to deploy discriminating values and to mark aspects of their self and others."[103] So it is with the material culture of green burial. However, it is striking to witness just how short a time it takes for the ecopiety and consumopiety of low-tech, low-consumption, "all-natural" burials to get "tricked out" with enticing new consumer technologies.

The Internet of Things: iSaplings and Smart Bios Urns

There is more than one way to become a tree, and smart technology for the eco-pious consumer can assist this process. Since 2013, the company Bios Urn has been manufacturing an all-natural biodegradable urn that mixes a body's cremated ashes with soil fertilizer and a tree seed. The main slogan used in the company's marketing materials is "Back to Nature. Back to Life."[104] One of the company's promotional videos features Bios Urn purchasers telling their stories of these products as "catalysts for growth." These stories have also been curated by the company under the Twitter hashtag #LifeAfterLife. In one story, a young woman speaks of how she is in her twenties now but one day she will be in her fifties, and her loved one (now in the form of a tree) will continue "growing with me just as a person would." Another woman speaks movingly of her loved one, via the Bios Urn, "growing into another lifeform that God made," as she smiles in a reassuring way. A daughter speaks about her mother's journey to tree form in the Bios Urn and exclaims, "The first day of seeing the sprout is like, 'She's back!'" These stories are set to the soft vocals of a background song that contains reassuring lyrics of eternal return: "I'm coming home to you" and "Makin' my way home." A montage shows family members in gardens and other idyllic green spaces, peacefully watering and caring for their loved ones in tree form.[105] The images are poignant and captivating, and they make me cry each time I review the promotional video.

In 2017, responding to consumer desire to remain closer to their loved ones' cremains during the sapling growth process, and to have the guesswork taken out of the care and successful nurturing of the new tree, the company launched a new feature to the Bios Urn—the iUrn digitally controlled, wi-fi-enabled, smart tree incubator accessory with USB port charger. The Bios Incube Kickstarter.com crowdfunding campaign identifies the product as "the world's first system to help the remains of your loved ones grow into trees," and it makes an even more suggestive claim: "Bios Incube. The world's first incubator for the afterlife."[106] Suddenly, "Getting back to nature and back to life" involves connecting the Bios Urn filled with cremains to the ever-expanding "Internet of things."[107] The Bios Incube smart incubator is a sleek white holder for the Bios Urn that monitors water, soil nutrients, temperature, and light levels in order to ensure the optimum growing conditions for the tree seed. Understandably, it can be upsetting to plant a loved one's remains in the conventional Bios Urn only to under- or overwater and thus have the sapling die or never germinate. Loved ones who care for and monitor the tree planted in the Internet-connected urn receive regular push notifications on their smartphone with technological updates on the tree's conditions and needs, with text messages such as, "Hey, your room temperature is too hot. Move your Bios Incube to a colder place."[108]

Here dimensions of participatory culture, already part of green burial in DIY forms such as washing the loved one's body or hand-digging his or her grave, take on a tech manifestation. Billy Campbell, a doctor and founder of Memorial Ecosystems, one of the first green cemeteries in the United States, frequently talks of the therapeutic aspects of digging a grave and crying, as bereaved family members get the opportunity to be useful and to fulfill the strong urge to "do" for their loved one in death as they would in life.[109] Similarly, in her book on caring for the dead, Lisa Carlson affirms the healing nature of personally washing the loved one's body, combing his or her hair, and tending to dressing the loved one yourself, rather than outsourcing these intimate opportunities to carry out "your final act of love."[110] In the case of the Bios Incube, the consumer is offered the opportunity to keep caring for the loved one's cremains and, subsequently, the emergent sapling right in his or her own home. If you would like to receive regular growth updates and care notifications about how your loved one is doing in the #LifeAfterLife as a tree, there is an app for that. In the digital moment, we are used to managing and taking care of many tasks via our cell phone, itself a kind of prosthetic extension of the human body, and so caring for our transmogrifying loved one in the form of a tree sapling via our smart phone is not such a stretch.

In the nineteenth century, the seemingly magical technology of telegraphy, with its series of rappings as dots and dashes, provided a model for spiritualist communication with the dead.[111] The spiritualist "medium," usually a woman, became the "spiritual telegraph," transmitting messages from the spirit world and relaying questions from the living to the "other side." Radio and television technologies similarly provided the languages for New Age "channeling," in which, again, certain especially receptive people with acute spiritual antennae serve as "instruments" in transmitting messages or signals from unseen sources.[112] In this case, the Bios Incube sapling incubator and its corresponding wi-fi connection and user app provide a this-world interactive medium with which to connect with the departed, thus suggestively narrowing the separation between the living and the dead. The Bios Incube is also designed to support an "interactive grief process," which appeals to consumers who want more contact with a loved one after death than is feasible through conventional burial. Although certainly this pertains to human cremains kept in the house, this is especially true of pets, whose owners, in order to have the deceased pet in closer proximity, often retain the cremains rather than burying them.[113]

The Bios Urn, the company claims in its web marketing materials, "is much more than an urn—it's a catalyst for life."[114] More than simply an eco-pious consumer choice, in its iUrn Incube form, it has become a medium for an experience of the "more." It mediates communication (in some sense also a kind of communion with the dead), and through wi-fi, connects the loved one to an ever-expanding cybernetic network of "smart" products and virtual intelligences. You and your loved one, rather than being separated by death, are *both* connected to the Internet, and by implication share in a kind of unified technological noosphere. And yet, this eco-virtuous product in effect links the isapling to a whole network of greenhouse-gas-producing server farms with considerable carbon footprints.[115] Even more complicated is the question of whether connecting to one's deceased loved one through smartphone-enabled electronic products enables us to affirm death or simply further to deny it.

Facing Death, or Mortality Salience and the Urge to Splurge?

In his 1973 Pulitzer Prize–winning book on "death denial," cultural anthropologist Ernest Becker theorized that much of human behavior is driven by a fundamental "terror of death."[116] Scholarly fascination with the denial of death attracted further public attention that same year, as French medieval historian Philippe Ariès traveled to Johns Hopkins University to deliver a series of lectures on history, political culture, and national consciousness. Ariès's history of death outlined in the series, later published and widely popularized, argued that Western attitudes toward death had not always been so fearful, shaming, and characterized by a kind of pathology of avoidance. For Ariès, who engaged in a sweeping historical survey of death from the Middle Ages forward, intense terror of death was a fairly recent development that one could trace to the changes concomitant with the developments of modernity. Prior to the seventeenth century, death in the West had been more a member of the family, a commonplace household regular—visible, accepted, and faced. In medieval times, Ariès argued, death was more "tamed"—whereas in contemporary times it has become more characterized by terror, shame, and invisibility. "The old attitude in which death was both familiar and near, evoking no great fear or awe," wrote Ariès, "offers too marked a contrast to ours, where death is so frightful that we dare not utter its name."[117]

In theorizing the denial of death, Becker argued that to deal with our own fears, we engage in "heroic" acts in a desperate scramble to assure our own immortality. Such acts include "collective aggression, the control of nature, the creation of personal fortunes, political empires, and great art."[118] Becker's theory has been criticized for being overly broad and less than earth-shattering news. That is, people fear death? How is this a scholarly breakthrough? However, his legacy continues in both popular and academic discourses. Death denial is sometimes argued to be one of the factors that motivates climate change denial and prompts fantasies that, through acts of pious consumption, we can shop our way out of global climate crisis and other environmental problems.[119] Founder of Terror Management Theory (TMT), social psychologist Sheldon Solomon, has made his career on testing Becker's death-denial theory, demonstrating through a variety of studies the ways in which terror of death exacts powerful coping mechanisms from research subjects—mechanisms Solomon characterizes as not reflecting "humanity's most sterling qualities." One such death-denying strategy is to distract ourselves with consumption. Solomon observes that "Americans are the best in the world at burying existential anxieties under a mound of French fries and a trip to Walmart to save a nickel on a lemon and a flamethrower."[120] Does choosing and pre-planning an eco-pious green burial then provide a process by which humans (at least those in the mostly developed world) can begin to learn how to die in the Anthropocene?—that is, as climate philosopher Roy Scranton envisions, "to die to a civilization" or way of life based upon extractive economies that continue to write checks that the earth cannot cash?

Stoic philosophers contended that contemplating death and one's limited time on earth can make one more appreciative of one's time here and more intentional about what one does with that time. Contemporary philosopher William Irvine finds much to agree with in their approach. Irvine's *Living a Good Life* argues in favor of the mentally salubrious benefits of contemplating one's own death and those of the ones we love. He tells us that Stoics contemplated death not because they were obsessed with it but because they longed "to get the most out of life." Concurring with the Stoics' strategies and techniques, Irvine contends that learning *not* to fear one's death makes space for more joy. Those who do not develop a coherent philosophy of life that accepts death may try to delay death: "[T]heir improvised philosophy of life has convinced them that what is worth having more of in life is everything, and they cannot get more of everything if they die."[121]

Irvine may source his strategies for dealing with death from Hellenic philosophy in the West, but contemplating death via the practice of meditating on corpses, including one's own, and indeed meditating on bodies in a progression of advanced and gruesome decomposition, is a powerful technique used by some Buddhists for cultivating mindfulness of impermanence. Forty objects comprise Theravada Buddhist "*kammaṭṭhāna*" (objects to meditate on), and ten of these consist of "objects of repulsion" in the form of human corpses: "swollen corpse, bluish corpse, festering corpse, fissured corpse, gnawed corpse, dismembered corpse, worm-eaten corpse" and so on, until finally one meditates on a skeleton. Meditating on corpses, in theory, assists one in regarding and accepting the human body as itself a live corpse.[122] This kind of meditation on death, decay, and impermanence aims at cultivating a life of fearlessness not only of death but of material insecurity. It renders Becker's death-denying "heroic acts" and Solomon's shopping trips to Walmart meaningless and moot.

With all due respect to both Stoics and Buddhists, consumer behavioral psychologists who study and apply Becker, Solomon, and other research on "mortality salience" (thoughts about one's own death) are more skeptical that contemplating death actually dissuades consumption, or at least that it does so when it comes to (mostly) Western consumers. A number of studies within the realm of marketing science point to mortality salience actually *stimulating* the urge to splurge. Some studies suggest not only that consumers purchase *more* when they are asked to think about death and/or presented with images of death but that they also eat more.[123] Marketing behavior researchers Naomi Mandel and Dirk Smeesters's study of European and US consumers, published in the *Journal of Consumer Research*, found that participants asked to write about their own deaths and then asked to check off items they needed on a grocery list checked off significantly more items. They also ate more cookies offered them than did participants who were merely asked to think and write about a "painful medical procedure."[124] Mandel and Smeesters term this reaction "escape from self-awareness" and found that key ways to deal with this state of discomfort are to overeat and/or to overspend. They observed that "[p]eople want to consume more of all kinds of foods, both healthy and unhealthy, when thinking about the anxiety-producing idea that they will die someday."[125]

Does encountering scary information about climate change, rising oceans, Frankenstorms, doomed agricultural crops and cities, increased threats of wildfires, and the imperiled life-sustaining systems of the planet—in essence, what has been termed "ecocide"—shame or scare us into better behavior and make us more eco-piously green? Or, does it merely stimulate our appetites to consume even *more*? If theories of mortality salience are correct, it may be that greater consciousness of the stark realities of the Anthropocene makes consumers fatter and fills our homes and self-storage rental spaces with more gratuitous junk. A study of the impact of mortality salience on Canadian buying patterns conducted at Concordia University's School of Business, however, yielded mixed results. When Canadian participants were reminded of their own mortality, increased awareness of death made those "with lower self-esteem more likely to purchase prestige items." Thoughts of mortality made those with higher self-esteem and those who were older less likely to purchase prestige items.[126] It may be that thoughts of mortality produce less anxiety in those who have lived longer, experienced more loss, and thus for whom death is not a stranger.

It is an open question then as to whether the practices and rituals of green burial planning work to redirect a culture of death denial that, at least in part, drives the desire to consume. Or, does the act of thinking about and planning one's death, even eco-piously, subconsciously whet the appetite to consume *more*? Legitimated by notions of ecopiety and consumopiety, in some respects, green burial has become subsumed into another form of consumer/marketplace-based "moral licensing" that legitimates business-as-usual consumerist behavioral patterns and volumes of consumption. And yet, as we have seen, there is still a demonstrably strong streak of anticonsumerism among green burial advocates who take pride in what they do *not* buy. There are also established pathways in the culture of green burial that connect practices of personal, individual ecopiety to more collective activist practices of public civic engagement and involvement in ecopolicy. Experiences of green burial and its planning cultivate a sense of "more" in many consumers that makes them contemplate their ultimate disposition upon death not simply in terms of pious duty but in gratifying and even joyful mystical terms of "becoming one with the earth," or of "living on" as a vital contributor to an ecosystem.

One must also consider whether contemplating one's death and associating that death with serene and satisfying visions of harmonious "communion"—a mystical return to a welcoming planetary whole—might elicit a considerably different reaction from that triggered by random subjects told to think about their deaths in a research lab. The planning process inherent in making arrangements for green burial not simply involves facing death but, in many cases, reflects a coming to terms with and even welcoming of the transformative processes of death and decay. Instead of green burial taking on a sensibility of grim denial and sober restraint, advocates of eco-friendly disposition express great delight in "feeding the microbes," becoming "food" for other beings, and also a sense of peace when contemplating "returning" to the earth and "merging with nature." Green burial marketing is still steeped in messages of individual ecopiety and consumopiety, not policy. The pronounced mystical dimensions expressed in green burial narratives, however, are hard to miss, and their potential impacts on consumer behavior merit further study.

Marketing and advertising are "implicated in the ways our culture speaks to itself about death and dying."[127] The applied psychology of marketing strategies stokes fantasies of evading mortality via performed acts of consumption. Green corpses, in effect, communicate a counternarrative that resists these messages. In a culture where Botox injections and facelifts send strong messages about impermissible mortality and aging, green corpses' decidedly unapologetic, uncosmetic, unadulterated state of unmitigated decay intervenes and says that it is okay to be dead and to look dead. However, the story the bodies and material culture of green burial tell about our relationship as humans to the planet, to mortality, and to both the forces of nature and the power of digital new media technologies is complex, conflicted, and, as we have seen in previous cases, contrapuntal.

"If we want to learn to live in the Anthropocene," advises Roy Scranton, "we must first learn how to die."[128] New/old directions in funerary media implicitly introduce that message to a whole new generation of consumers, but this process is clearly not without its rich and evocative narrative and rhetorical contradictions. This chapter has wrestled with these questions but ultimately leaves them to further exploration as the green burial movement continues to grow and evolve. One thing we can conclude, however, is that green corpses as "media," if nothing else, do model a more graceful acceptance of transmogrification from one form to another, a challenge that the entire earth community faces on a global scale as we meet our current and future environmental challenges.

7

Expanding the Scope of Justice

Tattooing and Hip Hop as Ecomedia Witnessing Tools

When social psychologists speak about "the scope of justice," they evoke this image as a way to talk about "our psychological boundary for fairness." In short, the scope of justice is the story we tell ourselves of who is (and is not) included within the category of "we" and thus worthy of moral consideration.[1] This chapter is a tale of two different varieties of activist ecomedia deployed as tools to *restory* who constitutes "we" in ways that expand the scope of justice. To lesser and greater degrees, and with markedly different priorities in mind, they propose alternate models to privatized acts of ecopiety and consumopiety as moral responses to environmental crisis. Both activist ecomedia tools also involve mediated expressions of bodily suffering and pain, although in one case self-inflicted and in the other externally imposed. Two environmental nonprofit organizations, based in the San Francisco Bay Area, offer particularly rich media archives for understanding contrasting approaches and investments involved in activist moral witnessing to various publics on environmental crisis. Both organizations work in their own unique ways to expand the scope of justice and moral concern to include underrepresented and neglected populations, both human and nonhuman. Both organizations also harness media tools from youth culture to expand the imagined and implemented boundaries of fairness by strategically adopting popular media forms designed to engage a younger demographic in environmental concerns.

The first organization featured in this chapter is an organization called "Tatzoo." This is a group that has made use of tattoo body art and its circulation through the digital mediasphere as an activist witnessing tool. Traditionally, the act of "witnessing," as it is known in the context of evangelism, occurs when evangelists draw upon scripture and share personal experiences of faith and their personal connection to such things as a higher power, sacred texts, mystical experiences, and so forth, for the purposes of persuasion and, ultimately, conversion. Witnessing tools are material objects, visual culture, and various media traditionally deployed in evangelism as openers to spark conversation about paths to conversion for the uninitiated. Such tools can span the spectrum from more conventional tracts or cards to religious works of art to radio and television programs, billboard advertisements, consumer kitsch like scriptural-verse golfballs or "Jesus Walks with Me" beach flip-flops, message-bearing t-shirts, license plates, and products like video games, mobile applications, popular music, and, of course, tattoos.[2]

Endangered-species body art, when consciously chosen and indelibly inscribed, bears public witness to the suffering and peril of nonhuman species, while also making a correction to what is regarded to be an anthropocentric or human-focused culture that establishes a specist hierarchy of human priorities. Those who engage in endangered-species tattooing and activist media outreach make common cause with endangered species, effectively donating their bodies as public devotional media to communicate the suffering of the vulnerable and to agitate for their compassionate protection. In inscribing their bodies in this way and then sharing the designs digitally, endangered-species advocates make use of the corporate marketing media strategy known as "skinvertising" but do so in a way that subverts its consumerist objectives.

The second organization analyzed in this chapter, like Tatzoo, is also a nonprofit organization focused on environmentally activist work. Cofounded by eco-justice advocates and community organizers Van Jones and Majora Carter, Green For All addresses the pressing interlinked problems of poverty, crime, hunger, crumbling infrastructure, and environmental health hazards in low-income, mostly minority, US urban neighborhoods. The organization unapologetically promotes not personal piety or consumopiety but collective public investment in green jobs and community revitalization, while pressing for the transition to a sustainable green-energy economy. Where Tatzoo has employed tattoos as activist media, Green For All has produced, sponsored, and circulated hip hop music performances, rap videos, rapper contests, and hip hop festivals in an effort to focus environmental verbal and visual rhetoric squarely on the plight of polluted and frequently dangerously toxic black and brown communities. The organization's media outreach consequently shifts its mediated visual and verbal rhetoric away from what has been characterized as the environmental movement's historical preoccupation in media campaigns with the welfare of whales, polar bears, wolves, and wildlife. Green For All's programs and media making challenge unmarked elitism in environmentalism.[3] The organization attends specifically to dynamics of ecoracism and refocuses media attention on the survival and health of vulnerable minority human communities disproportionately exposed to the output of toxic industries.

Urban environmental ethnographer Amanda Baugh's work provides context for Green For All's media interventions. Baugh explains, "Despite a growing body of literature suggesting an expansive understanding of environmental history that includes working-class industrial struggles, American environmentalism in the popular imagination continues to be associated with a legacy of white, middle-class efforts to protect nature for white, middle-class enjoyment."[4] One of Baugh's urban environmental activist informants notably voices a different set of environmental priorities than has traditionally been the case. She explains, "[F] or me, this isn't really about polar bears and narwhals. Frankly, where I live, down on the south side of Chicago, those are pretty hard things to argue for. At least until we stop shooting at children."[5] It is against this background that Green For All's ecomedia making intervenes in dominant environmental media narratives, placing at its radical center the suffering of overlooked, ignored, or willfully dismissed minority communities, particularly the health crises suffered by black and brown children in urban neighborhoods.[6] The organization does not champion consumopiety as an adequate response to environmental injustice.

Endangered-species tattooing and green hip hop are just two of many examples of the ways in which notions of moral inclusion, activism, and the scope of justice get more narrowly or broadly drawn in media forms.[7] For endangered-species advocates, expanding the scope of justice involves intimate but public bodily devotional suffering in solidarity with endangered species. For Green For All eco-justice activists, the scope of justice encompasses the communal body in its suffering, a capacious orientation that intersectionally includes communities of color, urban habitats, systems of healthcare, community safety, wellness and support, networked economies, employment opportunities, actions for social justice, and a prophetic vision of the co-creative linked futures of all earth's peoples.

Creatively using tattooing and hip hop, respectively, as activist media, Tatzoo and Green For All, in distinct ways and with different aims, push environmental activism toward a larger, encompassing "we" of moral inclusion. The story of who constitutes that "we" and the scope of its boundaries of inclusion and exclusion is a tricky and contested business. Examination, for instance, of Green For All's moral messaging evinces the ways in which its media making champions not individuals but collective "people power" as the "alternative energy" needed to address our environmental crisis. In ways largely absent in the individualistic consumerist-dominated rhetoric of environmental marketing, Green For All imagines a future in which organized public support and civic demand for green energy and green economic policies can "lift all boats," creating a more "just, verdant, and peaceful world" for all.[8]

Endangered-species tattooing efforts also use media interventions to expand the sphere of moral inclusion and the boundaries of fairness. These broader boundaries extend to encompass neglected nonhuman others. Activists engage in public suffering and endure pain on behalf of endangered species, while partnering with them in various symbolic, representational, and even spiritually connected ways to draw attention to the imperatives of conservation. The visual rhetoric of endangered-species tattooing also suggests prophetic dimensions of the storied body as short-form storytelling. Fusions of human skin in solidarity with the lives of endangered animals evoke a more peaceable world—one of collaborating human and nonhuman animals, and one where humans, who have clung to orienting stories that separate them from or elevate them above the biotic whole, finally let go of that model. The suffering of tattoo activists is passionate and devotional in nature, spawned in many cases by deep moral commitments. As self-inflicted and voluntary suffering, however, it provides considerable contrast to the suffering inflicted upon the communities ministered to by Green For All. Those who subject their bodies to tattooing choose to story their bodies in ways that help give voice to the voiceless, the vulnerable, and the oppressed. Those who donate skin space to help imperiled species choose those stories carefully and with deliberate discernment, deploying corporate "skinvertising" tactics but disrupting their associated corporate goals.

Black and brown bodies subjected to polluted urban "sacrifice zones" are also, in effect, storied bodies, but these stories, of such things as asthma, cancer, and lead poisoning, are forcefully and violently inscribed upon them.[9] These are not willing, noble, or devotional sacrifices. Eco-rappers use their art to bear witness to the forced suffering of communities of people unwillingly made into human sacrifices on the altar of corporate industrial profits. When low-income and poor minority communities are rendered acceptable sacrifice zones for toxic industries, the people in these communities are effectively treated as subspecies, their endangerment posing little or no concern for moral exclusion when weighed against financial gains.

Paralleling the "digital divide" that excludes lower-income minority students and job candidates from opportunities that necessitate greater digital literacy and access, Amanda Baugh's urban ethnographic informants refer to the disparity of environmental protection for white and black communities as the "eco-divide."[10] Baugh details the nature of this divide: "The environmental movement historically has been associated with the interests of white elites, and despite best efforts of white environmental leaders, most mainstream organizations have had difficulty attracting minority audiences. . . . [M]ainstream environmental organizations fail to attract minority involvement, studies suggest, because of their wilderness-focused agenda and overwhelmingly white leadership, membership, and image."[11] Identifying how each of these activist groups uses endangered-species tattooing and green hip hop, respectively, as activist ecomedia interventions to expand the scope of justice provides a more granular portrait of the eco-divide while demonstrating mixed responses to a consumopiety model.

Mediated Devotional Suffering and the Storied Body

The video camera's frame centers on a Tatzoo program fellow, who is draped, her belly facing down, over a portable tattooing chair, in the middle of a crowded nightclub setting.[12] The venue is actually the NightLife Party Lounge at the California Academy of Sciences, a San Francisco green-roofed research institute and natural history museum that attracts a hip urban crowd to its NightLife party events, concerts, and open-mic nights, which have a distinct club-like vibe to them. NightLife-goers mill around as the live tattooing takes place, curiously checking out what is going on beneath the "buzz buzz" of the tattoo machine. Electromagnetic coils move the mechanism's armature bar up and down, forcing ink into the skin on the lower part of the fellow's torso. The spectators stand and watch the young activist being stuck repeatedly, as, microphone in hand, she witnesses both to her surrounding audience and to those watching the tattoo video livestreamed online about the plight of the endangered sea turtle, the likeness of which gradually begins to appear on her skin.

As the turtle emerges on the woman's torso, it does so visibly below a prior tattoo, which reads simply "Unless" in a stylized script. "Unless" is a reference to *The Lorax*, Dr. Seuss's 1971 iconic fable of hope amidst environmental destruction. In the story, a greedy factory owner called the "Once-ler" despoils the local ecology until no animal species are left. The Lorax, an advocate for the local flora and fauna, is ultimately also forced to flee the factory pollution amidst wholesale biotic diasporic crisis. He leaves behind one single word engraved on a monument that stands where the Lorax had repeatedly issued warnings about the dire consequences of ecological habitat destruction. "Unless," as the Once-ler later figures out, is the Lorax's shorthand for "unless someone like you cares a whole awful lot, nothing is going to get better. It's not."[13] Lorax tattoos are popular tattoos in the wider world of environmental tattooing and in the digital online sharing of these tattoos on Pinterest.com and other social media platforms. Endangered-species tattoos are a subgenre in the larger genre of environmental tattooing, which includes renewable energy tattoos, earth activist organization tattoos (Greenpeace and Earth First! tattoos are popular), environmental lifestyle tattoos that advocate bicycling and veganism, Mother Earth tattoos, and, of course, the ubiquitous three-arrows recycling symbol tattoo as the icon of environmentalism.[14] I define environmental tattoos as those that, via content, messaging, public display, and digital circulation, function as *intentional* media advocacy tools for humans who have dedicated their bodies to promoting environmental activism. A number of tattoos of animals or earth images might be read as being environmentally themed tattoos, but for the purposes of this study, the designation "environmental tattoos" is defined by the tattoo bearers' conscious intent to raise environmental awareness, support an environmental cause, and use their bodies to make an environmental statement.

During the streaming of the "live tattooing," the sea turtle advocate was serving as a fellow in a year-long activist training program developed by Tatzoo. Funded in part by the Audubon Society and a grant from Toyota, Tatzoo's programming has been dedicated to training advocates for endangered species in California. Fellows in the sea turtle advocate's cohort went through a ten-week "Tatzoo bootcamp" program in which they were taught direct-action strategies, activist organizing, and DIY guerilla media making and marketing skills. They also gained other valuable tools that enable them to advocate for a moral "we" that includes endangered species. As described on the Tatzoo web site, "Tatzoo trains biodiversity conservation leaders who use creative advocacy to protect local endangered species. They believe our generation has an unprecedented ability and responsibility to pass on a world rich in biodiversity. Their tattoos represent a commitment to confront and solve the extinction crisis."[15]

After a period of training, study, and thoughtful discernment, each fellow chooses a species to commit to as an advocate. Each fellow also designs and embarks upon a Tatzoo activist project to serve and advocate as an ally for his or her partner species. At the end of the program, as a marker of constancy and enduring dedication, the fellows receive a free tattoo of the species. The tattoo is a symbol not only of indelible commitment but of the willingness to sacrifice and endure pain in order to make this commitment. Much as anthropologists Victor Turner and Arnold Van Gennep identified symbolic markings that accompany the culmination of rites of passage, the endangered-species tattoo marks the initiate's new status once he or she has come through the sacred ordeal of "Tatzoo bootcamp" training.[16]

This chapter draws data for media analysis from an online archive that includes tattoo media and testimonials posted on Tatzoo's public Facebook, Instagram, and Vimeo sites, accompanied by viewer comments, public testimonials from Tatzoo fellows published on Tatzoo's main web site, posted videos of fellows' projects on a variety of video streaming platforms, conversations about Tatzoo and Tatzoo events on publicly accessed social media sites, press coverage of Tatzoo, news media interviews with the organization's founder, and the organization's own press releases and other media making. The sum of this archive suggests multiple functions served by Tatzoo endangered-species tattoos, but two of these are (1) deployment as a witnessing tool to evangelize the need for species conservation and (2) adoption as a meaningful marker of suffering, devotion, and commitment. That is, the Tatzoo fellow's storied body itself becomes both powerful *medium* and *model* for suffering on behalf of, in solidarity with, and in devotional commitment to, nonhuman animals in peril.

The name of the organization, "Tatzoo," and its implications are worth noting here, especially as a key function of an endangered-species tattoo is that someone sees the tattoo on the storied medium of the body, stops, looks, wonders, and then asks the tattoo bearer about it. In an evangelical context, here is the hook that then enables the evangelist to bear witness to the uninitiated and subsequently to, as evangelists phrase it, "bring them the story." The deployment of endangered species tattoos as activist ecomedia ideally works in a similar fashion. Much as with zoos, the human host for the endangered animal tattoo (or tattoos) invites the public to come and have a "good look"—either in person or via digitally circulated media. Many zoos, active in animal conservation efforts, set up contained spaces to facilitate this human voyeuristic gaze in exchange for greater financial and public support for saving endangered species. The animal contained on the canvas of human skin also fascinates the spectator, as it elicits comments and questions, perhaps even a sense of awe, and thus provides the endangered-species activist an opportunity to educate, elicit compassion for the species, and gain more supporters for the conservation cause. In his now-classic essay, "Why Look at Animals?"—on humans, nature, and animal spectacle—art critic John Berger observes, "The public purpose of zoos is to offer visitors the opportunity of looking at animals. Yet nowhere in a zoo can a stranger encounter the look of an animal. At the most, the animal's gaze flickers and passes on. They look sideways. They look blindly beyond."[17] But the spectacle of the endangered-species tattoo often *does* look its public in the eye, holding the voyeuristic gaze. In this way, the viewing of the tatzoo becomes a seductive invitation into conversation on the media audience's own terms. It does not avoid the human gaze but welcomes it. This accessibility and invitational dynamic may contribute to its value as a witnessing tool.

Back at the NightLife Lounge, the Tatzoo fellow embarks upon another bodily inscription beneath her "Unless" tattoo, only this time the inscription process is public, confessional, and didactic. She instructs her audience, "The big thing with tattoos is that they are forever." Like the Lorax, the fellow establishes her constancy as an endangered-species advocate, proclaiming, "This [sea turtle] tattoo to me represents my commitment to this species—that I will forever act as a storyteller, as an advocate for the sea turtle. . . . I will give this animal a voice." Her vow to give the sea turtle a voice is suggestive on a number of levels, including her verbal testimony not simply to the turtle's plight but also to her storied body as witness—as corporeal media, testifying to the species' suffering, the visual marker of which is now artistically and dermally fused to her own devotional suffering.[18]

As the tattoo gun (sometimes called an "iron") continues to puncture her skin, she steadfastly directs NightLife-goers to pick up the attractive reusable cloth bags she has hand made from recycled materials and is distributing that night as gifts. She tilts the microphone up, gestures to a video screen of images, and explains how plastic bags look just like jellyfish to turtles, who then consume them. (Sea turtles can choke on these plastic bags, or their stomachs fill with plastic, which they think is food but which their bodies cannot digest.) She continues to educate her audience about the trials of the sea turtle as the continuous buzz from the machine never falters in the background. When the fellow testifies to the endangerment of sea turtles from plastic bags, she points out our direct moral culpability as consumers for causing the suffering and deaths of these creatures: "That's our fault," she says into the microphone, the needle buzzing in the background. "That's our responsibility. No one put plastic in the environment but us," she admonishes.[19]

This live devotional bodily inscription was then injected via social media into digitally networked flows of environmental civic engagement and endangered-species activism. Video of the live tattooing, along with still images of her turtle tattoo and its story, then circulated the digital currents. To display the turtle for the still camera shots posted online, the fellow lifts up the end of her shirt to reveal the image. In her day-to-day encounters, depending on clothing choices, climate temperature, and her workplace, very few people might actually encounter this singular witness to sea turtle suffering and endangerment. Once this image is entered into a variety of overlapping digital media ecosystems, however, the reach and impact of the medium that is her storied body, dermally fused representationally with the turtle's plight, is made manifold.

Media theorist Marshall McLuhan famously referred to media as the "extensions of man"—tools that extend or project our bodies in space and time.[20] Media make us more than we are. They augment our mere human corporeal capabilities and limitations, amplifying our impact and reach. And yet, in this video of an incredibly devoted, well-intentioned woman, who is, to her immense credit, doing so much more than most people do, the story communicated is incredibly limited in what it asks of viewers. The tatzoo fellow shames her audience for using single-use plastic, emphasizing that the suffering of plastic-ingesting turtles is "our fault," and then she supplies her immediate audience with reusable bags so that from then on they can make an eco-pious choice. What if instead she had taken this incredibly powerful, arresting moment of mediated corporeal suffering, spectacle, and display of skin-media inscription to witness to her audience about civic engagement steps to take in order to ban single-use plastics?

This civic and more collective approach to dealing with plastic is one that other countries are enacting across the globe, but the approach has been much slower to gain ground in the United States. In 2017, Costa Rica announced its comprehensive national strategy to ban all single-use plastic by 2021.[21] Before that, in 2016, France became the first country to ban all plastic cups and plates.[22] At least fourteen countries as of 2018, from Zimbabwe to Taiwan, had enacted similar bans, and as of July 1 of that year, Seattle became the first US city to ban all plastic straws and single-use plastic cutlery.[23] Numerous local ordinances in coastal cities and beyond have enacted laws banning plastic bags just like the ones turtles choke on, *not* leaving plastic use up to the voluntary ecopiety of the individual consumer but enacting public policy that requires businesses not to supply plastic bags in the first place. These actions are affirmations of the positive power of collective action and legislated change, deflating libertarian fantasies that avow noninterference and absence of all regulation as the best approach. A policy model is strategically, qualitatively, and practically different from one that shames citizens for not practicing virtuous abstinence amidst a consumer marketplace and supply system concertedly structured to inundate consumers with plastic bags at every turn.[24]

What is perhaps more effective in the visual culture of the fellow's messaging is that endangered-species tattooing itself is arguably a representational collapsing of the human/animal divide, where human skin symbolically fuses with endangered species. The inscribed body itself becomes a powerful medium testifying to human/nonhuman intersectionality. The devotional act of partnering as "advocate" for the species suggests the embodiment of a kind of morally "extended self," a phenomenon that social psychologists Marilyn Brewer and Wendi Gardner describe as a redrawing of boundaries of self such that the "content of the self-concept is focused on those characteristics that make one a 'good' representative of the group or relationship."[25] In this visibly avowed solidarity with the partnered species, there is public testimony to humans' being, despite our illusions to the contrary, animals, *too*. This arguably fosters greater feelings of empathy and compassion based upon a shared identity of "us" instead of "them."

Animal studies theorist Margo DeMello observes that "the divide between humans and all other animal species is neither universally found nor universally agreed upon. It is neither an exclusively behavioral nor biologically determined distinction but has, at times, included biology, behavior, religious status, and kinship. Ultimately, we will see that this divide is a social construction. It is culturally and historically contingent; that is, depending on time and place, this border not only moves but the reasons for assigning animals and humans to each side of the border change as well."[26] In the case of the sea turtle live tattooing described above, as the visually and verbally inscribed story is released into the mediasphere, it crosses multiple platforms, in effect offering the endangered sea turtle tatzoo virtual habitat in which to roam. The messaging about ecopiety and plastic bag consumption may be woefully limited, but the digitally shared public political act of dermally fusing with a kindred animal species is a profound statement in and of itself.

In analyzing tattoos and communication, the work of media theorist John Durham Peters once again proves insightful for reading bodies themselves as media. Recall from the discussion of green corpses in this book that Peters argues that the body is after all "our most fundamental infrastructural medium."[27] As living, organic media, then, Tatzoo fellows' animal-inscribed bodies effectively function as politically activist media interventions in a world in which habitat destruction and species extinction are frequently marginalized as peripheral concerns. Sociologist of media and culture Kevin Howley defines such media interventions as "activities and projects that secure, exercise, challenge, or acquire media power for tactical and strategic action."[28] Engaging the cultural politics and symbolic power dimensions of such interventions evokes media theorist Nick Couldry's insight that media are not *things* so much as "something we *do*."[29] Getting tattooed, sharing that tattoo in person, and of course digitally circulating it, accompanied by personal story or testimonial, are all very much in a Couldrian sense *doing* media and media *as practice*. Indeed, the devotional and witnessing practices of Tatzoo-inscribed activists engender the reality of humans themselves as "media animals" *in action*—in this case, *intervening* in the endangerment of species. As mentioned, to accomplish this, they use a corporate marketing method known as "skinvertising," but they do so innovatively and in unorthodox ways.

Skinvertising and the Spirit of the Tattoo

Tatzoo is just one of a number of organizations and projects that have utilized endangered-species tattooing as an activist media practice. A UK-based program called "extInked," for instance, celebrated the two hundredth anniversary of Charles Darwin's birthday in 2009 by soliciting one hundred local volunteers to get inked with an endangered plant or animal species of the British Isles. Key to this project was that many of these species were not cute, eye-grabbing, or even particularly interesting to a wildlife nonspecialist. Nonetheless, the human commitment to be dermally inscribed with these "unexciting" species spoke to the species' intrinsic existential worth and the value of their survival. Following a three-day tattooing marathon, extInked's advocates gave public talks, handed out educational pamphlets, met with conservation groups and government leaders, displayed their body art in endangered-species exhibitions, raised funds, and generally deployed their bodies as activist media to educate and advocate for the conservation of their partnered species.[30] The brainchild of an activist art collective, extInked was organized into a traveling photographic display of all one hundred advocates, each bearing his or her designated endangered British plant or animal species. Featured at galleries and museums across the United Kingdom, the exhibit traveled for some three years and is archived in the "Species" and "Ambassadors" galleries at the project's web site.[31] "At a time of alarming biodiversity loss and unprecedented climate change," declares a conservationist group leader on the project's web site, "it is more important than ever that innovative approaches are used to engage with people and reconnect them to their natural environment."[32]

In its most recent form, the "innovative approach" of using human bodies as billboards is a marketing tool that has been pioneered by corporations and businesses and is more technically referred to in the world of advertising as "skinvertising." Human billboards have long existed in the advertising world in the form of signholders, known as "human directionals," as well as in the form of "sandwichmen"—those who walk the streets wearing an advertisement attached to their bodies on front and back. Skinvertising is the phenomenon in which people either contact companies and offer to sell space on their skin for a price, or companies reach out to potential skinvertising influentials, offering compensation for skin-based brand exposure. The most famous skinvertiser is a husband and father of five children, known as "Billy the Human Billboard." He has supported his family since 2009 by selling sections of his skin to dot-coms and prior to that sold space on his back to pay for kidney donation surgery. A forehead tattoo alone can go for fifteen thousand to twenty thousand dollars.[33] Various athletes competing in volleyball, weightlifting, skateboarding, and running have readily accepted corporate sponsorship in return for skin space. Corporate sponsorship of NBA players in the form of skin billboarding has come to pose a thorny legal issue for debate.[34] Replacing the traditional bake sale for children in financially strapped US public schools, wearing six-week temporary tattoo logos has become a way to fundraise in order to replace money slashed from education and athletic program budgets, while a number of businesses now offer their employees a certain percentage raise if they agree to a permanent tattoo of the company's logo.[35]

Environmental tattooing practices reclaim and decommodify the corporate sponsorship of skin billboards by deploying skin media as tools for environmental activism in a "gift economy" that bleeds into the realm of sacrificial and devotional practice.[36] By using the term "devotional practice," I refer to acts of virtue and piety that are brought into daily life and usually involve some form of sacrifice in faithful observance. Many tattoo artists donate their services to this kind of work, and of course those who get inscribed donate their skin as canvases with no monetary compensation. Not only do they donate their time to the tattooing process itself but they also donate time to tending the new tattoo as it heals, while also dealing with any associated medical costs for complications.

Using digital media tools and social media networks, an online environmental tattoo-focused conservation organization called "Bush Warriors" has used skinvertising to focus greater public attention specifically on conservation and antipoaching work in Africa. The Bush Warriors Clan wildlife tattoo archive of their "Tattoo of the Day" contest showcases brightly colored tigers and African elephants splayed across men's backs. Zebras and wizened gorillas occupy sturdy calves. Lions and tigers roar across shoulders, torsos, and lateral muscles. A notably large tattoo in the archive shows a powerful dual-horned rhino on a cracked swath of savannah, surrounded by the skulls of its murdered brethren.[37] An anthropological-sounding explanation for the significance of skinvertising as activist tool issued by Bush Warriors' founder, Dori Gurwitz, an African-born, now US-based tech business innovation strategist, accompanies the tattoo archive:

> Throughout the centuries, tattoos have served as religious or spiritual symbols; represented rites of passage, status symbols, sexuality or punishment; and more recently a personal statement by its bearer. Regardless of the reason, humans have sacrificed part of their body and endured pain to prove a point, create change, and emerge as a new being. It is a transformative declaration of power, an announcement to the world of control of the body, life and destiny.
>
> Wildlife and nature tattoos go hand in hand, dating back to the beginnings of civilization. The mummy of a Scythian Chieftain from over 2,500 years ago is tattooed with an extensive range of animals, fish and monsters. Other civilizations have tattooed wildlife as a spiritual way of possessing their qualities and traits while modern portraits of wildlife usually depict their majestic beauty or raw power.[38]

In the comments section of the tattoo archive, an enthusiastic Molly Tsongas (founder of none other than Tatzoo) introduces herself and invites Bush Warriors founder Dori Gurwitz to head over to Tatzoo's web site to see how her organization is similarly combining tattoos and the conservation of local species. Here we are able to observe digital media networking in action, as Tsongas points out, "We have a lot in common!"[39] Both Tatzoo and Bush Warriors tap into a larger uptrend in tattooing, especially among millennials, and a more specific emergent trend in environmental tattooing as activist skin media, of which endangered-species tattooing is merely one subset.

The point Gurwitz makes about how "civilizations have tattooed wildlife as a spiritual way of possessing their qualities and traits" is an evocative one when considered in the context of contemporary environmental activism. Ethnographic research on radical environmental activists has pointed to the mystical dimensions of such activist work: for instance, some activists working to defend habitat for endangered species perceive their spirits as merging with the spirits of those species and even being assisted by their powers. An Earth First! activist saboteur in Bron Taylor's ethnographic work on "dark green activism" merges with the spirit of her "totem animal," a ring-tailed cat, in order to camouflage herself and evade federal law enforcement chasing her through the desert.[40] Forest activist and old-growth redwood tree sitter Julia Butterfly Hill writes in her memoir of her having experienced interspecies communication and spiritual oneness with Luna, the fifteen-hundred-year-old redwood tree she spent nearly two years defending from loggers. As Hill inhabited the tree, she sensed herself partnering with Luna, receiving guidance and even spiritual power from the tree in order to continue their collaborative protracted protest.[41]

Activism and advocacy are exhausting, draining work, and the kind notoriously prone to burnout. What if that burnout can be offset for activists, though, by partnering with an endangered animal ally they have ritually inscribed on their body in an avowed ritual of dedication and devotion? This is a connection that embodies more than mere ecopiety. Once symbolically fused with the endangered species, activists are no longer one small, overwhelmed individual but now fundamentally linked in a very tangible way: not only to other species tattoo activists but to the individual animal they bear on their body and representationally to the species' "clan." Here, social science theoretical discourses on the power of totems and totemic representation come to mind, a discourse of which Bush Warriors' tattoo activist and founder Gurwirtz appears to be self-consciously aware.[42] A number of tattooing cultural traditions also recognize a living soul or spirit endemic to a tattoo. Among the Maori of New Zealand, for instance, the *tā moko* facial tattoo (originally done with a chisel and thus technically a carving but now usually executed with a tattoo machine) traditionally needed to be conserved as a dead person's body received mortuary ritual processing. Such conservation is necessary because the *moko* is still alive and inspirited.[43]

When Tatzoo activists speak superlatively about the tattoo being "forever," they could be right. As with the Maori, conserving tattoos is increasingly a popular practice in the United States, where efforts to preserve tattoo artwork after death, or to retain the spirit of a deceased loved one by keeping a memorable piece of him or her has launched a new service industry in tattoo retrieval and curation. Businesses like Save My Ink Forever process the skin canvas bearing the tattoo with a preservative polymer and then frame the cutting under archival quality UV-protective glass.[44] Laser tattoo removal technologies aside, the perceived "eternity" of the tattoo, at least while its host is alive, and now through conservation perhaps well beyond the grave, lends weight to its inspirited and inspiriting power for the activist. There are also committed activists who get inked with "extinction empathy tattoos" to memorialize species that have already gone extinct. The tattoo includes an illustration of the species with the approximate dates under it in which the species lived on earth.[45] With conservation technology, those tattoos could now live on well beyond their human host. In the Tatzoo web site section labeled "Testimonials," a Tatzoo fellow displays her shoulder tattoo of the endangered Metalmark Butterfly and reflects on the tattoo: "It's a story that will live on in our bodies forever." Thus, the tattoo has both a living quality and a narrative quality that enable the storied body both to be read by others and to communicate its messaging across space and time.

Doing Media, Doing "Whatever It Takes"

A Tatzoo Facebook-posted online video shows a curator at a wildlife education center as she receives her endangered-species tattoo. Again, this is a public viewing and live tattoo exhibition, and the wildlife center that employs the curator Internet livestreams the video of her tattoo inscription. She lies face-up on a hospital gurney, in the middle of the bustling nature center, while families with children who have come to visit the center surround her inquisitively. As the tattoo artist zaps away at her arm, a wildlife handler steadfastly stands by her side, handling Helios, a live (endangered variety of) screech owl.[46] The curator turns her head, avoiding looking at the ongoing needle work, as she answers in a determinedly steady voice questions posed to her from the families who have stopped to watch this spectacle. "Is that a real owl?" asks a child. "Is that a real tattoo?" asks a parent. "Does it hurt?" asks another child. "It's not bad," she reassures. She explains to the immediate spectators and to viewers online, "Each person has a story behind why they are putting that piece of art [a tattoo] on their body. This is mine." She then proceeds to discuss her commitment to the screech owl, indicating the small owl who is standing by her side blinking at her as its likeness begins to materialize on her arm. Within the visual rhetoric and compositional tableau of the video, Helios the owl appears to be reciprocally bearing witness to his human advocate's devotional suffering. As she lies prostrate on the gurney, arms outstretched, receiving the tattoo, it is hard not to wonder at this sort of public offering of the storied body.[47]

Essential to media productions such as the Tatzoo activist's sea turtle tattooing recounted at the beginning of this chapter and the nature center activist's live-streamed owl tattooing is the very conscious public nature of the sacrificial act—performed in front of a live audience, recorded on video, and then posted to the Internet for more to view. Self-inscription of the body may seem like a very individual, personal, and intimate act, but these endangered-species tattooings are most definitely not private acts of piety. They are pronounced public acts of devotional suffering and witnessing to the sins of humans as perpetrators of species' endangerment, even as a visual rhetoric of human contrition, penance, and eco-pious redemption is displayed.

The Tatzoo official web site in its design and content also reflects a very public evangelism. It specifically dedicates a section to "Testimonials," in which fellows speak to the ecologically salvific dimensions of their tattoo inscription. Content and design also communicate a redemptive message about humans who, through acts of piety, devotion, and suffering, may work to redeem the error of their ways. Offering testimony is a regular practice in evangelical churches, by which those who have been saved get up in front of the congregation and speak of

God's grace in their lives and how they have been saved from their own sinfulness by giving their lives over to God. In the case of posted Tatzoo videos, however, especially videos of bodily devotional suffering, these media reflect more of a quasi-Catholic sensibility. Redemption comes from mending human ways and partnering with those whose lives we endanger, offering voice to the voiceless, and rendering activist bodies as testimonies to solidarity with those who suffer.

The content and semiotics of these videos evoke similar themes as those presented in Roman Catholic social justice media and Catholic Worker rhetoric—that is, expressions of solidarity through acts of voluntary suffering with the least among us.[48] Of course, in Catholic Worker literature, suffering with the least among us is a pronounced human-focused effort, particularly directed to the poor who suffer in America's slums. It is one that unmistakably argues for the moral inclusion of marginalized sectors of human society. In the case of activist tattooing, the "extended self" suffers with an endangered animal, making an implicit argument for the moral inclusion of nonhuman animals in the sphere of justice, but the material, corporeal nature of this suffering still bears resemblances to Roman Catholic sensibilities of devotional suffering.[49]

Some of the testimonial photos show raw-looking red tattoos that have not yet healed. A photo, for instance, of a fellow's inflamed shoulder and back, covered with tattoos of a school of scalloped hammerhead sharks—still red and weepy from the needle—bears the caption, "They're amazing creatures, and I will do whatever it takes to save them."[50] "Whatever it takes" is clearly displayed on the activist's inflamed skin. These are images of sacrifice but markedly those of willing, joyful, at times even ecstatic, sacrifice of committed devotees.[51]

Medieval self-flagellants joyfully whipped themselves and mortified their flesh in public parades, in public squares, and in other public demonstrations as a testimony to their piety and willingness to self-inflict suffering as penance. Their suffering bodies also became witnessing tools to inspire suffering, piety, and devotion in others. This practice continues today, most notably in Roman Catholic countries, such as the Philippines, but also in traditional groups of flagellants such as the Penitentes in the American Southwest.[52] The piety and pain of the devout self-flagellant is often expressed as ecstasy. The body of the tattooed conservation activist likewise becomes a media platform for an enduring story of suffering, loss, conservation, life-long commitment, public piety, and redemption.

In her work on the cultural politics of body modification, sociologist Victoria Pitts characterizes tattoos and tattooing in terms of "subversive bodies and invented selves."[53] Tatzoo fellows similarly evoke the subversive potential of their endangered-species inscriptions for bringing about a different future, indeed a prophetic one—an envisioned world of mutually enhancing human/nonhuman species relations. On the testimonials page, a Tatzoo fellow's account offers a narrative sense of his tattoo that goes beyond the mere material: "This tattoo is a vision of another future, one where threats to these butterflies are gone, one conducive to the process of life."[54]

Purity and Danger in Devotional Skin Sacrifice

Perhaps fittingly, the very composition of the environmental tattoo has become fraught with consumopiety concerns within the environmental community. Human and surrounding environmental health impacts from sourcing and utilization of mercury, heavy metals, including lead, and other toxins in the ink have prompted eco-pious questions and moral apprehension about the practice of tattooing. *Environmental Health News* has circulated a series of articles that raise potential problems with the aforementioned mercury and lead in tattoo ink, in addition to other toxic substances like cobalt nickel, chromium, manganese, and even arsenic and phthalates. Autopsies have shown that tattoo ink and thus the environmental toxins it contains migrate to the body's lymph nodes.[55] Environmental and food safety advocates have been active in getting phthalates (known hormone-disruptors implicated in cancer) removed from beverage containers, children's toys, and other household items, only to have phthalates injected directly into bodies via tattoo ink.[56] Both popular "all-natural" lifestyle publications and the World Health Organization have pointed out to consumers that just because an ink is marked "organic" does not mean it is not still toxic for human bodies and for the environment, citing studies showing that with sun exposure, or just simply with time, deposits of toxic tattoo ink break down in the skin and travel, clogging the lymphic system.[57] The categorizing of tattoo ink as itself an environmental toxin thus poses a tricky ecopiety and consumopiety quandary for those wishing to participate in environmental tattoo activism.

Vegans who wish to be inked face additional ethical challenges, since inks usually contain the ash of charred animal bones, and thus are not vegan. Specially formulated alternative vegan inks that leave out the animal bones often replace animal ash with an increased composition of heavy metals, thus increasing health concerns and environmental toxicity. Vegan "purity and danger" consumopiety concerns in this instance go beyond the mere symbolic dangers anthropologists observe with anxieties about culturally defined notions of purity and pollution and cross into the realm of documented biological health dangers.[58]

Although one of the powerful things about a tattoo is the commitment to a (mostly) permanent inscription of the body, ecopiety-minded individuals who wish to remove their toxic tattoos face a subsequent dilemma since the laser process most used to remove tattoos necessitates the use of tetrafluoroethane, a particularly potent greenhouse gas. Choosing eco-pious alternatives to remove these signifiers of commitment and green identity has become almost as important as the initial commitment itself. All this, while ongoing debates raise questions about the efficacy of outcomes from such self-inflicted acts of public sacrifice.

Green Hip Hop, Suffering, and Sacrifice Zones

Unlike those who willingly subject themselves to acts of eco-pious sacrifice and devotional suffering, the suffering experienced by those in urban minority communities subject to toxic environmental contamination is not self-inflicted, chosen, or agential. Neither is this suffering culturally recognized or valued as noble or redemptive. Often consigned as industrial pollution "sacrifice zones," these communities endure externally inflicted and nonredemptive suffering. The very term "sacrifice zone" is an atomic one, originating from the Cold War. It originally referred to National Sacrifice Zones—areas of the United States that would be served up as sacrifices so that nuclear weapons could be manufactured and stockpiled. Communities surrounding the mining of radioactive materials, or places closest to areas where atomic bombs are tested, were designated National Sacrifice Zones.[59] The Nevada Test Site, the bomb test site about an hour's drive from Las Vegas, is designated a permanent national nuclear sacrifice zone.[60] In Chernobyl in northern Ukraine, the site of a 1986 nuclear reactor meltdown, and, in Japan, the twenty-mile area established around the Fukushima nuclear power plant severely damaged in 2011 are both referred to as nuclear "exclusion zones." They are essentially sacrifice zones to nuclear energy but are also restricted areas that humans are not permitted to inhabit.[61] Here in the United States, the Navajo reservation and other Native American reservations where radioactive uranium mining has taken place, and where radioactive tailings still remain, or where nuclear reactor waste is stored, are peopled sacrifice zones.[62] Toxic low-income minority neighborhoods located near or next to polluting industries or waste hazard sites are also peopled sacrifice zones.

In environmental justice terms, the designation of sacrifice zone has gone beyond its original atomic associations to apply to low-income and minority communities whose damaged health and well-being becomes a morally acceptable exchange for corporate polluter profits. Wealthier and more socially advantaged communities, understandably, howl with cries of "NIMBY" (not in my backyard) when offending industries try to locate in or nearby their homes and children's schools, threatening the health and safety of their communities. Unlike residents in poor minority neighborhoods, NIMBYs possess the political clout, the financial resources, the legal access, and the litigative savvy to send those industries somewhere else. Somewhere else is usually a community of color, comprising what environmental justice advocates have termed "ecological others," who are then served up as sacrifices to city or state revenues and corporate interests.[63]

The activist group Green For All has innovatively used hip hop music and its messaging as an activist ecomedia witnessing tool to focus attention on environmental racism, toxic neighborhoods, and food deserts (areas with little or no accessible fresh food).[64] The mainstream images of environmentalism that continue to predominate in the mediasphere are those of wealthy, white, elite, lauded celebrity green experts and exemplars. These models of environmental virtue appear on the covers of *Vanity Fair*'s special "green issues" (Madonna, George Clooney, Leonardo DiCaprio, Julia Roberts, Al Gore). Green For All's media making *intervenes* in this predominant narrative of privileged eco-heroism as practiced through consumopiety and concertedly redirects our attention to brown and black peoples in decidedly unglamorous urban neighborhoods. The visual and verbal messaging of Green For All's rap music videos, in particular, bear witness not to the suffering of whales and polar bears but to the suffering of minority children who experience some of the worst asthma rates in the country.[65] "The environment," as conceived and framed in Green For All's media works, is not about saving pristine wilderness but squarely about things like jobs, health, and justice. The organization's activist media making engages in a conscious reorientation of environmental priorities to foreground neglected human populations that are too often vulnerable, taken advantage of, and rendered expendable. In effecting change, Green For All also concertedly and powerfully promotes policy solutions, not piety solutions. In the university courses I teach in Northwestern University's Program in Environmental Policy and Culture, I have a repeated motto that by the end of the term all the students can circle multiple times in their notes: "Policy, it's what's for breakfast." At Green For All, policy, not piety, is what is served up.

When used as witnessing tools, both mediations of endangered-species tattoos and hip hop music/performances intentionally speak to a younger demographic. There is also a degree of overlap between the two artistic expressions. As hip hop scholar Monica Miller points out, tattooing culture and its signifying components are integral ingredients of hip hop culture.[66] Both media practices also bring into public consciousness images of suffering, marginalized populations—those overlooked or excluded from moral concern and the scope of justice. But, whereas endangered-species tattoos embody a biocentric ethic and aesthetic that make more space in the human moral universe for nonhuman species, green hip hop squarely and unapologetically focuses on human struggles for survival, pressing for the urgency of marginalized-human moral inclusion. Green For All, as the organization's name suggests, in its work to expand the scope of justice, seeks to close the persistent eco-divide.

East Oakland rapper Doo Dat (aka Markese Bryant) grew up living next to a Chevron refinery. He recounts, in a video about the making of his rap music, how he and his friends would wade and play as children in a toxic ditch of chemical run-off that streamed through their backyards. As he grew older, he and his friends played basketball under high-voltage power lines and spent recess inhaling the fumes from the freeway located adjacent to their school playground. After a stint in jail for selling crack cocaine, Bryant straightened out his life, took classes at a community college, and made his way to Morehouse College, the alma mater of Dr. Martin Luther King Jr.[67] Bryant has since emceed the Bay Area's solar-powered hip hop concert Grind for the Green, and appears in Green For All–funded rap music videos about pollution in minority neighborhoods and their lack of access to basic things like clean air and fresh produce. Images of the toxic Chevron ditch and the school playground with high-voltage powerlines flash through his video as he raps, "Got a message for the 'hood: It's time to go green. We gotta go green. The food ain't fresh and the air ain't clean." Titled "The Dream Reborn," the video includes clips of Dr. King's famous "I Have a Dream" speech, as King declares, "Now is the time to make justice a reality for all of God's children." King's sound bites and speech clips are juxtaposed with clips of then-president Obama calling for green jobs.[68]

In the context of the video, "green" means *justice* and a pathway to better economic conditions. Rapper Doo Dat emphasizes, "We need green jobs / We don't need no mo' jails." Starting with a voiceover that says, "Let hip hop lead the way into a great future," the video brings lyrics and images together to evoke what that future might look like. The hook of the song, "My President is Black, but he's goin' green" (a nod to Jay-Z's "My President Iz Black"), is punctuated with photos of Michelle and Barack Obama digging and planting vegetables in inner-city community gardens, advocating policy and urban planning changes to create access to clean food and better nutrition as a key component of a greener economy and thus a more just future.[69]

Doo Dat's video, like a number of eco-rap videos, shows inner-city minority children with asthma inhalers, reflecting the disproportionately high rates of asthma in minority neighborhoods. To raise awareness of the eco-divide in terms of clean air, Green For All has produced and circulated a chilling video that portrays "the sounds of the old economy"—the clang of industrial mechanisms combined with the sound of a little Latina girl coughing and wheezing, and the "pssssst" noise of her desperately pumping her inhaler as she suffers asthmatic distress.[70] The inhaler also shows up in other eco-rap videos, such as a video called "Global WarNing" from the album *Earth Amplified* that rapper Stic.man (from dead prez) made with rappers Seasunz and J. Bless. The video, which is produced by Green For All, is about the global war on poor people, particularly minorities, waged by oil company executives and industrial chemical barons like the Koch Brothers. A street sign in the video reads, "Caution: Koch Brothers Profiting." Sinister figures of vulture capitalism loom over the city in Orwellian animated oversized billboards dominating the city, while down below on the street, African American children are sucking on inhalers, gasping for air just to survive.

Whereas the recycling symbol of revolving arrows provides the ubiquitous symbol of the environmental movement more generally, in the world of eco-rap, the *asthma inhaler* occupies that iconic space and reframes more general conversations about the environment to discussions of eco-justice and imperatives that demand collective moral responsibility. Seasunz raps,

> Uncle Sam making bets write his name by the Ex-xon Valdeez
> Probably asthma's next
> Little Johnny got a little funny knob in his neck so what the problem with that?
> You just a corporate exec who extract that oil from the boils of the backs of Nigerians
> Gimme that!

Images in the video juxtapose and link the suffering of US black and brown people economically confined to hazardous neighborhoods with the exploitation of black and brown peoples around the world.

In the same "WarNing" video, stic.man advocates resistance to companies that exploit vulnerable communities:

> Look, Look, Look
> Go green, Go solar, Go agriculture
> FYA [For Your Action] burn MacDonald's and Coca Cola
> All corporate vultures should burn in the toxic sulfur they emit for every dollar they get
> They exploit the culture
> And take resources
> Control the media and censor the voice of the voiceless

> Sitting back fat in they office they don't see people, just profits and losses.[71]

Images depicting diabetes and heart disease as killers in poor urban communities accompany the call to action ("FYA") to burn McDonald's and Coca Cola, two companies that aggressively target poor minority communities and are a pervasive presence in urban food deserts.

Adopting a homiletic didactic quality in media making, Doo Dat's "Dream Reborn" rap video conjures the moral authority and memory of Dr. King as his video reframes environmental discourse away from the tropes of a historically white-dominated environmental movement—wilderness, nature appreciation, species and habitat conservation, and nature as critical safety valve *and* sanctuary *from* the problems of the city.[72] Seasunz, J. Bless, and stic.man's video similarly casts issues of the environment as being about corporate policies and marketing that preys on the weak, while destroying the extended earth community in the process. Collaborating with Green For All, these artists restory environmental issues as fundamentally *justice* issues and place them directly into the practical realm of what Anthony Pinn has called the "nitty gritty hermeneutics" of rap music and hip hop culture.[73] In the realm of the "nitty gritty," in the specificity of the urban pavement, the environment is not an aesthetic abstract but is grounded in things like jobs, asthma, cancer, inaccessibility to community healthcare, and food scarcity. What is more, instead of a conventional back-to-nature environmentalist narrative that cast cities as problems, eco-rap reframes cities as potential *solutions* for our most pressing environmental problems. In the lyrics and narrative aesthetics of eco-rap, sustainability is about fair urban community planning, investments in fresh food sources and urban gardens, and access to healthcare, green jobs programs, clean air, water, and safe playgrounds, while building a just economy with opportunities for all.[74] Using rap as a tool of urban media resistance, Doo Dat's video repeatedly evokes the spirit of Dr. King to drive home the point that eco-justice is fundamentally a civil rights issue.

In other Green For All–sponsored eco-rap videos, artists paint the very real picture of "eco-apartheid," as when East Coast rappers Tem Blessed, First Be, and Outspoken address food deserts and the segregation of poor urban minority communities. The rappers won an eco-rap contest hosted and funded by Green For All, and the prize was funding to get their rap song, "Green Anthem," produced as video. The lyrics here address food and health inequities in terms of structural racism:

> If you're living in Chi or South Bronx, NY
> You're a 40-minute drive from a grocery store.
> Therefore, diabetes and obesity common,
> From eating processed garbarge, Lil' Debbies, and ramen.
> We need green for all,
> Not eco-apartheid.

Later in the same anthem, the group raps,

> I can't figure out why I can't breathe
> Toxins in the air just makes me OD
> While the whole world is begging. Please
> Stop destroying me from these factories.

The hip hop group does not simply bear witness to suffering, though; it offers a solution. Green power is fundamentally "people power"—organizing and speaking out. In the video, as the group paints a colorful mural of unified community action, they rap,

> The Future is art, the music the blueprint . . .
> Everybody shine—be solar-powered.
> Speak you mind—that's wind power.
> Move your H2O—wave power.
> Organize!
> People power—we got that soul power!

In an interview about "Green Anthem," rapper First Be talks about the need for music artists to be the "heart and soul" that engage young activists: "We know from history that youth have been an integral part of a lot of social movements. . . . If we don't get youth involved in the movement, it's not going to succeed. We've got to green the youth if we want to green the world."[75] For these rappers, collective people power, not personal, individual piety, is the alternative and renewable energy that will lead the way into a great future.

Pollution, Prophecy, and Resistance Media

In the wake of moral panics over rap music lyrics and videos in the 1980s and early 1990s, sociologist Theresa Martinez built upon and extended the then-current literature on popular culture as oppositional culture to make the case for rap music as a creative and *important* tool for minority political resistance, community empowerment, and substantive social critique.[76] For the most part, recent scholarship on rap music and the wider designation of hip hop culture assumes this premise, but at the time, Martinez and (less than) a handful of other scholars attending to the then-new subfield of hip hop studies were doing "heavy lifting."[77] In step with hip hop photographer/ethnographer Brian Cross (aka B+), Martinez places rap and other elements of hip hop, such as graffiti and dance, on a historical "continuum of minority resistance in the U.S." and places it within the tradition of such figures as Marcus Garvey and Malcom X.[78] Intervening in the dominant media discourse that feared and reviled rappers as mere "thugs and hooligans," scholars such as Tricia Rose recast groups such as NWA (Niggas With Attitude) as "prophets of rage" who—through their music—"bridge the gap between popular culture and social criticism by means of a potent oppositional culture."[79] Embracing African American literary critic Houston Baker's assertion that the work of political rap groups such as Public Enemy was not only critically interpretive but "homiletic" in nature, Martinez argues that rap is not simply *an* important form of minority oppositional culture in the United States but, because of the reach and impact of its moral, social, and political critique, it is *the* form of minority oppositional culture in the United States.[80]

Whereas hip hop scholars have focused on political issues in rap music—issues such as police brutality, high minority incarceration rates, poverty and unemployment, drugs, violence, racial disparity in healthcare and emergency services, and disillusionment with the US government—both rappers themselves and those who study them have been slower to call out dynamics of environmental racism and to take up the environmental crisis in minority neighborhoods.[81] This absence is one of the reasons Green For All, as early as in 2008, began sponsoring multiple eco-rap contests, producing eco-rap activist videos, and actively seeking working collaborations with rap artists both at the local and at the national level, relationships it continues to cultivate today. Much as with the eco-divide between socioeconomically advantaged communities and low-income communities, there has been an eco-divide in rap, whereby the environment was not identified as a pressing racial injustice issue by major players in the rap music world but instead consigned to the minor voices of local artists and groups.

That divide has begun to narrow. Think of Beyoncé's 2016 video "Formation" and its critical visual depiction of environmental racism displayed in scenes of Hurricane Katrina–flooded, largely abandoned, minority neighborhoods in New Orleans. She visually links the expendability of black lives during the Katrina disaster, vis-à-vis anemic and delayed federal emergency response, to the lethal effects of police shootings of black people around the nation—situating herself on top of a mostly submerged police cruiser in front of a flooded black neighborhood.[82] Within about a year after "Formation's" release, off-screen, Beyoncé found herself volunteering time and aid to desperate Hurricane Harvey victims in her hometown of Houston, where the US Census–based government Social Vulnerability Index (SVI) maps directly onto the metropolitan area's lowest-lying ground. Low-income minority communities are the most vulnerable to flooding in the city, the least likely to have resources to evacuate, and the least likely to receive disaster aid.[83] In the national "Hand in Hand" telethon concert to raise relief funds for Hurricane Harvey victims, participating rappers directly criticized the lack of federal response to climate change. They prescribed political action, not personal ecopiety, in response. As Chicago rapper Vic Mensa bluntly put it, "You can't come out and say we pray for Houston, we support Houston, and then deny climate change. . . . We need to take preventive measures, and we need to remove anyone in our government that doesn't think we do."[84]

Both Grammy Award–winning rapper Common and Billboard-chart-topping artist Big Sean have teamed up with Green For All's #FixThePipes campaign to raise money to address Flint, Michigan's lead-contaminated water crisis—made public in 2016 and, as of 2018, still not completely resolved, with pipe replacement not scheduled to begin until 2020. In a 2017 fundraising video message circulated by Green For All, Common raps, "Instead of educate, they rather convict the kids / As dirty as the water in Flint the system is," while urging rap fans and others to contribute to "getting the lead out of Flint."[85] More than twenty years before the Flint crisis, a self-identified eco-rapper in *Eco-Rap: Voices from the Hood*, a 1993 documentary on local Bay Area eco-rappers, reminds his audience of the fundamental intertwining of gun violence and environmental violence in many urban minority communities, commenting that there is "more than one way of getting filled with lead."[86] Black children are still roughly three times as likely as white children to test positive for elevated lead levels in their blood, and are ten times more likely than white children to be shot and killed by a gun.[87] In some cities the life-expectancy gap between black residents and white residents is as wide as twenty years, and the eco-divide plays a culpable role in these figures.[88]

Like Common and Big Sean, rappers like Drake, Ludacris, Wiz Khalifa, and Will.i.am, who are not self-identified as eco-rappers, work with organizations and programs like the Hip-Hop Caucus and Green the Block, all of which work to make inner-city living healthier and more sustainable through a variety of local initiatives. Green the Block, rap artists, and Green For All have, for example, teamed up in the civic work to promote the Keep It Fresh Campaign, which advocates for access to "real food" in minority communities, the eradication of inner-city food deserts, and healthy reforms in school lunch programs. In 2010, Drake traveled to his concerts in a bio-diesel concert bus, played at carbon-offsetting venues, and called for policies to invest in a clean-energy economy as vital for the survival of urban minority communities. His tour of college campuses was billed as "half rock tour/half environmental campaign," and his performances keyed up audiences to participate in green-jobs rallies around the nation. In talking with audiences and student groups about "greening the block," Drake reframed environmental issues as being squarely about jobs, poverty, food access, toxic exposure, and community health, all things to be addressed through collective civic action.[89]

Getting rid of toxics, greening the block, and improving community health all sound great. What would be the reason *not* to work to accomplish all of these positive changes? Unfortunately, as I have given research presentations over the years on eco-rap in a variety of conference and professional settings, African American clergy who live and work in toxic neighborhoods have pointed out a big concern with inner-city greening efforts. That is, once you clean up the cancer-causing agents and the lead contamination and make the water and air cleaner, while establishing accessible sources of fresh and healthy food, immediately, "fancy condo developers" and "Yuppies" move in. Gentrification takes place and prices out local residents of what has then become a hip, "up and coming neighborhood." Poor people are then pushed out to the next, perhaps even worse, toxic place to live. Despite this very serious and daunting challenge, eco-justice efforts continue, working toward an envisioned future in which low-income people can actually live in neighborhoods that are both healthy and affordable.

In his essay on rap music and the "art of theologizing," William Banfield echoes hip hop scholars before him (e.g., Cross, Rose, Martinez) in casting hip hop artists not as breaking *from* but as powerfully "continuing in the Black Arts prophetic vein and moving us to self-reflective listening and action."[90] More specifically, says Banfield, hip hop culture "provides a viable road map showing where contemporary culture is heading . . . [and] provides a road map for certain truths that are relevant to social, cultural, and theological interest and inquiry."[91] If so, the prophetic road map that hip hop artists construct in the genre of eco-rap looks considerably *different* from many of the standard byways commonly associated with the environmental movement. These byways have traditionally included scenic images of nature, such as those chronicled in environmental historian Roderick Nash's *Wilderness and the American Mind*, American studies scholar Leo Marx's unpacking of pastoral fantasies in *The Machine in the Garden*, and conservation historian Douglas Strong's idealized portraits of the moral wholesomeness of the great outdoors, its benefactors, and its patrons.[92] In striking contrast, West Coast local Latino eco-rapper 7-Stone tells us that "environmentalism" is about benzene and leukemia, race, and poverty.[93]

Eco-rappers conjure prophetic visions of a more just and sustainable future that includes, as Green For All puts it, "good jobs, clean energy, and opportunities for all." In short, eco-rap's environmentalism is fundamentally and uncompromisingly public, activist, and intersectional.[94] In Doo Dat's "Dream Reborn" video, former Green For All director Phaedra Ellis-Lamkins reminds us that eco-justice is "the unfinished business of the Civil Rights movement."[95] Addressing environmental crisis and tackling global climate change in a way the earth "will actually notice" necessitates taking a whole-systems approach that jointly addresses social and economic injustice. It is an immense project requiring major public investment—an investment that Green For All argues can create green jobs for those struggling in or near poverty, while revitalizing health and well-being in neighborhoods where people's bodies suffer the scars of cancer, the effects of malnutrition, and the damaging impacts of lead (from both paint and guns) as inhabitants struggle merely to breathe.[96]

Most Americans first became aware of the concept of a "Green New Deal" in 2019 when New York congresswoman Alexandria Ocasio-Cortez famously introduced a resolution to Congress that contained this phrase, but Green For All cofounders have marketed and advocated this kind of Green New Deal policy-based approach to the environment for more than a decade. In 2006, the Global Green Party called for a comprehensive public investment plan to address climate crisis by putting into place policies that would achieve 100 percent renewable energy by 2030. The party labeled and promoted this plan as a "Green New Deal." Soon after the Green Party's 2006 declaration, Green For All cofounder Van Jones began calling for the enactment of a Green New Deal in his speeches and subsequently devoted an entire chapter of his 2008 book on *The Green Collar Economy* to advocating for what he specifically termed "The Green New Deal."[97] Jones's repeated emphasis on social/economic reforms and the powerful role for change played by public works (like that historically demonstrated by President Franklin D. Roosevelt's New Deal) has shaped Green For All's civic engagement approach to intertwined environmental and economic justice. The organization's media works, in turn, dissuade audiences from relying on dubious market magic or a small, elite band of nobly eco-virtuous citizens to make real and lasting change, bearing witness instead to the effective changes that are possible when implemented through collective action and civic collaboration.

Restorying the Myth of Disposability

"The power to change the world comes from the people" and "the power of the people is more powerful than the people in power," raps teen eco-rapper Xiuhtezcatl Martinez, who is a cofounder of the global youth activist group Earth Guardians.[98] Martinez, who is indigenous Mexican American and speaks English, Spanish, and the native Nahuatl of his Aztec culture, is part of a group of twenty-one American youths who, in 2015, brought a climate suit against the US government for failing to protect the atmosphere for their generation. The suit charges that the government, knowing full well the man-made causes of climate change, is guilty of negligence in perpetuating fossil fuel policies that worsen the warming of the planet. The suit also charges the government with violating the "public trust doctrine," violating the plaintiffs' constitutional rights to life, liberty, and the pursuit of happiness, and violating the equal protection clause by denying them the fundamental rights enjoyed by past and present generations.[99] As of 2019, after two years of the US government being denied multiple motions for the case to be dismissed, plaintiffs awaited a Ninth Circuit Court of Appeals oral argument on the most recent appeal. While waiting, the youth filed a preliminary injunction to prohibit the granting of all new government leases for fossil fuel extraction in the time leading up to a trial.[100]

An eco-rapper since he was eleven, Martinez has rapped from the TED stage, at the Dakota Pipeline Water Protectors Youth Concert, and at the international Bioneers conference, has addressed the UN's General Assembly, and has written a how-to guide book called *We Rise* for youth earth activist organizing.[101] Frustration with protests that have fallen on deaf or indifferent ears, in conjunction with an ineffective "micro-changes" approach to addressing the enormity of climate change, prompted Martinez to help organize the youths' government lawsuit: "The marching in the streets, the lifestyle changes haven't been enough, so something drastic needs to happen. The change that we need is not going to come from a politician, from an orangutan in office, it's going to come from something that's always been the driver of change—people power, power of young people."[102] Punctuating Martinez's forecast of youth action on the way, in the early months of 2019, student climate strikes swept across Europe, in which school children walked out of their classes and took to the streets demanding immediate climate action from their governments.[103]

Both tattooing and rap are media art forms and modes of storying that engage the "power of young people" and call them to action and civic engagement as a way of life. Bronx rapper KRS-One famously expounded that "[r]ap is something you do, hip hop is something you live."[104] Likewise, Doo Dat's very name implies that it is the *doing* that reactivates "the dream" and puts prophetic visions into practice. Both endangered-species tattooing and eco-rap media making are media as *practice*, performed, captured or recorded, and posted to the Internet for more to view and circulate. Both are also forms of media intervention, in Howley's expanded sense, creatively overturning and contesting dominant media narratives, in their own unique ways, and supplanting them with stories of possibility and greater planetary partnership.

It is critical to observe here that both tattoos and rap are, in rhetorical terms, "short forms" of storying that employ liberal use of cultural quotations. Martinez's generation, Generation Z, the generation after the Millennials, makes up an estimated 25 percent of the US population, making them larger than either the Baby Boomers or the Millennials. Generation Z is populated ubiquitously with digital natives raised on new-media short forms and sampling that flourish in the digital era. The visual quotability (and Pinterest pinability) of tattoos, inscribed on bodies and then injected into online media flows, and of hooks, rap rhymes and/or samplings that repeat and stick in the listener's head, provide *tout court* mediums perfectly tailored to target and engage a younger activist audience.[105]

Tattoo activists' biocentric orientation and prioritizing of nonhuman species contrast sharply with eco-justice activists' people-centered focus. This kind of gap, or eco-divide, between activist interests, especially when so many communities of color are struggling for survival, understandably, can be divisive in the realm of environmental activism and a source of friction. Perhaps surprisingly, then, these activist groups may have more common cause and overlapping stories than imagined. When Van Jones recounts the early inspiration for cofounding Green For All, he cites a series of pivotal conversations he had at an environmental event with redwood-conservation activist Julia Butterfly Hill—the woman who experienced mystical interspecies communication with Luna, her fifteen-story-high redwood tree-sit host. Hill's major activist concern was with conserving old-growth forests and endangered species. Jones's was with saving human lives in desperation—organizing to, as Majora Carter puts it, "green the ghetto." With such different priorities, Hill and Jones struggled to understand one another, were frustrated, and made little headway. Finally, says Jones, they bonded on "the myth of disposability."[106] This is the story we tell ourselves that we can simply throw stuff away with no consequences, and what is tossed out won't come back to haunt us and cause problems in other ways. Some examples of this would be clear-cut forests that then result in massive mudslides, single-use plastics that have now come to form floating islands (twice the size of Texas and three times the size of France) choking our oceans, the nuclear waste that—despite our best container technology—eventually leaks, medical waste that washes up on our beaches, prescription drugs flushed down toilets only to end up in our drinking water, and, of course, toxic e-waste (the United States alone throws out 130,000 computers a day) that gets melted down for lead, mercury, and gold content by trash-pickers in now-poisoned third-world "recycling" communities.[107] But the *myth of disposability* extends beyond examples such as these, says Jones. "It's disposability," he contends, "that's at the root of this [earth crisis and economic failure]—the myth of disposability. We can reinvent . . . [by having] no throwaway species, no throwaway resources, and no throwaway children or neighborhoods or people either."[108]

Jones's morally inclusive prophetic vision is of what he describes as a "green wave that lifts all boats," involving not only our *deconversion* from carbon-based capitalism to a green economy but the very *reinvention* of civilization to reflect a radically expanded scope of justice.[109] It is also a vision reflected in the media slogan of the 2014 People's Climate March: "To Change Everything, We Need Everyone!"[110] Green For All continues to work toward this radically restoried vision of who "we" are, expanding the "scope of justice," protecting and defending humans in the most vulnerable communities threatened by injustice, exploitation, and poisoning, transitioning the economy, and in doing so, helping to heal and protect the larger ecosystems that support nonhuman species and foster holistic biotic well-being. Hip hop, defined once again by William Banfield, is a prophetic cultural "road map to the future." As such, it offers a tool, both efficient and poetic, for restorying the earth—activating, in its eco-rap form, a story not of personal piety but of common destiny—an expansive "we" living collectively into a "great future."

Conclusion

Storying the Future: Becoming Green Scheherazades in the Anthropocene

"We're wired for story" instructs qualitative psychological researcher and storyteller Brené Brown. Referencing neuroscience research that studies "your brain on story," Brown relates how hearing a story, "a narrative with a beginning, a middle, and an end—causes our brains to release cortisol and oxytocin. These chemicals trigger the uniquely human ability to connect, empathize, and make meaning. Story is literally in our DNA."[1] When Brown was invited to consult with the directors, producers, and writers at Pixar Animation Studios, a display on the wall of Pixar's Story Corner caught her eye. Three sentences on the Pixar wall read, "Story is the big picture. Story is process. Story is research." A crown at the top of the wall represented Pixar's guiding motto—"Story Is King."[2] That is, what matters and makes a true difference in cultural production is the power of story and how it grips, moves, transforms, and transports. This book has examined multiple sightings in contemporary mediated popular culture of stories of ecopiety, including stories from such varied sources as popular erotica and its fan-fiction responses, green consumer product marketing, social media and streamed video, reality television, digital mobile software applications, fashion manuals, corporate greenwashing, vampire media, gravestones and caskets, tattoos and hip hop music.

As demonstrated in a number of the examples featured here in *Ecopiety*, stories contribute to shaping social energetics, edging those who are engaged by story in one direction or another. And yet, the directional motion of stories can be more complex, contrapuntal, and contrary, edging us in different directions at once. Stories, their crafting, and their circulation cannot be separated from frameworks of power, control, and domination, which are adept at quickly identifying resistant stories and assimilating or swallowing them to use for their own purposes. The affective power of story and the power of *who* controls the frame of the story and its medium of communication have been and continue to be operational factors in the effectual use of fascist propaganda—factors to which the critical theorists associated with the social theory and philosophy of the Frankfurt School were keenly attuned.[3] Even resistant media are almost invariably coopted into replicating the form and style of dominant media's cultural forms.[4] In our current marketplace, these are predominantly corporatized media forms. As evidenced from examples in this book, what appears to be a story designed to dislodge, disrupt, and intervene, upon closer examination, can be one that actually reinscribes the very narrative or messaging it purports to challenge. Though faced with the narrative-digesting dynamics of late carbon-based capitalist consumer culture, environmental advocates, activists, artists, scientists, and humanist media makers from myriad backgrounds continue to "do media" in ways that restory earth for the present and future.

Philosopher Peter Sloterdijk observes, "It is a characteristic of *humanitas* that human beings are confronted with problems that are too difficult for them and that nevertheless cannot be left unaddressed on account of their difficulty."[5] When faced with such problems, advises Sloterdijk, the philosopher's function is that of both "provocation" and "interruption"—in the latter case, interrupting habituated loops, flows, and escalations. Rather than pass these on, the task is allowing them to die, or, should this strategy not work, "retransmitting [them] in a totally metamorphosed, verified, filtered, or recoded form."[6] Intervention, remixing, remaking, reinventing, and restorying, via the tools of media in conjunction with systems of networked collaboration and civic action, are the mechanisms of social transformation. Rescripting the future of planet earth so as to turn from its current crisis trajectory is a tall order and, perhaps, as Sloterdijk suggests, may prove too difficult for humans. This is especially so when prevailing narratives, such as those based upon extractive economics, appear to us to be fixed or even "factory installed."

A central question raised in this book has been how we approach the challenges of intervention. Dour prescriptions of ecopiety are both unappealing and misleading in their asserted scale of environmental outcomes and degrees of efficacy. Green consumer messaging often champions a superficial consumopiety as the vehicle for practicing earth-saving ecopiety in ways that generate little impact that the planet "would actually notice." Stories of ecopiety too often reinforce an individualistic, voluntary, libertarian model of environmental virtue that demands little or no public investment, regulatory affordances, government involvement, policy enactment/enforcement, collective action, and/or structural change. I have suggested that this particular combination—marketing green virtue in terms of the unappealing notions of duty and obligation associated with piety, while offering a way around this grim duty by cheerfully shopping our way into a greener future—can produce a "worst of both worlds." Self-congratulatory moral offsetting of feel-good consumption encourages pacifying practices that supplant more substantive environmental engagement, while also legitimating resistance to structural change.

This book has also made the case that mediated popular culture can act as both a powerful mirror and an engine for moral engagement, and that, as we tell our stories in and through popular culture, they in turn tell us. In turning, then, in the conclusion of this study, toward storied visions of earth's future, conflicting narratives emerge of "[w]ho we are, where we are we going, who 'we' includes, and what we are to do."[7] In framing conflicting narratives at the start of the book, I identified it as useful to think with economist David Korten's model of the "sacred money and markets story" and the "sacred life and living earth story." These stories appear entangled with one another in visions of human planetary or extra-planetary futures. Two juxtaposed subplots characterize the fate of earth in these visions: (1) stories of heroically escaping a disposable earth; and (2) stories of collectively rebooting a reinhabitable earth. Both of these are powerful orienting and motivating visions of the future and ones that, in the examples below, astutely embody the energetics of play over piety.

As we experiment with ways to live into the Anthropocene, making different life community choices than we have in the past, play, not piety, is a recipe for getting humans to, as Sloterdijk says, deal with problems that are too difficult for them. "Every game ever invented by mankind," points out poet John Ciardi, "is a way of making things hard for the fun of it."[8] Here, Ciardi, as a teacher of poetry, is talking to students about "poetic play" and the time and practice needed to develop the skill of both reading and writing poems—skills that require less attention to *what* a poem means than to "*how* a poem means." A poem, like a human, "has difficulties it imposed on itself."[9] But there is delight in the struggle and play of puzzling with a poem, and a deep satisfaction in getting better at it. Kindling our delight in difficult things, if not our fascination with problems too difficult for us, and harnessing the energy of our big-brained primate love of storied play, problem solving, and puzzles, afford ways of enticing rather than shaming humans into crafting a more life-giving, sustainable future on the planet.

Make no mistake: a shift in approach toward the poetics of play in meeting our daunting environmental challenges is not at odds with hard work as critical to what we as a species must accomplish in the Anthropocene. Nor does a strategic turn in emphasis from piety toward play suggest a Pollyanna vision of the sobering realities ahead. What *is* enormously adaptive about play, however, as developmental psychologist and play theorist Brian Sutton-Smith asserts in his classic book, *The Ambiguity of Play*, is that play is "the willful belief in acting out one's capacity for the future." Sutton-Smith carefully clarifies that the opposite of play is *not* work. The opposite of play is "vacillation, or worse, depression." To play, then, is "to act out and be willful, as if one is assured of one's prospects."[10] Play can, in essence, offer us a powerful antidote to the feelings of fear, fatalism, and futility that keep us shut down, paralyzed, and passive—ineffectual deer in the headlights of climate change and environmental disaster. Imaginative storied play, its mediations, and its role in social transformation might then be our species' most important work on behalf of the planet.

Shifting Social Energetics

Models of dutiful ecopiety, thus far, have not been effectual in motivating or inspiring collective shifts in human "social energetics" to the degree needed adequately to address our environmental crisis.[11] Perhaps counterintuitively, the mystical writings of theologian Matthew Fox and atheist political philosopher Jane Bennett both suggest modes of mystical connection, love, and delight as constituting the most promising energy to fuel political and social change in the Anthropocene. Granted, this is an odd couple to match up but a surprisingly synergistic one. Both figures are, perhaps not coincidentally, former Roman Catholics, having been shaped early on by a similar worldview formation.[12] In Matthew Fox's work, we encounter an empathetic indictment of the inadequacies of the duty-oriented environmental "stewardship" framework or mindset—one that I would argue fuels grim messages of ecopiety. Fox urges,

> The stewardship model tells us that God is "out there"; it is theistic. God is the absentee landlord, and we are here to do God's dirty work and steward the earth. Therefore, we have a duty-oriented morality—but you cannot arouse people by duty. You can make them feel guilty, you can pressure them, but the idea of duty morality only goes back to Kant. It is part of the Enlightenment. Let it go.
>
> Aquinas said that you change people by delight, you change people by pleasure. The proper model for theology for an ecological era is not stewardship, which reinforces theism. The proper model is mysticism . . . where we realize that God is the garden, that God is expressed in the word, and the plants, the trees, the animals and the soil.[13]

Though Jane Bennett is a self-identified atheist, her articulation of a world of "vital materiality"—in which we as humans are not separate but always located in webs of relationship to, and working within what she calls "intimate assemblages" of human and nonhuman "actants"—suggests a kind of nontheistic materialist mysticism that harmonizes with Fox's expressed delight motivator. For Bennett, it is a mixture of "delight and disturbance" that moves us from ethical notions into ethical action by providing "the impetus to act against the very injustices that you critically discern."[14] Bennett's call for "working with energetic flows," in conjunction with her call for a "course in careful anthropomorphism," bespeaks a world that is alive, intelligent, inspirited, and agential—one with which we as humans are enmeshed in sensual relationship. Such a world is one that inspires our attentive action, not out of dutiful moral obligation but out of delight, interconnection, love of the world, and, indeed, "enchantment" with a world of vital materiality.[15] Along these lines, rather than duty-identified piety, imaginative storied play provides a potent recipe for cultivating affective experiences of delight and connection that move us into action.

The research of neuroscientists Sam Wang and Sandra Aamodt on "Play, Stress, and the Learning Brain" suggests further reasons for cultural works aiming to elicit environmental moral engagement to emphasize the delight of ecoplay over ecopiety.[16] The researchers' focus is on the brain science of developing children and teens, but they make illuminating observations about play, neurobiologically, as critical practice and work more generally for all humans throughout the lifespan.[17] The researchers also pay keen attention to the effects of stress on the human brain and the ways in which play can offset some of these effects. At a time of environmental crisis, not to mention concomitant cascading political, economic, and social crises, stress on the human brain is in no short supply (if ever it has been). In performing a number of its functions, play offers a powerful source of lubrication for shifting social energetics in the Anthropocene. Here are just three to consider:

1. *Play opens human brains to learning*: Wang and Aamodt observe that "[t]he conditions of play—the generation of signals that enhance learning without an accompanying stress response—allow the brain to explore possibilities and to learn from them." One of the things humans need to cultivate, to lean into, in the Anthropocene is a *learning mindset*, in which we feel it is safe to explore new possibilities, try out new ways of doing things, and extend our comfort zone, without being paralyzed by anxiety and stress. Play creates the neurobiological conditions in our human bodies that let us know it is okay to explore and try new and comparatively strange things.[18]
2. *Play makes us want to do something more*: Wang and Aamodt find that "[p]lay activates the brain's reward circuitry."[19] Much as when humans engage in consensual sexual activity, when we engage in activities in the context of play, our brain's pleasure centers reward us and make us want to do those activities some *more*. We do not want the play to end but wish to extend it and keep it going. In terms of developing more sustainable modes of being in relation to both the immediate and the wider planetary community, dopamine rewards for play send signals to our human brains that encourage *more* connection, more empathy, more collectively engaged play. This, once again, is a far cry from being asked to do something out of duty or grim, guilt-induced obligation.

3. *Play is good practice, enhancing problem solving and fostering creativity*: One of Wang and Aamodt's most compelling findings is that "work in adult life is often most effective when it resembles play. Indeed, total immersion in an activity often indicates that the activity is intensely enjoyable; this is the concept of flow, or what athletes call being in 'the zone.' Flow occurs during active experiences that require concentration but are also highly practiced, where the goals and boundaries are clear but leave room for creativity."[20] With all the challenges our planet faces, we are going to need what Elizabeth Kolbert identifies in *The Sixth Extinction* as humans' big "problem-solving" brain.[21] Our creativity and curiosity as a species, Kolbert recounts, have driven us to move on and explore, often enacting destruction in our path. And yet, these same qualities hold the potential for leading us in positive game-changing directions. Imaginative storied play is also an excellent virtual simulator, allowing us to practice, role play, and experiment with different possibilities in the world before we enact and live them. Like a flight simulator, in which pilots develop muscle memory in how to handle aircraft calmly and skillfully in emergencies, storied play and its rehearsals have the potential to help us keep our heads, while acting with greater deliberation and wisdom during environmental crises.

Not all forms of play are storied, of course, and not many take up the task of restorying earth and rescripting the future. What follows is a comparative analysis of content and approach to speculative stories—popularly mediated visions of earth's future. Environmental prophecy, the trope of apocalypse, and eco-catastrophe can be adopted as a kind of "kaleidoscope for cultural self-consideration."[22] That is, speculative visions, like the contrasting ones below, tell us much more about ourselves today in this cultural moment than they foretell the future.

Earth as a Disposable Planet

On February 6, 2018, billionaire tech mogul Elon Musk posted to Twitter live video feed and selfies from Starman in space. Musk's space-travel company, SpaceX, had dressed a dummy (called "Starman" in an homage to David Bowie's song of that title) in a spacesuit and strapped him into the driver's seat of a convertible cherry-red Tesla roadster before rocket launching the Tesla into orbit. Musk, who owns both SpaceX and Tesla, the electric car manufacturer, hailed the demonstration (the more jaded dubbed it a publicity stunt) as a validation that SpaceX rockets would be capable of successfully carrying the heavy payloads of cargo needed to colonize Mars.[23] As Starman braved the galactic highway solo, images of the space hero in his eco-friendly Tesla, heading into the frontier, covered newspapers, television, and online news services across the globe.

Musk has argued vehemently for urgency in pursuing the colonization of Mars, promising that his private company will make human settlements a reality sooner than we might imagine. "We need to leave earth as soon as possible!" decreed billionaire Musk in the fall of 2015, shortly before meetings began for the United Nations Framework Convention on Climate Change (the Paris Climate Summit).[24] Citing possible asteroid collisions and climate catastrophes, among other apocalyptic disasters such as killer AI robots and outstripped resources, the Silicon Valley entrepreneur has repeatedly made public statements declaring that "humanity's survival rests on its ability to move to the red planet."[25] Comparing the earth to a computer "hard drive" that eventually will crash, Musk markets Mars colonization as an essential "backup copy" to preserve the human species.[26]

Digitally circulated selfie tableaus of Starman and his red Tesla silhouetted against planet earth, and heroically conquering space, have provided iconic visual representation for Musk's protracted campaign to "cut and run," leaving the "crashed hard drive" of earth behind. Media coverage of the Starman launch lauded the event as a "new day for American space travel," a "new day for exploration," a demonstration of "new capabilities," the dawning of a "new entrepreneurialism," a "new space race," and a "rekindling of dreams" of greatness that had faded back in the 1960s. For all this acclaimed newness of Starman in his new space suit and shiny new convertible, this frontier fantasy is not so new. Despite Musk's involvement in more environmentally focused industries such as solar energy and electric cars, SpaceX's public relations messaging taps into embedded Western cultural narratives of salvation and redemption via escape to some *other off-site home*—an old story made new again that conveys temporary flawed survival here on earth, while finding ultimate salvation elsewhere. Whether intentionally or not, the marketing of Starman's exit from earth, free at last to conquer space, leverages unmarked culturally embedded Christian fantasies of human dominion, otherworldly promise, and individual salvific escape, all intersectionally entwined with the colonial promise of moving on to "the next great place."[27]

The story of fouling one's nest and then, amidst disaster, leaving it behind to colonize anew so as to stick one's flag in virgin territory, though an old story, still clearly holds popular public resonance. Both the visual rhetoric and the verbal rhetoric displayed in Elon Musk's pissing contest with billionaire Amazon.com CEO and rival rocket developer Jeff Bezos, betting on whose will get there first to colonize Martian territory, read like parodic takes on nineteenth-century expressions of manifest destiny.[28] What is more, unsurprisingly, salvation can be bought for a price. When Musk insists, "We need to leave the earth as soon as possible," who is "we" in that declaration? Mars colonization provides a bug-out plan for the 1 percent in a time of planetary crisis, with profits handsomely going to SpaceX. Salvation is only afforded to the specially chosen—those who can pay SpaceX prices and afford SpaceX equipment, and then real estate once they get there. The rest of earth's denizens are presumably left behind in this narrative. A disposable earth is left to disposable persons.[29] The myth of disposability, embraced as a point of common ground by both urban eco-justice and forest activists working together, described in the last chapter, is no myth at all for space barons like Musk. It is, instead, as with the myth of manifest destiny, a business plan. As Musk's private company has received to date billions of dollars in US government subsidies, the 99 percent are ironically funding public investment toward colonizing Mars and contributing to their own and earth's disposability.[30]

The "sacred money and markets" story is in full display at the SpaceX online store, where consumers can purchase "Occupy Mars" terraforming mugs, SpaceX silver flasks and tumblers for the journey, SpaceX luggage tags, an "Occupy Mars" bug-out backpack, Mars jewelry, and all sorts of tie-in merchandise to celebrate humans' abandonment of an obsolete earth.[31] Ten different SpaceX flags are available for purchase and planting, two of which celebrate the slogan of "Occupying Mars."[32] The clothing and other merchandise have an alluring fan cosplay feel to them, providing the props needed for colonization preenactments.[33] Musk's salvific storying of moving humans off a doomed and disposable earth to a newer, more exciting, and radically libertarian retail space also benefits from cross-media free tie-ins, from popular mediated storylines about noble Mars-traveling astronauts such as the hit novel and film *The Martian*, and from Mars exploration–themed digital games, which make earth apocalypse engaging and *fun*.[34] Like his maker, Starman is hip, exciting, affluent, playful, well-dressed, digs David Bowie, and drives a cool car. Who *wouldn't* want to go party on Mars with him, or at least play along with the fantasy?

What has been most troubling to those doing the heavy lifting of addressing enormous global challenges like climate change is that Musk's charismatically communicated manifest destiny of earthlings bugging out to Mars becomes a self-fulfilling prophecy of earth's impossible redemption from its current problems.[35] Critics, including scientists from NASA, engineers in the aerospace science community, and economists, argue that Musk's plan for human migration to the red planet is factually flawed and dangerously misleading. Even in worst-case scenarios (global pandemic, asteroid hit, AI takeover, thermonuclear war, climate chaos, etc.), planet earth is *still* exponentially more inhabitable and hospitable at its very poles to supporting human life than any other planet in our solar system.[36] When asked about Mars, even Musk's rival Jeff Bezos admits, when compared to Mars, Antarctica is a "garden paradise."[37] Following astrobiologist David Warmflash's evisceration of the science of Musk's Mars-based colonial fantasy in *Discover Magazine*, the comments section is peppered with readers—some Warmflash's scientific peers—incredulously questioning the science of what could possibly make the earth any less inhabitable than other planets. One notes, "No matter how bad we treat earth, we can never make earth as screwed up as the other planets in our solar system." Another weighs in that even if earth did crash like Musk's prophesied computer hard drive, "Post-apocalyptic Earth would still be an absolute paradise in comparison to Mars, the Moon, or hellish Venus. None of these places will ever be more welcoming than Earth, no matter how bad Earth gets. Even if we have to live and farm under pressurized domes, we'll have it a million times easier than the people who try to do the same on Mars."[38]

Economically and practically, "rebooting" and repairing what Musk foresees as the "crashed hard drive" is wildly cheaper than tossing it into the trash and colonizing Mars, where the *average* temperature is minus eighty degrees Fahrenheit, or Venus, where lead melts on the surface and it rains sulfuric acid. The costs of colonizing another planet are indisputably astronomical, even when compared with the prospect of enacting a full-scale global Green New Deal here on planet earth. What is more, launching thousands of fifteen-ton payloads would require the extraction and burning of tons of fossil fuel, and resources drawn from earth for Mars colonization are resources diverted from investment critically needed to heal and repair this planet—a planet that already possesses, not insignificantly, a breathable atmosphere.

In effect, Musk hawks the promise of a new heaven and a new earth in an envisioned great migration. This promise capitalizes on, once again, what economists term "heuristic System 1 thinking" and the availability bias of mutually symbiotic narratives of apocalyptic otherworldly escape and divinely destined colonial conquest and progress. These Christian-inflected narratives are so culturally embedded in American storied national sensibilities and popular consumer messaging as barely to garner public notice as such.[39] Max Levchin, Musk's former cofounder of PayPal, has observed, "One of Elon's greatest skills is the ability to pass off his vision as a mandate from heaven."[40] Indeed, Musk's wildly successful branding is driven, in part, by the perception of his virtually superhuman powers. When actor Robert Downey Jr. researched his role of Marvel Comics billionaire superhero Tony Stark for the film *Iron Man*, the actor set up a meeting with Elon Musk to observe him so that he could model the character after him. But, in March of 2007, as Downey Jr. was sitting down with Elon Musk at SpaceX headquarters in Segundo, California, to mine material for Tony Stark, a launch of a very different kind was taking place.

"Play It before You Live It": Reinhabiting and Rebooting

Hatched in Silicon Valley and announced to the world in March of 2007, the alternate reality game *World without Oil* (*WWO*) was a historical preenactment game in which eighteen hundred players/coauthors "real-played" the story of humans' reinhabiting and rebooting earth's systems in a post-carbon future. Players hailed from twelve countries, ranging from Iraq to Japan and Australia to France. By the end of the year, there had been over 110,000 visitors to the game's Internet platform. Apocalyptic narratives are popular with gamers.[41] *World without Oil*'s speculative story of the future, however, was different from the typical apocalyptic gaming fare. This was not a story of earth as a disposable wasteland, a story of grim embattlement, or a story of a human "cut and run" to the next great planet. Nor was this a story that centered on a lone rugged hero. It instead engaged a globally networked collaboratory of media-making ordinary citizens figuring out together how to live imaginatively and successfully carbon free. Described as "part serious game, part collaborative storytelling, part social network, part multimedia experience," *WWO* was created by game writer and designer Ken Eklund, in conjunction with Gaming for Good cofounder Jane McGonigal, and sponsored by the Corporation for Public Broadcasting–funded ITVS (Independent Television Service).[42] The game challenged players to live their own lives as if dealing with the first thirty-two days of a massive global "oil shock." In the game world, world oil reserves have been outstripped, fossil fuel is dead, and players need to figure out how to deal with the reality of the crisis and press on with their lives.

Instead of *WWO* enacting a story of how it all ends for the earth, it explored and played with ideas for how it all begins, inviting players to be active co-creators in imagining earth's future. The tagline for the game was "World without Oil: Play It before You Live It." In restorying earth and rescripting the future, this kind of imaginative storied play has a powerful contribution to make. Story becomes a virtual simulator to try out new plotlines and possibilities, rewrite outcomes, and experiment with new ways of being in the world. "You've opened our eyes and infected our minds with realities and possibilities," wrote player Organized Chaos, as the game came to a close.[43] Without getting dully didactic, *WWO* created pathways to move play into policy, as play and accompanying game materials tied into six major state school curriculums on civics and public policy, teaching students a variety of "civic levers" with which to shape or influence public policy, such as attending political and governmental meetings, demonstrating, contacting public officials, writing letters, boycotting, community organizing, petitioning, and picketing.[44]

Different from other digital play, alternate reality games (ARGs) are not played in a simulated space with characters. Instead, as Jane McGonigal explains, "You are yourself," and players play the game in the physical world of their own lives. "It's not thinking about change, it's making change," asserts McGonigal, "but in a really fun context where lots of other people around the world are making changes, too."[45] Because ARGs are played and practiced in everyday life, they provide opportunities for taking direct action in the physical (or "natural") world. As such, ARGs are more than merely a call to action; they *are* action. Tangible changes made in playing beyond the platform in the practice of everyday life were documented by the game's eighteen hundred players in over fifteen hundred posted blogs, vlogs, images, e-mails, voicemails, and journal entries. *World without Oil* successfully integrated improvisation, experimental solutions to conflicts, and the work of imaginative storied play as applied to real places on earth.

Although some players, as predicted, retreated to relative isolation and armed themselves, the vast majority of the eighteen hundred players instead got to work, reached out and shared their involvement in the game with others in their community, started building mutual relationships of assistance to cope on a zero-carbon earth without oil, and ended up rewarded for collective projects and collaborative solutions. Play ranged from, as McGonigal explains, "immediate, personal changes (public transportation instead of driving) to perhaps inspiring more active citizen participation in crafting new or existing public policy and, most importantly, by developing better case scenarios. Or, as Stefanie Olsen of CNET says, 'The best way to change the future is to play with it first.'"[46]

There is a palpable delight in many of the players' postings, as they slow down their de-carbonized lives to a different pace, bike to work, convert their cars to run on recycled restaurant vegetable oil, run their laptops on solar chargers, and get rid of lawns to create home vegetable gardens, while collaborating with neighbors, family, and other ARG players both locally and around the world to find civic solutions and workarounds to the challenges of a life lived beyond oil. On May 7, 2007, "Morie" posted the following update: "We grow our own food now, and eat it every night together at the dinner table. Fresh, wonderful, healthy and gorgeous! Lettuce grows outside the back door, beans take over the middle of the yard. No more driving to the store, fighting the traffic, finding a parking space. Birds have come back, and the bees. The kids run barefoot. The sky is endless blue. I use old dipsticks to mark the plant rows."[47]

More dicey was when players in northern climates, where spring was still chilly, turned off the heat in their houses with some trepidation but went on to reassure and encourage other players to follow suit. In the posted journals of players, now archived online as a resource for "teachers, policymakers, and ordinary citizens," there are unmistakable anxieties expressed about shortages, but there is also an almost audible collective exhale in living life on a more human scale and setting down the frenetic—and often isolating—dimensions of a carbon-based economy. For all the uncertainty and apprehensions expressed, there is also a collective, cooperative excitement in the preenactment that this can be done. Although players are responding to an externally imposed crisis, implicit in the game is that communities and countries could, if they marshal the public will to do so, decide to do this for real—of their own accord, live without oil. Rather than hightailing it to Mars being the only option, there would be ways to pull together and make a post-carbon future on earth work.

In *WWO*, play and its pleasures generated a desirable *invitation* to players into creative resilience as they work to craft a more ecological culture. At some point, the game makers portend, we will not simply be playing oil shortage. In the meantime, the "as if" world of "real play," serious play, and ARGs holds promise and potential for pulling us into, as Jane Bennett says, "a topography of becoming" in which we co-join with complex human/nonhuman assemblages to play with lively animate forces in order to co-create more ecological cultural sensibilities and ethical action. ARGs by definition enter into a lively engagement with "as ifs," where working groups become playing groups, steeped not in a grim, duty-bound morality but in what cultural historian Michael Saler calls the imaginative "modern enchantment" of energetic flow, delight, and connection.[48]

It is telling that, even after *WWO* officially ended, some players continued to play on, no longer dependent on the gaming platform but working their own networks and continuing to journal their lives beyond oil. This kind of ownership and coauthorship of the game in players' lives was concertedly engineered into the design of what McGonigal describes as "the first collectively puppet-mastered game ever."[49] On June 8, 2007, posting from Chicago, player "Fallingintosin" wrote, "Just because *WWO* is 'officially' over, it doesn't HAVE to be OVER. I'm going to keep my journal going. I'm going to keep documenting ways I'm saving oil and other oil-related information/items I come across."[50] Elsewhere in her live journal, Fallingintosin exclaims, "This has been one of the most amazing experiences of my life. . . . I end this post with HOPE. Because, that is what I'm filled with in this moment. I have HOPE for our future and what a powerful feeling that is." The player then gives out her social media contact information so that players can continue to stay in touch with her and shift play over to that platform. Other players urge each other to stay in touch and to play on, continuing to live the game in their lives. As neuroscientists Wang and Aamodt point out, the powerful dimensions of play, its affective delights, dopamine rewards, and feelings of connection, make us want to do the actions associated with play *more*.

"Video games could be the greatest storytelling medium of our age—if only the worlds of art and technology would stop arguing and take notice," contends novelist/game designer Naomi Alderman.[51] In the case of *WWO*, which is not a video game but made use of a multimedia participatory online digital platform with many video-game-world-like storytelling aspects, its premise had both a post-apocalyptic and a post-millennialist feel. That is, by living more sustainably and organizing for greater sustainability measures in their communities, players worked via play to bring about a more sustainable world. Or, as the adage goes, "Act the way you want to become, and you'll become the way you act." *World without Oil* creates a time of tribulation over the sacrifices and chaos generated by the end of oil, while at the same time realizing, if not appreciating, a world released from the "bondage" of oil addiction. Players thus express taking *pleasure* and *delight* in exploring new kinds of post-oil community and rediscovering an intimacy in human/earth relationships, including seeing their communities in different ways as they bike and walk. Of course, the mood would likely be very different if there actually were no more oil, and if hunger and desperation reached dystopian proportions. Even so, the "living as if" experience of working together in real time and cooperating to make the best of the best, while players transformed their physical communities, living spaces, and in some cases sites of employment, appears to open players' experiences to be more accepting, or at least less fearful, of what environmentalist Paul Hawken and colleagues would later publicly outline as "Project Drawdown"—a comprehensive plan for arresting and reversing global warming.[52]

It is also worth noting that, in the *WWO* story world, players' actions are not prompted by stipulations of noble virtue or demands of ecopiety. Their actions are practical exigencies in a world where a carbon-free lifestyle becomes *not* a point of virtue but the only option and each successful workaround a triumph. In reviewing the enormous time put into crafting *WWO* postings, which document time spent implementing real-world *WWO* actions, including those that collaborated on moving play into public policy, an aggregate promulgation emerges from the archival opus. It is one that parallels ecotheologian Catherine Keller's declaration on eschatology and ecology: "We need no new heaven and Earth. We have this Earth, this sky, this water to renew."[53]

As advocates of Mars colonization rationalize that our planet is used up, outstripped of its resources, and that it is time to move on to the next great place, players of *WWO* seem to channel the modus operandi employed by the characters in the popular post-apocalyptic/ecotopian novel, *The Fifth Sacred Thing*. The novel tells the story of a post-oil society in northern California, where revolutionaries have de-paved the roads to build community organic gardens. Healthcare and fresh water are shared in collectives, and the community has worked diligently to reinhabit and restore damaged ecosystems. Resources are scarce, fossil fuels inaccessible, and yet, as one of the revolution's coleaders reflects, "You'd think we had plenty of everything, plenty of land, plenty of water. Whereas we've simply learned how not to waste, how to use and reuse every drop, how to feed chickens on weeds and ducks on snails and let worms eat the garbage. We've become such artists of unwaste we can almost compensate for the damage."[54] This speculative vision of a future in which humans creatively become "artists of unwaste" is the antithesis of the fatalistic "disposable planet" story.

Media activist projects like Sustainable Human TV that have emerged since *WWO* echo its collaboratory model of multiple coauthor storytelling, the crowd-sourcing of collective and networked wisdom, and the civic involvement of, as Jane McGonigal puts it, "normal people, ordinary citizens." An independent collaborative video and content creation community, with contributor-volunteers around the world, Sustainable Human TV provides a laboratory for citizen storytelling in the Anthropocene. Founded by a former IT consultant for Accenture, the project operates along the lines of a gift economy, so citizen media makers donate their time as video/audio editors and producers to create short video provocations that explore new stories and possibilities for earth's future. All of the media produced through Sustainable Human is licensed as creative commons, making it freely downloadable and uploadable to any site. Most of distribution is through social media, like Sustainable Human's Facebook page, which has more than 1.7 million subscribers. Like a physical maker space, Sustainable Human's web-based collaboratory allows media makers to brainstorm and play with different ideas, experiment with developing different script possibilities, get help with production from volunteers, and then receive assistance with strategies for viral launch plans.

The community's tagline is, "By Changing the Story, We Change Our World." A button on the "Participate" page on the platform invites, "How do you want to change our story? Click now." Video titles cowritten and coproduced by the community include "Earth Trek: Economics of the Future," "The Myth of Human Nature," "There Is No Away," "Earth Democracy," and "It's Not an 'Investment' If It's Destroying the Planet." Sustainable Human TV's "Why We Exist" section provides a window into its mission: "If life on Earth is to not only survive but *thrive*, we must change the narrative of modern civilization. The world we have inherited is built on an unsustainable story of infinite resources, limitless waste and endless competition, degrading both our planet and our humanity. It is time to collectively re-imagine the story of the world we wish to live in—one that enables each of us to develop our unique gifts and give them to the world."[55] This theme, familiar from David Korten's work and sounded by a number of futurists, is addressed directly in a voiceover for a Sustainable Human TV collectively produced video called "To Change the World, We Must Change Our Story: A New Story of the People." The video is narrated by sustainability and gift-economy advocate Charles Eisenstein. Set first to visuals of eco-destruction, tents from the Occupy Movement, homeless people in the streets, and despondent-looking youth confined to corporatized classrooms, the video then moves into majestic scenes of humans communing with nature—a woman joyfully greeting the dawn, a man walking on the beach and wading in the surf, people marveling at trees and interacting peacefully with animals. The voiceover tells a story of restorying:

> As our ecosystems are falling apart, as our political systems are falling apart, as our educational systems fall apart, things are not working out so well anymore, and it is harder to fully believe in our stories. We are moving into a different story—a different story of self, a story of the world, a different story of people. . . . By acting from a different story, we disrupt the psychic substructure of our mythology. And we offer an alternative. This is something eminently practical. Any time we give someone an experience that does not fit into that old story, it weakens that old story. It disrupts it.[56]

Similar disruptive and interventionary themes undergird Sustainable Human TV's diverse array of video provocations. Some repeated themes are as follows: everything we do to this world we do to ourselves; being in service to something larger than ourselves is what makes life meaningful; none of us can go it alone; our world is built on a misguided story of human separation that we need to change; and earth is home. As with the players in *WWO*, Sustainable Human's story making is cooperatively participatory, cowritten, coproduced, and fueled by delight, improvisation, play, and connection. The "change the story" mantra that appears throughout its web site copy, social media presence, and video productions avows that, in the media activist work of Sustainable Human organizers and volunteers, "Story Is King."

When compared to "big media," Sustainable Human is of course small and dependent on donations and a gift economy of media-maker volunteers. If story *is* king, it cuts both ways. While those who work for greater eco-justice, more equitable social transformation, and planetary reinhabitation operate from the premise that "story is king," so do commercial marketers and political interests that would oppose such directions. These makers have much deeper pockets to produce and promote their own stories. Big media and their mega franchises can play an enormous role in social transformation and shifting social energetics with narratives that move and transform, and many storyteller/creators who work in that realm care about doing just that. A prime example of this is *Black Panther*'s restorying of Africa, the ecotopian narrative aesthetics of which engender provocative political messages and models of colonial resistance, the value of public investment, and the primacy of collective moral engagement in creating a more fair, equitable, just, and green world, where no one is left behind. The challenge is in somehow keeping bold restorying at the big media level from being coopted, swallowed, and digested by contrapuntal corporate marketing and merchandising agendas.

What Do We Do Once We Know?

This book opened with the haunting question of Drew Dellinger's video, "Planetize the Movement," in which the children of the future stare each of us in the eye and demand to know, "What did you do, once you knew?" On the very first day of a course I teach called "Media, Earth, and Making a Difference," I show Dellinger's video to my students and ask them to imagine what they might reply to the children of the future if asked this question. One student raised her hand and commented that she is "just one person," and that sometimes she feels completely overwhelmed by the enormity of all that we are facing. It *is* overwhelming. Rather than causing students to shut down, feel despondent or disheartened by the environmental material we engage in the course, my goal is to empower students to act. When I say this, they look at me a bit skeptically. Drawing from Marshall McLuhan's theory of media being "extensions of man," we talk about how, as humans, we are amazing story-spinning, tool-creating creatures, or, as Jonathan Gottschall says, "the great ape with the storytelling mind."[57] When we don't have enough torque in our fingertips to turn something, we invent and use a screwdriver or a wrench as their extension. When our hand is not strong enough to pound something flat, we create a hammer, again, as our hand's extension. And when our fingernails or teeth are not sharp enough to rip or tear something on their own, we polish stone and make a knife we can use to cut. Yes, this creative tool-making ability ultimately gets us in trouble, especially as the broader category of "extensions of man" now includes fossil-fuel-extraction technologies, the burning mechanisms of the greenhouse carbon economy, and nuclear warheads to boot. But combining those big storytelling primate brains with our social abilities to work together and form networks, and then adding in the tools of media as extensions of our physical communicative bodies, it is like having superpowers. Our impact can quickly become manifold. A bystander with a smartphone and access to the Internet can circulate powerful stories around the world at a speed that is now taken for granted.

On the darker side of this capacity exist the Orwellian aspects of surveillance culture: the corporate colonization of social media, screen space, our eyeballs, and our attention. What is more, the mediasphere now legitimates, perpetuates, and celebrates conspiracy and confabulation. The manipulative power of a bot-creating cyberflunky, writing code to make computer programs talk like humans, can relatively easily fabricate the appearance of a surge of online populist support for a demagogue candidate. To be perfectly honest, the odds are terrible and the deck is stacked in the favor of forces of domination, control, greed, and exploitation. When has it ever not been so in human history? Even with the enormity of all this in mind, I *still* want to empower students with the tools of storytelling and media making to act out their capacity for the future.

What do we do once we know? There are many options, but for the purpose of the course, we focus on one: *we make media*. We make media "as a bridge to social action" and public policy.[58] We also make media to interrupt and intercede in existent media loops, reframing and reformulating them. First, I have students identify, critically analyze, and evaluate case studies of environmental media interventions, appraising what proves, or not, to be effective strategies in each case. Then I have the students choose a medium (video, film, podcast, web site, social media project, short graphic novel, etc.), research an environmental issue that speaks to them, and then create their own storytelling media interventions. I remind them that when they pick up that camera, or audio recorder, or just put their cell phone or iPad to work, craft a story, and then release it into media circulation, it is like putting on "the Iron Man suit," with powers that extend their bodies and persons in myriad ways. Granted, not all Iron Man suits are fancy, durable, or have as many sophisticated bells and whistles as those that have been financed by wealthy mega-media conglomerates. Citizen "suits" may be borrowed, used, shared, last year's model, worn, cheap, and/or jerry-rigged to hold together, much like Han Solo's precarious ship the *Millennium Falcon*, as it rattles its way toward the Empire. When "networked" with other "suits," on the cheap, or in collaboration and exchange, they have the potential to challenge even the best that Tony Stark, or his human avatar Elon Musk, may have to offer.

Countless instances of citizen video streaming and viral social media making—from air travel incidents to fleeing refugees to African American individuals and families ousted from public coffeehouses and swimming pools, separated immigrant families at the border, footage of children in cages, and #MeToo social media declarations—demonstrate that media-making makeshift Iron Man suits *do* generate much more impact than those in big-media corporate and/or political authority anticipate or bargain for.[59] Are these media-making extensions of the citizen—these DIY or independent digital tools—sufficient to counter the mammoth corporate algorithmic powers that, as former Google ethicist Tristan Harris chillingly puts it, can shift a billion minds around the world at the turn of a dial?[60] I fully concede that this is bringing the proverbial knife to a gunfight. Omnipresent in our cognizance needs to be Nick Couldry's admonitory counsel on the troubling politics of voice and the power dynamics involved in silencing and repressing the voices of society's most resistant and vulnerable.[61]

Even so, like Drew Dellinger, as we face the "specters of children of the future," not to mention children in the now, who call us on us to explain ourselves, won't we want to know that we used all the tools we had at our disposal and acted—as Henry Jenkins and his research collaborators put it, "by any media necessary"—to make interventions while we could?[62] In the moral question that haunts Dellinger's video, the emphasis is unmistakably on the "doing." The children's question to us is not "What did you *think*, once you knew?" but "What did you *do*, once you knew?" If we engage and understand media as *action*, that media are something *we do*, creatively, imaginatively, strategically, it expands the possibilities of how we do media, what forms those media practices take, and the impacts they have—all critical resources to make use of as we live imperfectly but resiliently into the Anthropocene.

Becoming Green Scheherazades

Finally, it is both a terrible and a spectacular time to be reinventing the world as we know it—a time when, as speculative fiction writer Ted Mooney puts it, we are experiencing "[e]verything on the verge of becoming something else."[63] For better or worse, story does have the potential to be that algorithmic dial that shifts the world's course. "Humans evolved to crave story," Jonathan Gottschall reminds us.[64] Because of its crave-inducing, addictive quality, story can also be used deliberately, tactically, to make change and, no less, to stave off what appears to be inevitable demise. Like Scheherazade, the heroine in *The Thousand and One Nights*, who evades a death sentence by telling stories, night after night, to her king and captor, those who set about restorying earth do so with life on earth in the balance. All of Scheherazade's predecessors had been subject to certain obliteration. She is determined to interrupt that cycle of ordained demise—determined to shift dynamics in another direction and hell-bent on changing existing policy. Her stories, like those found in today's mediated popular culture, have the potential to interrupt and intervene in what appears to be a foregone conclusion. If clever, like Scheherazade, environmental storytellers captivate not only with adventure tales that include plenty of play, romance, sex, and bawdy details (think slinky green vampires), but they infuse these stories along the way with moral insight and wisdom that teach and transform (think green hip hop), even as they delight and entertain. As the complex, multidirectional sightings analyzed in this book demonstrate, in this life-saving scenario, media makers restorying earth are simultaneously Scheherazade, while at times, frustratingly, still serving as her captors.

Even so, using the power of story and, by extension, storied play, edging us toward new directions and urging our human "hive" toward social transformation, becomes not a grim act of dutiful ecopiety—sober moral responsibility reluctantly shouldered—but something we actually take pleasure in doing—something we are drawn to, night after night, day after day, and even crave. As we participate in the restorying of earth, like being drawn to participate in a favorite TV series online fan forum, delighting in taking a walk with a podcast that makes us laugh out loud, or sharing a juicy novel with our local book group, our motivation to expend energy digging in, organizing, healing, restoring, disrupting, reinhabiting, remaking, and reinventing is fueled not by the draining energy sources of guilt and shame, duty and deprivation, but by sustaining storied sources of delight and play. St. Francis may lay claim to the title of "patron saint of ecology," but it is Scheherazade's clever captivating narrative strategy that holds the passcode for "renewing" the earth and rescripting the future.

ACKNOWLEDGMENTS

To write this book and to address in a substantive way the range of issues and subject matter it entails, I realized that I would need to go back to school to get an advanced degree in media studies. It is somewhat unusual for a tenured professor in midcareer suddenly to become a graduate student again, but that is what happened. Taking a flying leap, I embarked upon a graduate track in Media History, Philosophy, and Criticism in the Graduate School of Media Studies at The New School for Public Engagement. It is one of the best decisions I have made. Energized by being a student again and quickly obsessed with new theorists, concepts, and texts, I have been deeply enriched by this experience as a scholar. I hope that sense of energy and passion at exploring and forging further connections between religious studies and media studies is apparent to the reader and reflected in this work.

Making the move to learn a new field, however, necessitated extreme patience on the part of a number of parties, not only my family and department chairs at Northwestern University, but especially NYU Press intrepid editor Jennifer Hammer. Teaching and grading my students' term papers while also submitting my own seminar final projects for graduate school meant more extensions on book deadlines than I had foreseen. Jennifer graciously continued to provide me with detailed keen feedback on drafts and urged me forward. She is, hands down, the best editor I have ever worked with. She is a credit to the profession, and I am profoundly grateful to have had the honor of working with someone of her caliber and insight.

Thank you to Stacy Floyd-Thomas and Anthony Pinn for their vision in directing NYU's series on religion and social transformation and for giving me the opportunity to contribute my voice to the expanding conversation in this area. I am deeply grateful to the anonymous reviewers of the manuscript for their careful reading of the book and their helpful feedback. Thank you also to my dedicated research assistant, Andrés

Carrasquillo, for tracking down countless books, articles, images, videos, and apps.

To patient departmental chairs Cristie Traina and Kenneth Seeskin, who also thoughtfully commented on the framing sections about piety in the book, I am truly grateful, as I am to Richard Kieckhefer, who talked Kurosawa and *Rashomon* with me, and to Sarah Jacoby, who consulted on Daoist energy vampires and Buddhist practices of corpse meditation. Thank you as well to Dean Adrian Randolph and Northwestern's "And" model for encouraging extra-disciplinary training that enables faculty to take risks, innovate, and explore new visions, while modeling the kind of research that transcends discipline-specific approaches in order to tackle pressing collective problems.

This book would not have been possible without the stellar teaching support and mentorship I have received from my media studies professors at The New School. Elizabeth Ellsworth's extraordinary seminar on Media in the Anthropocene was nothing less than life-changing for me. That pivotal learning experience introduced me to Roy Scranton's work, scholarship on "social energetics," and it encouraged me to engage Jane Bennett's philosophies of "vibrant matter," along with John Durham Peters's theorizing of "elemental media" in ways that have now reshaped my thinking as a scholar. I am also indebted to the teaching and guidance of my supportive advisor, Sumita Chakravarty. Anne Balsamo, Dierdre Boyle, Lauhona Ganguly, Dawnja Burris, and Aras Ozgun, and many others at the School of Media Studies, including SMS advisor Robbie Powers, have supported me every step of the way. The New School's Provost's Academic Scholarship for graduate study made it possible for me to pursue this degree. I cannot say enough about the learning experience I have had through The New School and how much it has expanded my horizons as a researcher, a teacher, and a person.

I am grateful to many others who have read parts of this manuscript or encountered my research for this book in scholarly presentations and offered constructive feedback. SCMS Media and Environment SIG cochair Hunter Vaughan kindly took the time to do a close reading of my Prius chapter and to give me extensive comments. Former ICA Popular Communication Division cochair Melissa Click generously read most of an earlier draft of the manuscript and offered feedback. Kathryn Lofton considerately responded to portions of the hip hop chapter at the

ISMRC in Canterbury, UK. The research in that chapter was also enhanced by the feedback I received from participants in Northwestern's Summit on Sustainability and participating Engineers for a Sustainable World, in addition to responses I received from the AAR program unit on Religion and Hip Hop. Middlebury College faculty in both environmental studies and religion also contributed helpful commentary to the hip hop material. Many thanks to the scholars who offered feedback on my research dealing with coal rollers and pollution porn at the UCSB media and environment "Power Dynamics" conference and especially for the encouragement from Janet Walker and Adrian Ivakhiv. Religion scholars at the University of Edinburgh were kind enough to invite me to lecture on my conceptualizations of ecopiety and consumopiety, which helped me hone my articulations of both in subsequent work. I am appreciative to my colleagues in Northwestern's Environmental Humanities Workshop who responded to portions of this manuscript, to my colleagues in the Program in Environmental Policy and Culture for their continued support throughout this project, to School of Communication colleagues Jake Smith, Mimi White, Bob Hariman, and Jan Radway for engaging exchanges on media topics, and to Northwestern's Center for Global Culture and Communication for so amiably welcoming a cultural study of religion scholar into their conversations.

Scholars in the University of Colorado's Center for Media, Religion, and Culture "Hypermediations" project on Public Religion in the Digital Age provided invaluable feedback for my thinking about Elon Musk and SpaceX. Dynamic and dedicated field builder Stewart Hoover has been an important mentor for me, has been supportive throughout my transdisciplinary journey, and through his unflagging efforts to support scholarship on religion and media, has given scholars like me an exciting and rapidly evolving place to work. I have learned a tremendous amount from Stewart and am grateful for his magnanimity and for his enthusiasm for this and future projects.

From the beginning of my first foray into media and popular culture studies, Henry Jenkins has been a guiding figure and inspiration. In fact, it was reading Henry's work that first prompted me to go back to graduate school. I am grateful for Henry's feedback on this book's content when I presented a section of it in a session dedicated to Henry's research at the San Diego AAR meeting and when he responded, in conjunction with the very gracious Diane Winston and Tok Thompson, to my presentation in the Civic and Social Media Group Symposium at USC. Henry's generosity to more junior scholars, such as myself, his unparalleled collaboratory skill, and his tireless work toward greater public civic imagination and engagement have motivated me throughout this project to press on and continue this work.

Sarah Banet-Weiser and Mara Einstein, in their innovative scholarship on consumer culture, branding, marketing, and advertising, have helped me by example to find my voice as a scholar and to own it, and I am grateful to both of these incredible women for their perspectives shared in discussions of my work. My dear mentor Wade Clark Roof has never ceased his caring assistance and involvement in my career since graduate school. Clark supported my applying to go back to graduate school in media studies, cheered me on, and gave me the reassurance to try something outside my comfort zone. I am also grateful to PCA/ACA colleagues Paul Booth, Brian Cogan, Bob Batchelor, Kathleen Turner, and Brendan Riley, who made me feel right at home in cultural studies and spurred me on to new challenges. AAR colleagues who have helped contribute to this work in ongoing conversations and exchanges are too numerous to name here, but I'm especially grateful to Chip Callahan, Elijah Siegler, Gary Laderman, Eric Mazur (my clipping service!), John Modern, Melissa Wilcox, Rachel Wagner, Erica Andrus, Kristy Nabhan-Warren, Bruce Forbes, Jeffrey Mahan, Lynn Schofield Clark, Curtis Coats, Marie Pagliarini, Katie Lofton, Laurel Kearns, Becky Gould, my fabulous Religion, Media, and Culture cochair and friend Jenna Supp-Montgomery, Debbie Whitehead, Anthea Butler, Heidi Campbell, Greg Grieve, Chris Helland, and to my doctoral mentees Amanda Baugh, Steph Brehm, Hannah Scheidt, and Myev Rees.

Special thanks to my Academic Ladder online authors' support group, who cheered me to the end. My local author group, North Shore Writers (working in Suburban Exile), shared motivation and mojitos. Wise mentor Susanne Simmons helped me make it through the "noodle gauntlet," while encouraging me to approach the very visual way I perceive the world as a dyslexic, intensified by my perpetual "on switch" for cultural pattern recognition, less as a liability and more as a strength. Christie Jordan's compassionate and intuitive presence has been indispensable to enabling me to run the "energetic marathon" that has been this project, reminding me to take it step by step. Ladies of Ghost Ranch, courage is contagious and your solidarity helped me make it to this point.

I am also grateful to Darla, who drove summer camp carpool so that I could finish chapter revisions, and to both Darla and Tina for bringing dinner over when I was most crunched. My soul sister Lauren Harrison kept me going with laughter, love, hiking, and *hygge*. Kent Place's remarkable posse of women offered strength, warmth, connection, and kinship.

My first and most important teacher, my mother, kindly proofread my chapters and gave me infinite moral support. Profound gratitude to my patient husband, trained in economics and a business school graduate, who was a great sport as I repeatedly drew him into conversations about behavioral economics and consumer psychology. He is a veteran of many years working in consulting and consumer research, and his repeated caveat—consumers do not behave rationally—was useful for me to keep in mind as I explored the affective dimensions of ecopiety, consumopiety, and their marketing to consumers. I am thankful for his emotional support, intellectual engagement, loyalty, love, and partnership, throughout the years. Finally, to my son, who is the great blessing of my life and whom I love more than words can ever express, it is with an enduring commitment to you, the children of the now, and the future, that I write this book as just one voice among many working toward rescripting the future.

NOTES

INTRODUCTION

1 See Drew Dellinger, *Love Letter to the Milky Way: A Book of Poems* (Ashland, OR: White Cloud Press, 2011), 1–4; and the "Planetize the Movement" video on Dellinger's web site: https://drewdellinger.org (accessed June 4, 2018).

2 "Planetize the Movement."

3 T. Christopher Hoklotubbe, *Civilized Piety: The Rhetoric of Pietas in the Pastoral Epistles and the Roman Empire* (Waco, TX: Baylor University Press, 2017), 1–4.

4 Ibid.

5 Earthworks Group, *50 Simple Things You Can Do to Save the Earth* (Berkeley, CA: Earthworks Press, 1989).

6 See "Commodity Activism in Neoliberal Times," in Roopali Mukherjee and Sarah Banet-Weiser, eds., *Commodity Activism: Cultural Resistance in Neoliberal Times* (New York: NYU Press, 2012), 1.

7 Models of "moral self-licensing" and "moral offsetting," and their implications, are discussed more fully later in this book.

8 Christian ethicist and former president of the Society of Christian Ethics Cristina Traina astutely unpacks what she characterizes as this "devastating shortcut." She observes, "[V]irtue theory tends to assume a universe in which virtuous personal behaviors advance and support the common good, in part because that is just how things work and in part because virtue is defined (based on wise reflection on experience) with respect to the common good, with optimism that these two are readily made coherent. This assumption can lead to a devastating shortcut: not in fact analyzing whether one's patterned acts are fulfilling obligation out there in the world, because one assumes that (in ways invisible to us) one's actions somehow will do so, because of the Order of Things." Personal communication with author, January 13, 2018. I am indebted to Dr. Traina for generously offering her expertise and comments on an earlier version of this chapter.

9 Plato, *Euthyphro*, trans. Benjamin Jowett (Hoboken, NJ: Pantianos Classics, 2016), 6–8.

10 Socrates points out to Euthyphro that "what is agreeable to Zeus" may be "disagreeable to Cronos or Uranus, and what is acceptable to Hephaestus" may be "unacceptable to Here, and there may be other gods who have similar differences of opinion." Plato, *Euthyphro*, 16–17. My thanks to philosophy professor and colleague Kenneth Seeskin for his generous consultation on notions of piety in

Plato's *Euthyphro* and for commenting on an earlier draft of this chapter. Personal communication, January 14, 2018.

11 For a history of the notion of the "citizen consumer," see Mukherjee and Banet-Weiser, *Commodity Activism*, 5–6.

12 For a more specific discussion of the "Protestant ethos of capitalist individualism," see Tracy Fessenden, *Culture and Redemption: Religion, the Secular, and American Literature* (Princeton, NJ: Princeton University Press, 2007), 186–87. For a broader cultural study of the outgrowth of individualism as the dominant force in Western philosophy and culture, its products, and contrasting philosophies/orientations, see Louis Dumont, ed., *Essays on Individualism: Modern Ideology in Anthropological Perspective* (Chicago: University of Chicago Press, 1986).

13 Like Martin Luther's concept of a "universal priesthood," by which laypersons have direct access to God without the mediation of clergy, so consumer narratives about the "direct power" of the consumer to make change by "shopping for good" implicitly argue that governmental policies, legislation, and regulation are irrelevant to substantive change. See also Mara Einstein's scathing moral critique of "social marketing" in *Compassion, Inc.: How Corporate America Blurs the Line between What We Buy, Who We Are, and How We Help* (Berkeley: University of California Press, 2012), especially chapter 5, "Shopping Is Not Philanthropy. Period."

14 I am indebted to Bishnupriya Gosh, scholar at UC–Santa Barbara of globalization in contemporary mediascapes, for her insightful comments in response to my presentation on ecopiety and "pollution porn," in which she highlighted instances and mediation of *ecstatic* piety in South Asia. Personal communication, Power Dynamics: Media and Environment conference, UCSB, Santa Barbara, California, April 30, 2016. See also Ronald Inden's discussion of the influence of early representations of "ecstatic piety" in Hindu devotionalism on contemporary notions of Hinduism in *Imagining India* (Bloomington: Indian University Press, 2001), 121.

15 Rachel Rinaldo, *Mobilizing Piety: Islam and Feminism in Indonesia* (New York: Oxford University Press, 2013), 3; Saba Mahmood, *Politics of Piety: The Islamic Revival and the Feminist Subject* (Princeton, NJ: Princeton University Press, 2011), 45–47.

16 Elaine Peña, *Performing Piety: Making Space Sacred with the Virgin of Guadalupe* (Berkeley: University of California Press, 2011), 67.

17 Peña, *Performing Piety*, 1. Emphasis is mine.

18 See Sarah McFarland Taylor, "Shopping, Religion, and the Sacred 'Biosphere,'" in Bruce Forbes and Jeffrey Mahan, eds., *Religion and Popular Culture in America*, 3rd edition (Oakland: University of California Press, 2017), 242. Here, I adopt Hine's term "the buyosphere" to probe the enmeshment of consumption and religion as something that takes place in a "sacred" sphere. On the "Capitalocene," see Jason Moore, ed., *Anthropocene or Capitalocene? Nature, History, and the Crisis of Capitalism* (Oakland, CA: PM Press, 2016), which implicates the mechanisms of unregulated capitalism in biospheric crisis (1–7).

19 Thomas Hine, *I Want That! How We All Became Shoppers* (New York: Harper Perennial, 2003), xiv–xv.

20 Laurence Iannaccone argues that in the American "spiritual marketplace," the Roman Catholic Church made a strategic error in getting rid of aspects of belonging and practice tied to Catholic identity and loyalty, while continuing to enforce the doctrines that Catholic laity were least willing to accept, such as prohibitions on birth control. Iannaccone writes, "The Catholic Church may have managed to arrive at a remarkable 'worst-of-both-worlds' position, discarding cherished distinctiveness in liturgy, theology, and lifestyle, while at the same time maintaining the very demands that its members and clergy are least willing to accept." See Iannaccone, "Why Strict Churches Are Strong," *American Journal of Sociology* 99, no. 5 (March 1994): 1180–1211.

21 Anne Lamott, *Plan B* (New York: Penguin Books, 2006), 145.

22 Stuart Hall, "The Work of Representation," in Hall, ed., *Representation: Cultural Representations and Signifying Practices* (London: Sage, 1997), 5.

23 See Stuart Hall's discussion of language as medium in representation, *Representation*, 1.

24 For religion scholars who trace, identify, and shine a light on these sorts of phenomena, see the following examples: Catherine Albanese, *Nature Religion in America: From the Algonkian Indians to the New Age* (Chicago: University of Chicago Press, 1991); Gary Laderman, *Sacred Matters: Celebrity Worship, Sexual Ecstasies, The Living Dead, and Other Signs of Religious Life in the United States* (New York: New Press, 2009); John Modern, *Secularism in Antebellum America* (Chicago: University of Chicago Press, 2011); Richard Callahan, "The Study of Religion: Looming through the Glim," *Religion* 42, no. 3: 425–37; Kathryn Lofton, *Consuming Religion* (Chicago: University of Chicago Press, 2017); Melissa Wilcox, *Queer Nuns: Religion, Activism, and Serious Parody* (New York: NYU Press, 2018).

25 Michelle Lelwica's work on shame and bodies similarly complicates this binary of "religious" and "secular," as does the aforementioned works of Lofton, Laderman, Modern, and Fessenden, among others. See Lelwica's *Shameful Bodies: Religion and the Culture of Physical Improvement* (London: Bloomsbury Academic, 2017).

26 This notion of the "sovereign individual" or "sovereign individualism" is most notably articulated by English political philosopher John Stuart Mill in his famous essay "On Liberty." Mill asserted that "over himself, over his own body and mind, the individual is sovereign." Mill argued that the individual should be "free from interference" of state or social control in matters concerning the self, and that such freedom from government interference, unless harm was being done to others, was an essential element of liberty. This kind of "liberalism" was associated with Mill's political party, the Liberal Party. See John Stuart Mill, *On Liberty, Utilitarianism, and Other Essays*, 2nd edition (New York: Oxford University Press, 2015), 13; Nicholas Abercrombie et al., *Sovereign Individuals of Capitalism* (New York: Routledge, 2014); and Mara Einstein's discussion of the proliferation of the increasingly "personalized economy" in *Compassion, Inc.*, 3–4.

27 The following chapter features greater explanation, characterization, and theorizing of "media interventions" and what it means to "intervene," as outlined in Kevin Howley, ed., *Media Interventions* (New York: Peter Lang, 2013).
28 Roy Scranton, *Learning to Die in the Anthropocene: Reflections on the End of Civilization* (San Francisco: City Light Books, 2015), 55–56. Scranton at first uses this term to refer to literal ways to move energy (oil, gas, coal, etc.), but "social energetics" also takes on a double-entendre to refer more sociologically to something closer to "social energies." It is in this sense that I use the term in this book, taking cognizance of the complex relationships between humans and energy sources and power dynamics of multiple sorts. An early definition of sociology was the "science of social energies." See John Henry Wilbrandt Stuckenberg, *Introduction to Sociology* (New York: A.C. Armstrong and Son, 1898), 129.
29 Scranton, *Learning to Die in the Anthropocene*, 19.
30 Scranton, *Learning to Die in the Anthropocene*, 55–56.
31 Henry Jenkins, *Convergence Culture: Where Old and New Media Collide* (New York: NYU Press, 2006); and Jenkins's "Avatar Activism," *Le Monde Diplomatique*, September 2010, https://mondediplo.com; Henry Jenkins, Sangita Shresthova, Liana Gamber-Thompson, Neta Kligler-Vilenchik, and Arely Zimmerman, *By Any Media Necessary: The New Youth Activism* (New York: NYU Press, 2016).
32 Henry Jenkins, "What Black Panther Can Teach Us about Civic Imagination," 21st Century Global Dynamics, May 22, 2018, www.21global.ucsb.edu; Jenkins et al., *By Any Media Necessary*.
33 In her book responding to the 2016 U.S. election, Naomi Klein similarly argues that resisting and saying "no" to the current state of our earth is not enough: "[W]e need to fiercely protect some space to dream and plan for a better world." See Klein's *No Is Not Enough* (Chicago: Haymarket Books, 2017), 257.
34 Alissa Walker, "Wakanda Is Where Every Urbanist Wants to Live," Curbed, February 19, 2018, https://www.curbed.com; Stephen Jorgensen-Murray, "What's Up with Wakanda's Trains? On Public Transport in Black Panther," *City Metric*, February 20, 2018, www.citymetric.com; William Snoeyink, "Wakanda: The Future of Urban Sustainability?" Planning for Sustainable Communities, February 20, 2018, https://planningforsustainablecommunities.wordpress.com.
35 Andrew Dobson, "Ecological Citizenship: A Defence," *Environmental Politics* 15, no. 3 (2006): 447–51; Dobson, *Green Political Thought*, 4th edition (Abingdon, UK: Routledge, 2007); Neil Carter, *The Politics of the Environment: Ideas, Activism, Policy*, 2nd edition (Cambridge: Cambridge University Press, 2007); and Tim Hayward, "Ecological Citizenship: Justice, Rights, and the Virtue of Resourcefulness," *Environmental Politics* 15, no. 3 (2007): 435–46.
36 See in particular James Hay and Nick Couldry, eds., "Rethinking Convergence Culture: An Introduction," *Cultural Studies* 25, nos. 4–5 (2011): 473–86; and Roy Scranton's discussion of the force of corporate logics writ large in the Anthropocene and their role in catalyzing and worsening climate crisis in *Learning to Die in the Anthropocene*, 19.

37 For a study of analogous dynamics with the feminist movement and how principles of feminism became watered down and neutralized within a depoliticized marketplace, see cofounder of Bitch Media Andi Zeisler, *We Were Feminists Once: From Riot Grrrl to Cover Girl, the Buying and Selling of a Political Movement* (Philadelphia: Public Affairs, Perseus Book Group, 2016). Zeisler critically queries what it means when a social movement becomes a brand identity and eviscerates the ways in which feminism has been strategically co-opted by fashion and advertising. See also Sarah Banet-Weiser's critical analysis of the Dove "real beauty" campaign for women in *Authentic: The Politics of Ambivalence in a Brand Culture* (New York: NYU Press, 2012), 15–49; Mara Einstein's work on the disturbing manipulation of feminist rhetoric for corporate profit in the marketing of breast cancer campaigns in *Compassion, Inc.*, 69–100; and Molly Bandonis with Paul Booth, "Branding Feminism: Corporate Blogging and the Shaky Relationship between Ideology and Profitability," in Amber Davisson and Paul Booth, eds., *Controversies in Digital Ethics* (New York: Bloomsbury Academic, 2017), 279–94.

38 Michael Pickering, "Introduction," in Pickering, ed., *Research Methods in Cultural Studies* (Edinburgh: Edinburgh University Press, 2008), 9.

39 David Korten, *Change the Story, Change the Future: A Living Economy for a Living Earth* (Oakland, CA: Berrett-Koehler, 2015), 23–36.

40 On moral foundations theory, see Jonathan Haidt, *The Righteous Mind: Why Good People Are Divided by Politics and Religion* (New York: Vintage, 2012); and on the subject of "encoding" and "decoding," which I will discuss and explain further, see Stuart Hall, "Encoding, Decoding," in Simon During, ed., *The Cultural Studies Reader*, 3rd edition (New York: Routledge, 2007), 477–87.

41 Theodor Adorno and Max Horkheimer, "The Culture Industry: Enlightenment as Mass Deception," in Simon During, ed., *The Cultural Studies Reader*, 3rd edition (London: Routledge, 2008), 405–15; and Jane Stokes, *How to Do Media and Cultural Studies*, 2nd edition (Los Angeles: Sage, 2013), 36.

42 Michael Saler, *As If: Modern Enchantment and the Literary Prehistory of Virtual Reality* (New York: Oxford University Press, 2012).

43 When incorporating online primary source data throughout this book, I have taken an archival research methodological approach, in which I view the Internet as my archive or repository of primary sources (commentary, blogs, opinions, comments, visual culture, etc.) related to various expressions and debates surrounding environmental virtue and ecopiety precipitated by prosumers' encounters with and circulation of cultural products. For the scope of this book, I have accessed only publicly accessible materials published in platforms that provide free and open access of content to all Internet users. I have not reviewed confidential or privileged information for this study, or information that necessitated my joining groups in order to access material. As much as possible, I worked to understand my primary sources in relationship to their particular context and in relation to surrounding sources, analyzing those sources within their particular situatedness and context. Primary sources' associational qualities to related media

were likewise recorded and considered, as was the timing of when a source was produced, and where, and in what context it was reposted, remixed, and repurposed elsewhere, if applicable. Rather than search for particular themes or key words on fan discussion sites, I printed out the entire listing of comments in each case and then went through these with a set of coding markers, color-coding different themes and marking the frequency of their appearance in discussions. I approach these published online comments and opinions expressed, especially those posted to blog entries, news articles, or media reviews, as cultural theorist Walter Benjamin might analyze publicly published letters to the editor, now collected in a repository or archive that is free and open to the public. See Benjamin's "Work of Art in the Age of Mechanical Reproduction," in Benjamin, *Illuminations: Essays and Reflections* (New York: Schocken, 2007), 232; on working with Internet sources and communications, Nancy Baym, *Personal Connections in the Digital Age*, 2nd edition (Cambridge, UK: Polity, 2015); and sociologist Erving Goffman's (pre–Internet era) work on the presentation of self in varied social contexts, which still retains remarkable salience when one is analyzing online interactions: *The Presentation of Self in Everyday Life* (Edinburgh: University of Edinburgh Social Sciences Research Centre, 1959).

44 Nick Couldry, *Media, Society, World: Social Theory and Digital Media Practice* (Cambridge, UK: Polity, 2012), especially chapter 2 on "Media as Practice," and 43–44 specifically on the varieties of media practice.

45 John Durham Peters, *The Marvelous Clouds: Toward a Philosophy of Elemental Media* (Chicago: University of Chicago Press, 2015), 6.

46 This term receives more detailed explanation in chapter 7. More immediately, see social psychologist and justice researcher Susan Opotow's seminal definition in "The Scope of Justice, Intergroup Conflict, and Peace," in Linda R. Tropp, ed., *The Oxford Handbook of Intergroup Conflict* (New York: Oxford University Press, 2012), 72–88.

47 For political theory that supports such inclusion, see Massachusetts Institute of Technology economists Daron Acemoglu and James Robinson's *Why Nations Fail: The Origins of Power, Prosperity, and Poverty* (New York: Crown, 2012), in which the authors argue, on the basis of their use of "big data" to record the economic history of every country, that *inclusive* political and economic institutions are the key to sustainable growth and a thriving economy. Inclusion strategies are just part of a larger conversation taking place about aims to replace extractive economic models with "regenerative economic models."

48 William Chafe, "Is There an American Narrative and What Is It?" *Daedalus: Journal of the American Academy of Arts and Sciences* (Winter 2012): 11. Historian William Chafe posits that if there does exist something we might call "the American narrative," it would include two conflicting paradigms or sets of values that have clashed for nearly four centuries of American history. Chafe sees a clash of values between *self* and *community* as critical to understanding American history and to understanding the political and social polarization of today. Parallel conflicts be-

tween a moral universe centered on an anthropocentric "I" versus an ecologically expanded moral scope of the "we" (the "community of life") factor significantly into environmental narratives in popular culture.

49 See Jenkins, *Convergence Culture*, 21.

50 In Couldry's afterword to Howley, ed., *Media Interventions*, 397; see also Couldry's *Media, Society, World,* especially 33–58.

51 See Andrew Bennett and Nicholas Royle's discussion of Chinua Achebe in *An Introduction to Literature, Criticism, and Theory*, 4th edition (New York: Routledge, 2009), 54.

CHAPTER 1. RESTORYING THE EARTH

1 David Korten, *Change the Story, Change the Future: A Living Ecology for a Living Earth* (Oakland, CA: Berrett-Koehler, 2015). The very title of Korten's book, *Change the Story, Change the Future*, bears rhetorical resemblance to the New Thought movement's optimistic adage, embraced by such figures as Charles Fillmore, Ernest Holmes, and Norman Vincent Peale: "Change your thinking, change your life" or "Change your thinking, change your world." See for examples, Catherine Albanese, *Republic of Mind and Spirit: A Cultural History of American Metaphysical Religion* (New Haven, CT: Yale University Press, 2008), 440.

2 Korten, *Change the Story*.

3 See, for instance, Frances Moore Lappe, *Changing the Way We Think to Create the World We Want* (New York: Nations Books, 2013) and Anthony Weston, *Mobilizing the Green Imagination* (Gabriola Island, BC: New Society Publishers, 2012); Buddhist eco-philosopher Joanna Macy's *Pass It On: Five Stories That Can Change the World* (Berkeley, CA: Parallax Press, 2010); and Macy's section on "Choosing Our Story" and "The Great Turning" in *Coming Back to Life: The Updated Guide to the Work That Reconnects*, revised edition (Gabriola Island, BC: New Society Publishers, 2014), 1–18. Eco-theologian Thomas Berry's work puts stock in the global adoption of the scientific story of evolutionary cosmology to effect change in contemporary human planetary relations. See Berry's "Emerging Ecozoic Period," reprinted in Ervin Laszlo and Allan Combs, eds., *Thomas Berry, Dreamer of the Earth: The Spiritual Ecology of the Father of Environmentalism* (Rochester, VT: Inner Traditions, 2011), 12. For a counterargument to Berry's thesis, see Lisa Sideris, *Consecrating Science: Wonder, Knowledge, and the Natural World* (Berkeley: University of California Press, 2017).

4 French structuralist anthropologist Claude Lévi-Strauss famously observed that totems (emblems of sacred beings, objects, or symbols representing kinship groups or tribes) were "not good to eat" but were "good to think." That is, they serve an intellectual speculative function within a larger structure of metaphoric thinking, particularly thinking about difference. Here, I make no universal claims about Korten's work, which is highly situated within specific cultural and economic discourses, but I do find that the structural opposition of his narrative schema provides insights into a particular sort of narrative dynamics at work

within the active "storying" of global capitalism and its discontents in media forms. For Lévi-Strauss's related discussion, see *Totemism*, trans. Rodney Needham (Boston: Beacon, 1963), 89.

5 Here, Korten and others take the stance that at some point in time, traditional societies, tribal societies, and indigenous cultures once had and (those that survived) still possess a "shared cultural lens." According to this narrative, this "shared lens" was an effective/functional one for living sustainably, and such a shared cultural lens today is possible (and needed) if we can only cultivate the right conditions to bring it about. I am more skeptical of this premise and see more of a multiplicity of contested lenses, as have sociologists such as Anthony Giddens who have contended that the nineteenth- and twentieth-century romantic notion of society as a "whole" is antiquated and a misreading. Were societies ever "wholes"? See Giddens, *The Constitution of Society* (Berkeley: University of California Press 1984), 164–65.

6 See Theodore Schatzki's literature review discussing the fallacy of conceiving societies as wholes in *Social Practices: A Wittgensteinian Approach to Human Activity and the Social* (New York: Cambridge University Press, 1996), 4–5. Couldry later echoes this theme and adopts it for theoretical discussions of media in *Media, Society, World*, 1.

7 See neuroscientist Alex Huth et al.'s brain imaging study, in which the Berkeley research team had research subjects listen to stories told by participants in *The Moth Radio Hour*. The researchers then "mapped" the subjects' brains with MRIs as they listened to the storytelling, thus creating a kind "semantic atlas." Their findings demonstrated that stories activate our brains in incredibly stimulating and powerful ways, exercising the entire brain, forging and developing various meaning-based systems of representation in the brain. See Huth et al., "Natural Speech Reveals the Semantic Maps That Tile Human Cerebral Cortex," *Nature*, April 28, 2016, 453–58; Benedict Carey, "This Is Your Brain on Podcasts," *New York Times*, April 28, 2016, www.nytimes.com.

8 The quotation, "We read to know we are not alone," appears in the teleplay version of William Nicholson's *Shadowlands*, which was produced for BBC Wales in 1985.

9 A classic example of religion doomsaying is Sigmund Freud's *Future of an Illusion* (New York: Norton, 1989). For a chronicling of the many figures who have historically predicted a demise of religion that never materialized, see Canadian sociologist Reginald Bibby's *Beyond the Gods and Back: Religion's Demise and Rise and Why It Matters* (Toronto: Project Canada Books, 2011). Pivotal secularization theorists such as Robert Bellah and Rodney Stark ultimately reversed their positions on secularization. See most notably Rodney Stark's "Secularization Theory, R.I.P.," *Sociology of Religion* 60, no. 3 (Autumn 1999): 249–73; and sociologist Peter Berger, "Secularization and De-secularization," in Linda Woodhead, ed., *Religions in the Modern World: Transitions and Transformations* (London: Psychology Press, 2002), chapter 13.

10 Leslie Marmon Silko, *Ceremony* (New York: Penguin, 1989), 1–2.

11 Leslie Marmon Silko, "Interior and Exterior Landscapes: The Pueblo Migration Stories," in *Yellow Woman and the Beauty of Spirit* (New York: Simon and Schuster, 1996), 30; and Silko's *Storyteller* (New York: Seaver Books, 1981).

12 Here, by "media," I am referring to communication tools for transmitting and storing information. However, I employ a broader and more expansive use of the term so that "media" refers to more than print, electronic, and so-called mechanical media, but also to "representational" media, which can include things such as body art and tattoos, which I will discuss later on in the book. Media theorist John Durham Peters's theorizing of "elemental media," especially human bodies as elemental media, also informs treatments of "media" in this book. Analysis of environmental tattoos, for instance, encompasses more than one category and definitional understanding of media, since I also attend to the digital display and circulation of tattoo-inscribed skin in space and time. See John Fiske's articulation of different categories of "media" in *Introduction to Communication Studies*, 3rd edition (New York: Routledge, 2011), 16–17; and Peters, *The Marvelous Clouds.*

13 Mircea Eliade, *Patterns of Comparative Religion* (New York: World Publishing, 1958).

14 Mircea Eliade, *The Myth of the Eternal Return; or, Cosmos and History* (Princeton, NJ: Princeton University Press, 1991), 3–11.

15 On the former point, see for example, Ursula King, "Historical and Phenomenological Approaches," in Frank Whaling, ed., *Theory and Method in Religious Studies: Comparative Approaches to the Study of Religion* (Berlin: de Gruyter, 1995), 123; and Russell T. McCutcheon, *The Discipline of Religion: Structure, Meaning, Rhetoric* (New York: Routledge, 2003); and McCutcheon, *Manufacturing Religion: The Discourse on Sui Generis Religion and the Politics of Nostalgia* (New York: Oxford University Press, 2003), 74–100.

16 Jonathan Gottschall, *The Storytelling Animal: How Telling Stories Makes Us Human* (New York: Mariner Books, 2012), xiii–xiv.

17 The scope of this book project, which engages in cultural study, is not sufficiently capacious for me to debate whether or not we are in environmental crisis. Certainly, though, people are responding in numerous ways to their experience of living in an age of environmental crisis, producing cultural works documenting and commenting on ongoing crisis, and those responses are key to my investigations and scholarly concerns. For a short list of accessible resources on this topic, however, see Pulitzer Prize–winning author Thomas L. Friedman's *Hot, Flat, and Crowded* (New York: Picador, 2009); James Lovelock, *A Rough Ride to the Future* (New York: Overlook Press, 2015); University College London Professor of geophysical and climate hazards Bill McGuire's *Global Catastrophes: A Very Short Introduction* (Oxford: Oxford University Press, 2014); University College London Professor of climatology Mark Maslin's *Climate Change: A Very Short Introduction* (Oxford: Oxford University Press, 2014); Bill McKibben, *Eaarth: Making a Life on a Tough New Planet* (New York: St. Martin's Griffin, 2011); Pulitzer Prize–winning author Elizabeth Kolbert's books, *Field Notes from a Catastrophe: Man, Nature,*

and Climate Change (London: Bloomsbury, 2006) and *The Sixth Extinction: An Unnatural History* (New York: Henry Holt, 2014). A more comprehensive and technical bibliography may be found on the King's College, University of Cambridge Global Warming Resources web site: www.kings.cam.ac.uk.

18 Bennett and Royle, *An Introduction to Literature*, 54.

19 The term "assemblage" is theorized by Gilles Deleuze and Félix Guattari as a constellation of sociolinguistic philosophical operational components that relationally interact and make up complex social systems. See Deleuze and Guattari's *Thousand Plateaus* (Minneapolis: University of Minnesota Press, 1993). In her "congregational understanding of agency," political philosopher Jane Bennett draws, in part, on this notion of "assemblage" in asserting what she calls "thing-power" or the "agency of assemblages" that characterize "vibrant matter." She also engages Bruno Latour's concept of "actants," which is discussed later in this book. See Jane Bennett, *Vibrant Matter: A Political Economy of Things* (Durham, NC: Duke University Press, 2010). In his cultural study of media, men, and sport, Clifton Evers summarizes, "Mediated assemblages involve communication technologies, such as mobile phones and the extensive network of other communication technologies they connect to. . . . These communication technologies are more than objects. . . . [M]ateriality, whether it is physiological, mechanical, or digital, is vibrant and has agency." See Evers's "Masculinity, Sport, and Mobile Phones: A Case of Surfing," in Gerard Goggin and Larissa Hjorth, eds., *The Routledge Companion to Mobile Media* (New York: Routledge, 2016), 376.

20 James Conlon, *Earth Story, Sacred Story* (Mystic, CT: Twenty-Third Publications, 1994), 11.

21 On the subject of media interventions, see Howley, ed., *Media Interventions*. On DIY (do-it-yourself) media, its trends, and its currents, see Clay Shirky's *Here Comes Everybody: The Power of Organizing without Organizations* (New York: Penguin, 2009); and Jean Burgess and Joshua Green, *YouTube: Online Video and Participatory Culture* (Cambridge, UK: Polity, 2009), especially Henry Jenkins's chapter, "What Happened before YouTube," which offers historical understanding for DIY media practices and various innovative precursors to YouTube, such as "garage cinema" and 'zines, web publishing, creative uses of VCRs and patch cords, and many more historical efforts to "democratize channels of communication" (109–25). Michele Knobel and Colin Lankshear's *DIY Media: Creating, Learning, and Sharing with New Technologies* (New York: Peter Lang, 2010) provides a guide to the range of DIY media practices with practical application and an emphasis on digital literacies and epistemologies.

22 See the product description of the Adobe "Magic Lens": http://lightfield-forum.com; and at www.techeblog.com (accessed January 15, 2018).

23 For a comprehensive study of the "*Rashomon* effect," *Rashomon's* legacy, and the film's aesthetic and cultural impact over time, see Blair Davis, Robert Anderson, and Jan Walls, eds., *Rashomon Effects: Kurosawa,* Rashomon, *and Their Legacies*

(New York: Routledge, 2015). I am indebted to my colleague, medieval historian Richard Kieckhefer, for our fruitful conversation about this concept.

24 Henry Jenkins draws attention to this grassroots "hijacking" in both *Convergence Culture* (2006) and (with coauthors Sam Ford and Joshua Green) in *Spreadable Media: Creating Value and Meaning in a Networked Culture* (New York: NYU Press, 2013), especially in the latter, 20–27. See also Julie Bosman, "Chevy Tries a Write Your Own Approach and the Potshots Fly," *New York Times*, April 4, 2006, http://www.nytimes.com; and emergent media and communication analyst Jenny Mizutowicz's online case study of Chevrolet's viral marketing campaign at https://jennydmizcapstone.wordpress.com (accessed February 11, 2019).

25 Ryan Martin, "Indiana GOP Asked Facebook for Obamacare Horror Stories: The Responses Were Surprising," *Indiana Star*, July 5, 2017, http://www.indystar.com/story.

26 Atmospheric chemist and Nobel laureate Paul Crutzen proposed the "Anthropocene" at a conference in 2000 as a way to characterize our most recent geologic era—one dominated by human-influenced or anthropogenic alternations. Crutzen later used the term in print in an article coauthored with Eugene Stoermer, "The 'Anthropocene,'" *Global Change Newsletter* 41 (2000): 17–18. For further cogitation in the field of Crutzen's concept, see J. Zalasiewicz et al., "Are We Now Living in the Anthropocene?" *GSA (Geological Society of America) Today* 18 (2008): 4–8.

27 Scholars, educators, and policy makers need to take seriously and address the digital divide. That being said, the greater accessibility of humans around the globe in the last decade or more to relatively inexpensive DIY digital production technologies is nothing short of astounding. For a comprehensive review of this issue, see Kim Andreasson, ed., *Digital Divides: The New Challenges and Opportunities of e-Inclusion* (Boca Raton, FL: Taylor and Francis, 2015), particularly 265–74. For a more sobering analysis of freedoms lost and power imbalances aggravated by the so-called digital revolution, see Mark Andrejevic's cautionary monograph, *iSpy: Surveillance and Power in the Interactive Era* (Lawrence: University of Kansas Press, 2007). Still, when comparing historically the relatively inexpensive, easily handheld, and widely adopted technologies of today to previous eras of "gatekeeping," elite, and expensive, not to mention bulky and difficult/costly-to-run-and-store media production equipment, the glass is arguably half full in terms of greater accessibility to tools of both media production and media circulation.

28 Diane Winston, ed., *Small Screen, Big Picture: Television and Lived Religion* (Waco, TX: Baylor University Press, 2009), 12.

29 For the "nothing is *ever* new under the sun" camp, I of course recognize that people have always interactively participated in cultural works of media production. In his discussions of "participatory culture," Henry Jenkins goes back, for instance, to the era of local ham operator hobbyists, Depression Era locally produced neighborhood newspapers, fan "'zines," and much more. See Jenkins's "What Happened before YouTube," in Burgess and Green, *YouTube*, 109–25; and

Jenkins's keynote address, "All Over the Map: What Oz the Great and Powerful Can Teach Us about World-Making," delivered to the Rethinking Intermediality in the Digital Age conference, University of Transylvania, October 25, 2013, https://www.youtube.com/watch?v=35VxoElyqpc. On the importance of maintaining historical perspective in the study of religion and media practices, see historian David Morgan's review essay, "Religion and Media: A Critical Review of Recent Developments," *Critical Research on Religion* 1 (2013): 347–56.

30 See, for instance, Wilcox, *Queer Nuns*; Laderman, *Sacred Matters*; Jeffrey Kripal's *Mutants and Mystics* (Chicago: University of Chicago Press, 2011); Kaya Oakes, *The Nones Are Alright: A New Generation of Believers, Seekers, and Those in Between* (Maryknoll, NY: Orbis Books, 2015); Rachel Wagner, *Godwired: Religion, Ritual, and Virtual Reality* (New York: Routledge, 2012); Stewart Hoover and Curtis Coats, *Does God Make the Man? Media, Religion, and the Crisis of Masculinity* (New York: NYU Press, 2015); Lynn Clark, *From Angels to Aliens: Teenagers, the Media, and the Supernatural* (New York: Oxford University Press, 2003); Gregory Grieve, *Cyber Zen: Imaging Authentic Buddhist Identity, Community, and Practices in the Virtual World of Second Life* (New York: Routledge, 2016); Heidi Campbell, ed., *Digital Religion: Understanding Religious Practice in New Media Worlds* (New York: Routledge, 2012); and Monica Miller's *Religion and Hip Hop* (New York: Routledge, 2012).

31 On our own existence as "stuff" and on our reciprocal relationship to the stuff we make, see Daniel Miller, *Stuff* (Cambridge, UK: Polity, 2010), 6–7; and Ian Woodward's *Understanding Material Culture* (Los Angeles: Sage, 2007), particularly discussion of anthropologist Mary Douglas and economist Baron Isherwood's *World of Goods*, which delves into the ways in which as "people construct a universe of meaning through commodities, they use objects to make visible and stable cultural categories, to deploy discriminating values and to mark aspects of self and others" (vii).

32 bell hooks, *Talking Back* (Boston: South End Press, 1989), 5.

33 Keith Hopkins and Mary Beard, *The Colosseum* (Cambridge, MA: Harvard University Press, 2011), 97 and 186.

34 For a mapping of the scope and scale of digital ubiquity, see George Beekman and Ben Beekman's *Digital Planet*, 10th edition (Harlow, Essex, UK: Pearson Education, 2013). For a more ominous look at these changes in scale, see Nicco Mele, *The End of Big: How the Internet Makes David the New Goliath* (New York: St. Martin's, 2013).

35 Benjamin, "The Work of Art in the Age of Mechanical Reproduction," 218, 223.

36 Benjamin, "The Work of Art in the Age of Mechanical Reproduction," 232.

37 Shirky, *Here Comes Everybody*, 55–80.

38 Mele, *The End of Big*.

39 Henry Jenkins, public lecture, "How Content Gains Meaning and Value in the Age of Spreadable Media," Center for Media and Communication Studies, Central European University, Budapest, Hungary, June 21, 2012, https://www.youtube.

com. In the same talk, Jenkins emphasizes that "media is not something that *happens* to us," reminding us of the active roles we each play as media outlets. "We are users, producers, and circulators of media, and those relationships are essential to understanding transmedia—the flow of media across media platforms."

40 Scott Shane and Mark Mazzetti, "The Plot to Subvert an Election," *New York Times*, September 20, 2018, www.nytimes.com; Issie Lapowsky, "The Real Trouble with Trump's 'Dark Post' Facebook Ads," *Wired*, September 20, 2017, https://www.wired.com; Massimo Calabresi, "Inside Russia's Social Media War on America," *Time*, May 18, 2017, http://time.com; Garret Sloane, "Adgate: A Step by Step Guide to Russian Interference," *AdAge*, October 9, 2017, http://adage.com.

41 Michael Saler, *As If: Modern Enchantment and the Literary Prehistory of Virtual Reality* (New York: Oxford University Press, 2012), 131.

42 Paul Booth, *Digital Fandom 2.0: New Media Studies* (New York: Peter Lang, 2017); and Melissa Click and Suzanne Scott, eds., *The Routledge Companion to Media Fandom* (New York: Routledge, 2017).

43 See David Croteau and William Hoynes, *Media/Society: Industries, Images, and Audiences,* 5th edition (Los Angeles: Sage, 2014), 1–31. Croteau and Hoynes argue that media are so ubiquitous in our lives that it makes sense to talk about the fused designation of "media/society," rather than discussing these as realms unto themselves.

44 Emily Nussbaum, "Must See Metaphysics," *New York Times*, September 22, 2002, http://www.nytimes.com.

45 This curation of apps is reflective of the ongoing self-curation of our own identities, points out Rachel Wagner in "You Are What You Install: Religious Authenticity and Identity in Mobile Apps," in Campbell, ed., *Digital Religion*, 199–206.

46 On dynamics of convergence, see Henry Jenkins's discussion of mobile phones in his introduction to *Convergence Culture*, 4–5. On the mechanics of consolidation and conglomeration, narrowing the number of corporations that own media and increasingly concentrating power among a select few, see "The Economics of the Media Industry," in Croteau and Hoynes, *Media/Society*, 32–55; and Ben Bagdikian, *The New Media Monopoly* (Boston: Beacon, 2004).

47 Jenkins, *Convergence Culture*, 2–3. This is an important distinction that Jenkins argues for in his work.

48 See, for instance, Richard Foltz, *Worldviews, Religion, and the Environment: A Global Anthology* (Boston: Cengage Learning, 2002); Roger Gottlieb, *A Greener Faith: Religious Environmentalism and Our Planet's Future* (New York: Oxford University Press, 2006); Mary Evelyn Tucker and John Grim, series editors of the *World Religions and Ecology Series* (London: Cassell, 1997–2003); John Carroll et al., eds., *The Greening of Faith: God, the Environment, and the Good Life* (Lebanon: University of New Hampshire Press, 1997); Sarah McFarland Taylor, *Green Sisters: A Spiritual Ecology* (Cambridge, MA: Harvard University Press, 2008).

49 "Religious Landscape Study," Pew Research Center on Religion and Public Life, http://www.pewforum.org (data from 2018 updated web site, accessed January 22, 2018).

50 Indiana University public policy professor David Konisky, who holds a doctorate in political science from Massachusetts Institute of Technology, in addition to master's degrees from Yale University in environmental management and international relations, respectively, is recognized as one of the most widely respected experts on environmental issues and public policy. See David Kosinsky, "The Greening of Christianity? A Study of Environmental Attitudes over Time," *Environmental Politics* 27, no. 2 (January 2018): 267–91; and Niina Heikkinen, "Christianity Is Not Getting Greener," *Scientific American*, January 16, 2018, https://www.scientificamerican.com. Konisky's study also provides more statistical basis for some of the skepticism about the veracity of the "greening of religion hypothesis" raised by Taylor et al. in "The Greening of Religion Hypothesis (Part Two): Assessing the Data from Lynn White Jr. to Pope Francis," *Journal for the Study of Religion, Nature, and Culture* 10, no. 3 (2016): 306–78. Additional research would be needed to determine whether or not Konisky's findings might extend to other religious affiliations.

51 Deanna Sellnow, *The Rhetorical Power of Popular Culture: Considering Mediated Texts*, 2nd edition (Los Angeles: Sage, 2013), xiii.

52 Sellnow, *The Rhetorical Power of Popular Culture*, 6.

53 Gary Laderman, *Sacred Matters*; Laderman, "Religion Is Dead . . . Long Live the Sacred," *Sacred Matters*, April 18, 2014, https://sacredmattersmagazine.com; Conrad Ostwalt, *Secular Steeples: Popular Culture and the Religious Imagination*, 2nd edition (New York: Bloomsbury, 2012); John Lyden, *Film as Religion: Myths, Morals, and Rituals* (New York: NYU Press, 2003); and Sellnow, *The Rhetorical Power of Popular Culture*.

54 Vatican digital technology interest and savvy predated Pope Francis. See Alex Kantrowitz, "Meet the Tech Company Bringing the Vatican into the Digital Age," *Forbes*, February 5, 2013, www.forbes.com.

55 Christopher Helland, "How the Pope's Climate Message Went Viral," *World Economic Forum*, July 31, 2015, https://agenda.weforum.org.

56 Kevin Dennehy, "Only Four in Ten Catholics Aware of Pope's Encyclical, Yale Poll Finds," *Yale School of Forestry and Environmental Studies News*, August 19, 2015, http://environment.yale.edu/news.

57 Patricia Miller, "Survey Finds Little 'Francis Effect,' Two U.S. Catholic Churches," *Religious Dispatches*, August 26, 2015, http://religiondispatches.org.

58 In 2018, this number had reached 34 million.

59 Jenkins et al., *Spreadable Media*, 6–36.

60 Observatório do Clima, "Pope Francis—The Encyclical," YouTube, June 11, 2015, https://www.youtube.com/watch?v=76BtP1GInlc.

61 This tagline, borrowed from the 2014 People's Climate March in New York City, is also identified as the motto of environmental journalist Naomi Klein and appears frequently in her work. See Klein's *This Changes Everything* (New York: Simon and Schuster, 2014).

62 Tierney McAfee, "The Pope's New Climate Change Encyclical Gets an Epic New Movie Trailer," *People*, June 18, 2015, http://www.people.com.

63 See YouTube's charted "Most Viewed Videos of All Time," https://www.youtube.com/playlist?list=PLirAqAtl_h2r5g8xGajEwdXd3x1sZh8hC (accessed February 11, 2019). On the mass success of E. L. James's erotica series and the record-setting success of the tie-in movie, which in 2015 crossed the $500 million mark for box-office sales, see Pamela McClintock, "Box Office Milestone: Fifty Shades Crosses a Sexy $500M Worldwide," *Hollywood Reporter*, March 5, 2015, http://www.hollywoodreporter.com. See also Jenkins et al., *Spreadable Media*.

64 Ray Browne, "Popular Culture as the New Humanities," reprinted in Ray Browne and Marshall Fishwick, *Symbiosis: Popular Culture and Other Fields* (Bowling Green, OH: Bowling Green State University Popular Press, 1988), 1–2.

65 Browne, "Popular Culture as the New Humanities," 1. See also Kathryn Lofton's argument for taking "popular culture seriously as a critical archive" in *Consuming Religion*, 168.

66 Marcelline Block, "Popular Culture Matters Because It Reflects, Expresses, and Validates the Spirit of Our Epoch," in *Popular Culture* (Bristol, UK: Intellect Publishing), www.intellectbooks.com, 15. Emphasis on "current" is Block's, and emphasis on "future" is mine.

67 See Sellnow's definition of mediated popular culture in her text, *The Rhetorical Power of Popular Culture*, 3.

68 Phyllis Japp, Mark Meister, and Debra Japp, eds., *Communication Ethics, Media, and Popular Culture* (New York: Peter Lang, 2005), 6.

69 Japp et al., eds., *Communication Ethics*, 9. Religion and media scholar Diane Winston takes an inverse approach to making the case for the importance of studying popular culture, purposefully *not* asking why we as scholars should take television seriously but instead pointing out that, as it is the most pervasive medium in American households (edged now closely by the Internet and often delivered through the Internet), why in the world *wouldn't* we? Winston points to various television series as being the "parables of our time," demonstrating how shows such as *The Wire*, *Battlestar Galactica*, and *House* become the cultural works through which we engage pressing moral questions of our age.

70 University of Southern California film and critical studies professor Marsha Kinder first coined the term "transmedia" in 1991 to talk about stories that extended beyond a single media platform. Henry Jenkins's development of this concept in a paper published on "Transmedia Storytelling" in 2003, his book on "convergence culture" in 2006, and his blog post, "Transmedia Storytelling 101," launched "transmedia storytelling" as a key and critical area of study. See Jenkins, "Transmedia Storytelling: Moving Characters from Books to Films to Video Games Can Make Them Stronger and More Compelling," *MIT Technology Review*, January 15, 2003, www.technologyreview.com; *Convergence Culture*; and "Transmedia Storytelling 101," *Confessions of an Aca-Fan*, March 22, 2007, http://

henryjenkins.org; Marsha Kinder, *Playing with Power in Movies, Television, and Video Games: From Muppet Babies to Teenage Mutant Ninja Turtles* (Berkeley: University of California Press, 1991).

71 On participatory digital culture, see both Jenkins's *Convergence Culture* and Jenkins, Mizuko Ito, and danah boyd, eds., *Participatory Culture in a Networked Era* (Cambridge, UK: Polity, 2016).

72 Jenkins, "All over the Map."

73 See Karen Hellekson and Kristina Busse, eds., *The Fan Fiction Studies Reader* (Iowa City: University of Iowa Press, 2014), 6.

74 See Maria Jose and John Tenuto, "Spockanalia: The First Star Trek Fanzine," October 20, 2014, http://www.startrek.com.

75 See Jenkins's chapter "Why Heather Can Read" in Jenkins, *Convergence Culture*, 175–216. See also MuggleNet.com's story award winners: http://fanfiction.mugglenet.com (accessed July 31, 2015).

76 Rachel Wagner describes the "cross-pollination of story" in transmedia storytelling as immensely "intricate and evolving story construction." For Wagner, this kind of storytelling also constitutes an interesting form of "rich religious modeling (and modding)." Rachel Wagner, *Godwired*, 213.

77 My playing with this concept extends the notions of "transmedia storytelling" beyond its strictest confines but does so in a way that stays true to many of its characteristics: participatory culture, debates over "core" texts, struggles for narrative extension and control, questions over whether a story of great mythic proportions can even be "owned," etc.

78 Note that the ethics of "culture-making" fans can be controversial when media companies profit from this free labor and/or arguably exploit the dynamics of fan "gift economies."

79 On "the art of world making," see Jenkins's *Convergence Culture*, 21.

80 In Leslie Marmon Silko's rendition of "The Witchery" clan story, the storytelling witch sets in motion the invasion and conquest of the Americas merely by telling the story of it. When asked to undo this witchery, the witch just shakes its head and says, "It is already turned loose. It is already coming. It can't be called back." In the instance of "The Witchery," a story of destruction and devastation cannot be called back, once set in motion, but the implication is that stories of healing and repair also carry this immense world-transforming power. See Silko's novel *Ceremony* (New York: Penguin Books, 1987), 138.

81 For a multidisciplinary collaborative work on the history of human-induced environmental change, including its contemporary consequences and implications, see Will Steffen et al., *Global Change and the Earth System: A Planet under Pressure* (Berlin: Springer, 2004).

82 Kevin Howley, ed., "Introduction," *Media Interventions*, 3–5. See, for example, Anita Howarth's work in the same volume on "Newspaper Campaigning in Britain in the Late 1990s," 37–54.

83 Victor Pickard, "The Postwar Media Insurgency: Radio Activism from Above and Below," in Howley, ed., *Media Interventions*, 249–50.

84 See Couldry's "Afterword" in Howley, ed., *Media Interventions*, 400.

85 Howley, ed., *Media Interventions*, 397.

86 In Jeff Jay and Debra Jay's popular-selling book, *Love First: A Family's Guide to Interventions*, 2nd edition (Center City, MN: Hazelden Books, 2008), they define "interventions" as being about loved ones revealing "the truth to the addicted person in such a way that he will hear the message and say 'yes' to treatment" (4).

87 Howley, ed., *Media Interventions*, 3.

88 Korten, *Change the Story*, 24.

89 The term "contrapuntal" is closely associated in cultural studies with the work of postcolonial theorist Edward Said, who employs a "contrapuntal reading" of texts as a way of reading together the simultaneities and subtleties of both imperialism and resistance in action. Here, I am using the word in its melodic and choreographic sense, especially as story is told through the medium of dance, but I am also indebted to Said's creative and insightful use of the term. See Said's *Culture and Imperialism* (London: Vintage, 1993), 66.

90 These very real ethical concerns are sobering and raised by Mark Andrejevic, *ISpy*; Christian Fuchs, *Social Media: A Critical Introduction* (Los Angeles: Sage, 2013); Mara Einstein, *Black Ops Advertising: Native Ads, Content Marketing, and the Covert World of the Digital Sell* (New York: OR Books, 2016); Mele, *The End of Big*; Vincent Mosco, *Becoming Digital: Toward a Post-Internet Society* (Bingley, UK: Emerald Publishing, 2017); and John Cheney-Lippold, *We Are Data: Algorithms and the Making of Our Digital Selves* (New York: NYU Press, 2017), among a growing literature on surveillance culture and the use of digital algorithm manipulations to exploitative and even criminal ends. For detailed discussion of the power dimensions of the "panopticon" as a tool of surveillance and control, as theorized in the work of Michel Foucault, see *Surveiller et Punir: Naissance de la Prison* (Paris: Gallimard, 1975). On the subject of contrapuntal movement and dance, I am indebted to New York City–based professional dancer/choreographer Stephanie Schwartz for her generous consultation.

CHAPTER 2. FIFTY SHADES OF GREEN

1 Belinda Luscombe, "The World's 100 Most Influential People: 2012," *Time*, April 18, 2012, http://content.time.com.

2 Emma Green, "Consent Isn't Enough: The Troubling Sex of Fifty Shades," *Atlantic*, February 10, 2015, www.theatlantic.com.

3 Saba Hamedy, "Fifty Shades Sets Record at Box Office," *Los Angeles Times*, February 15, 2015, www.latimes.com.

4 See Jenkins, *Convergence Culture*, 175–216. See also Karen Hellekson and Kristina Busse, eds., *Fan Fiction and Fan Communities in the Age of the Internet* (Jefferson, NC: McFarland, 2006).

5 Hellekson and Busse, *Fan Fiction Studies Reader*, 8–9.

6 James's extreme profiting from the novels is controversial in the fan fiction community precisely because her work was produced in collaboration with other writers, who did not benefit from the series' sales, and her *Fifty Shades* story was highly derivative of another fan fiction writer's story. See Gregory Miller, "Fan Fiction Writers Speak Out against *Fifty Shades of Grey*," *New York Post*, February 7, 2015, http://nypost.com; and a now famous "Reddit Origins Essay" posting (hosted and archived at Fanlore.org) penned by a member of James's former community of writers who tells the story of James exploiting and profiting from the free labor of other fan fiction writers. See "*Fifty Shades of Grey*: The Reddit Origins Essay," Fanlore, July 30, 2014, http://fanlore.org.

7 Northrop Frye, *The Great Code: The Bible and Literature* (Toronto: University of Toronto Press, 2006), 238–39. See also Neil Postman's explication of the importance of "resonances" to the study of media and culture in *Amusing Ourselves to Death: Public Discourse in the Age of Show Business* (New York: Penguin, 2005), 17–18. Postman explains that, for Frye, "metaphor is a generative force" and defines "resonance" as "metaphor writ large."

8 E. L. James, *Fifty Shades Darker* (New York: Vintage, 2012), 329. Note that, here, I am not denying the pleasure that many women receive in reading romance novels that include violence toward women. James's sales figures demonstrate clearly a market for her wares. For ethnographic and focus group study of women's pleasure in reading the *Fifty Shades* series, see Melissa Click, "Fifty Shades of Postfeminism: Contextualizing Readers' Reflections on the Erotic Romance Series," in Elana Levine, ed., *Cupcakes, Pinterest, and Ladyporn: Feminized Popular Culture in the Early Twenty-first Century* (Urbana: University of Illinois Press, 2015), 15–31. My intent is also not to dismiss or trivialize the value of the romance genre for women. Janice Radway has skillfully shown the real and important work this genre does in the lives of women in *Reading Romance: Women, Patriarchy, and Popular Literature* (Chapel Hill, NC: University of North Carolina Press, 1984). What I do find worth examining, however, is the "talking back" that readers do to James via new media as they challenge her in online forums, making it clear that resistant readers are not buying what she is selling in terms of abuse and stalking packaged as love story. The *Fifty Shades* series reached its apotheosis in the years before the ascent of "pussy-grabbing" Donald Trump to power and the emergence of the #MeToo era. I do wonder how enthusiastically the series would have been received if introduced to women at this later time period, when a broader media "lid" was pulled back on persistent societal dynamics of misogyny present all along, thus bringing them to the fore in wider public consciousness.

9 Susan Donaldson James, "BDSM Advocates Worry about 'Fifty Shades of Grey' Sex," *ABC News*, October 2, 2012, http://abcnews.go.com.

10 Green, "Consent Isn't Enough." BDSM educators also take issue with James's portrayal of Grey's inclinations toward BDSM as indicative of mental illness or a

pathology resulting from his childhood abuse. These educators stress that BDSM-related "kink" and "play" can be entered into safely and maturely and can be part of a fulfilling, healthy sex life, characterized by meaningful intimacy and good communication. However, this is not how BDSM is portrayed in the series.

11 Those who visit the web site, and sign up to become "interns" for Mr. Grey, end up signing onto an application that allows them to complete assignments for Mr. Grey and earn rewards. They also receive publicity notices and materials: http://www.greyenterprisesholdings.com. Home Box Office's series *True Blood* famously used similar alternative reality viral marketing, fostering and capitalizing upon fan participation in the story world by marketing real bottles of faux "Tru Blood" and creating a whole host of "in-world" reality vampire web sites, blogs, and vlogs to publicize the series and excite the curiosity of fans.

12 Melissa Locker, "7 Leadership Lessons from 'Fifty Shades of Grey,'" *Fortune*, February 12, 2015, http://fortune.com; David M. Ewalt, "#8: Christian Grey" from "Forbes 2013 Fictional 15," *Forbes*, July 31, 2013, http://www.forbes.com.

13 Julia LaRoche, "27-Year-Old Seattle Man Says That Women Fly across the Country to Be with Him Because He's a Real-Life Christian Grey," *Business Insider*, February 20, 2015, www.businessinsider.com.

14 Frye, *The Great Code*, 238–39.

15 See James Ridgway, "BP's Slick Greenwashing," *Mother Jones*, May 4, 2010, www.motherjones.com.

16 "BP Supports the Student Conservation Association," BP web site promotional video: www.bp.com (accessed January 25, 2016).

17 "Meeting the Energy Challenge," BP Global web site, www.bp.com (accessed January 25, 2016).

18 Article from BP's global corporate web site on Gulf of Mexico restoration. See http://www.bp.com/en/global (accessed September 16, 2015). Since I accessed this site, BP Global has shifted the positioning of the Gulf of Mexico restoration material out of the menu and to the very bottom of its "Sustainability" page, but now the web site actually shows a photo of the platform on fire while a rescue barge delivers "capping equipment" to address the problem. Their "Gulf of Mexico" restoration section, however, shows pretty photos of pelicans and affable-looking, cooperative fisherman conversing civilly with BP employees. See "Gulf of Mexico Restoration" under the subheading "Commitment to the Gulf of Mexico Restoration" at www.bp.com/en_us (accessed January 25, 2016). See also Amy Connelly, "Study: BP Oil Spill Left Millions of Gallons Buried in Gulf Floor," UPI (United Press International News Service), January 31, 2015, http://www.upi.com.

19 See, for instance, "The Problem Becomes the Solution," Chevron "Human Energy" print advertisement, featured in "Critical Thinking Resources: Ads," Education for Sustainable Development, September 6, 2010; and "Chevron Protects Sea Turtles on Barrow Island," Chevron YouTube channel, September 11, 2012, www.youtube.com (accessed December 3, 2018).

20 Chevron Corporation's official greeter page greets visitors with the headline "The Power of Human Energy: Finding Newer Cleaner Ways to Power the World," http://www.chevron.com (accessed January 25, 2016).

21 Ann Mulkern, "Oil and Gas Industry Sets Spending Record for Lobbying in 2009," *New York Times*, February 2, 2010, http://www.nytimes.com. See also "chevron rap sheet" at www.corp-research.org and senior legal analyst and international legal correspondent Michael Goldhaber's *Crude Awakening: Chevron in Ecuador* (New York: Rosetta Books, 2014). Due to space limitations, I have highlighted some of the critiques of Chevron and BP's greenwashing practices, but this is not to imply that other oil companies have been free of such criticisms. At the time of this writing, Exxon Mobil Corporation was under criminal investigation in two states for allegedly lying to the public and to its shareholders about what they knew and when they knew it with regard to the catastrophic effects of climate change and the oil industry's role in contributing to those effects. The company was also being investigated for its alleged actions to suppress and discredit climate change scientific research.

22 For a more detailed discussion of "greenwashing" public relations campaigns on the part of energy companies and other corporate polluters, see Fred Pierce, "Greenwash: BP and the Myth of a World 'beyond Petroleum,'" *Guardian*, 20 November, 2008, http://www.theguardian.com; Justin Elliott, "How to Sell Big Oil on the Web: A New Media Marketing Firm Creates a Novel Greenwashing Strategy for Chevron," *Salon.com*, July 19, 2011, http://www.salon.com; "Greenwash 101," http://thegreenlifeonline.org (accessed September 11, 2014). Each of these campaigns is predicated on trading heavily on the capital produced by cultivating the popular perception of ecopiety in corporate practices.

23 Deborah Whitehead's excellent study of religious "mommy bloggers" who become fans of *Fifty Shades of Grey* reveals how these women's interpretations of the novels are shaped by the fan culture surrounding *Fifty Shades* even as they themselves shape that culture, adding religious and moral meaning to women's godly submission to their husband's authority or headship as an extension of God's authority. Where some readers might read the book series and find a disturbing portrayal of an abusive relationship, some conservative Christian and Mormon women find affirmation of "male headship" in Christian households as both blessed and erotic. See Whitehead, "When Religious 'Mommy Bloggers' Met 'Mommy Porn': Evangelical Christian and Mormon Women's Responses to *Fifty Shades*," *Sexualities* 16, no. 8 (December 3013): 915–31.

24 Japp et al., eds., *Communication Ethics*, 2.

25 Alison Takeda, "E. L. James' *#AskELJames* Twitter Q&A Event Would Have Made Christian Grey Cringe," *Us*, June 29, 2015, http://www.usmagazine.com; Denise Florez, "E. L. James Event Backfires When 'Grey' Critics Air Grievances Using #AskELJames," *Los Angeles Times*, June 29, 2015, www.latimes.com.

26 Steve Schmadeke, "Prosecutors: UIC Student Charged with Assault Said He Was Re-Enacting 'Fifty Shades of Grey,'" *Chicago Tribune*, February 14, 2015. The ac-

cused student was later cleared by a judge, who upon determining that there had been *previous* consensual sexual relations between the two students, dismissed the young woman's charge that she had been forcibly tied up with two belts, a tie stuffed in her mouth that prevented her screams from being heard, beaten with a third belt, and then raped by the student in question. The judge ruled that this encounter had indeed been consensual despite the young woman's protestations to the contrary.

27 Of course the narrative of *Fifty Shades* clearly *does* appeal to many women, and we should take that appeal and the pleasure readers derive from the *Fifty Shades* fantasy seriously. See Radway's previously cited work on the genre of romance. In the case of *Fifty Shades*, however, it also bears keeping in mind the simultaneity of the "hate reading" phenomenon, in which readers check out a popular book to see what all the media buzz is about, despise the book, but derive pleasure from "hate reading" it, which offers the "pay off" later of being able to deride and eviscerate the work with greater specificity and sense of authority. On the phenomenon of "hate watching" and "hate reading," see Darren Franich, "The Rise of Hate Watching: Which TV Shows Do You Hate to Despise?" *Entertainment Weekly*, August 16, 2012, http://www.ew.com; Wendy Braun, "The Big Wedding: Why I Hate-Watch Films for Research," *Humanities and Social Sciences Online*, October 11, 2014, https://networks.h-net.org; Samantha Vincent, "Entering the World of Smut: Hate-Reading 'Fifty Shades of Grey,'" *Bust*, February 2012, http://bust.com.

28 Walter Benjamin, "The Storyteller," in *Illuminations*, 83–84, 100.

29 Shirky, *Here Comes Everybody*, especially chapter 13, "Everyone Is a Media Outlet," 55–80; Jenkins et al., *Spreadable Media*.

30 Jenkins, *Convergence Culture*, 186.

31 There are many examples of various "remakes" and "retellings" of *Fifty Shades* that have germinated in the ecologies of social media, including the Lego Movie version of *Fifty Shades*—think plastic figurines with blindfolds and a snap-together brick-constructed "Red Room of Pain." The satiric film *Fifty Shades of Black* highlights the absurdity of "rich white people problems," the self-congratulatory egotism of Christian Grey, and Ana's mind-boggling doormat docility as portrayed by James in the *Fifty Shades* novel. See *Fifty Shades of Bricks* at https://www.youtube.com/watch?v=S7AvZPTT4kU; and *Fifty Shades of Black*, www.imdb.com. There is a gender reversal depicted in the *Fifty Shades of Pink* Barbie video, in which Barbie is the dominant. A *Fifty Shades of Frozen* version and countless others engage in critiques of the simplistic notions of race (*Fifty Shades of Black*), gender and power (*Fifty Shades of Pink*), and class (*Fifty Shades of Broke*) engendered in the James novels. This does not begin to approach the Facebook discussion groups, fan sites, hate-consumption sites, blogs, vlogs, and now *Fifty Shades of Grey* ongoing fan fiction on the Internet. In short, there is no "Finis" to any of this in Benjamin's sense.

32 "Out Damned Logo! Shakespearean Flashmob," *BP or Not BP? Administrative News*, October 27, 2012, http://bp-or-not-bp.org.

33 "Surprise Theatrical Protest inside British Museum over BP Sponsorship of Shakespeare and the Olympics," *BP or Not BP? Administrative News*, July 22, 2012, http://bp-or-not-bp.org; Kevin Rawlinson, "Activists Occupy British Museum over BP Sponsorship," *Guardian*, September 13, 2015, www.theguardian.com.
34 For arguments that media are tools of anesthetization, pacification, and control, see Noam Chomsky and Edward Herman, *Manufacturing Consent: The Political Economy of the Mass Media* (New York: Vintage, 1995); and Neil Postman's cautionary *Amusing Ourselves to Death*. For a more optimistic view, see Jenkins's work on flashmob digital activism and the Harry Potter Alliance: "How Dumbledore's Army Is Transforming Our World," *HenryJenkins.org*, July 27, 2009. Analysis of the swift and spontaneous organizational power of the Black Lives Matter movement, the tools to publicize police brutality, and the greater mobility and reach of civil rights activism made possible by digital social media can be found in Bijan Stephen, "Get Up, Stand Up: How Black Lives Matter Uses Social Media to Fight the Power," *Wired*, November 18, 2015, www.wired.com.
35 Katherine Hayles, *How We Think: Digital Media and Contemporary Technogenesis* (Chicago: University of Chicago Press, 2012), 1.
36 Bennett and Royle, *An Introduction to Literature*, 54. Emphasis is mine.
37 "Conspicuous consumption" is a term coined by economist/sociologist Thorstein Veblen in *The Theory of the Leisure Class* (1899) to refer to the practice of spending more money on goods than they are worth, a practice often associated with the lifestyles and customs of the nouveau riche.
38 E. L. James, *Grey: Fifty Shades of Grey as Told by Christian* (New York: Penguin, 2015), 11–12.
39 Benoit Monin and Dale Miller, "Moral Credentials and the Expression of Prejudice," *Journal of Personality and Social Psychology* 81, no. 1 (July 2001): 33–43.
40 Ibid. Monin and Miller describe this more specifically as "self-licensing through moral credentials" and specifically examine how people more readily expressed prejudiced attitudes if they already established credentials as nonprejudiced people (34). The classic example of this would be the phrase, "Some of my best friends are black, but . . ."
41 Anna Merritt, Daniel Effron, and Benoit Monin, "Moral Self-Licensing: When Being Good Frees Us to Be Bad," *Social and Personal Psychology Compass* 4, no. 5 (2010): 349.
42 Merritt et al., "Moral Self-Licensing," 349.
43 Merritt et al., "Moral Self-Licensing," 348.
44 Matthew Harding and David Rapson, "Does Absolution Promote Sin? The Conservationist's Dilemma." Paper presented jointly at the February 2, 2011, Stanford Institute for Theoretical Economics; David Rapson, Ken Gillingham, and Gernot Wagner, "The Rebound Effect and Energy Efficiency Policy," *Review of Environmental Economics and Policy* 10, no. 1 (2016): 68–88.
45 Tom Crompton and Tim Krasser, *Meeting Environmental Challenges: The Role of Human Identity* (Cambridge, UK: Green Books, 2009).

46 See Robert Gifford and Reuven Sussman, "Environmental Attitudes," in Susan D. Clayton, ed., *Oxford Handbook of Environmental and Conservation Psychology* (New York: Oxford University Press, 2012), 65–80. In the same volume, see also Linda Steg and I. M. de Groot's review of the cumulative research conducted on environmental values, 81–91. For a study of the gap between pro-environmental attitudes as self-reported in quantitative measurement scales and actual behavior that would suggest the translation of values into action, see Anja Kollmuss and Julian Agyeman's "Mind the Gap: Why Do People Act Environmentally and What Are the Barriers to Pro-Environmental Behavior?" *Environmental Education Research* 8, no. 3 (2002): 240–60.

47 Roddy Sheer and Doug Moss, "Use It and Lose it: The Outsize Effect of U.S. Consumption on the Environment," *Scientific American*, September 14, 2012, http://www.scientificamerican.com.

48 "Strong Enough," copyright Sheryl Crow, 1994, published on her *Tuesday Night Music Club* album. Full listing of lyrics at http://www.azlyrics.com.

49 Samira Mehta, "Big Vampire Love: What's So Mormon about *Twilight*?" *Religious Dispatches*, December 3, 2009, http://religiondispatches.org; Jana Riess, "Yes, Robert Pattison, There Really Are Mormon Themes in *Twilight*," *Beliefnet*, July 2010, www.beliefnet.com; John Granger, "Mormon Vampires in the Garden of Eden," *Touchstone*, November/December 2009, www.touchstonemag.com; Ashley Fetters, "At Its Core *Twilight* Is about ———," *Atlantic*, November 5, 2012, www.theatlantic.com.

50 Theories as to derivative sources for the Book of Mormon are too numerous to address within the scope of this project. One source often mentioned is Reverend Solomon Spalding's historical novel, *Manuscript Found*, composed around 1810–1812. Whether this specific case is so or not, the larger point is that rewrites, remixes, "mashups," and even "fanfic" are certainly not *new* phenomena in the cultural history of storytelling. The "ownership" of story can be tricky, and the "viral" and transformative nature of storytelling is arguably endemic to its success.

51 See Edward Navas and Edward Gallager, *The Routledge Companion to Remix Studies* (New York: Routledge, 2014).

52 This point is debated within media history, with some historians arguing that there is nothing new under the sun and other observers failing to take cognizance of historical media precedents to the dynamics of new media. I find Clay Shirky's efforts to split the difference between these two positions useful, as he takes differences in speed, scope, and scale into account. He explains,

> The social urge to share information is not new. Prior to e-mail and weblogs, we clipped articles and published family newsletters. Recalling these old behaviors, it's tempting to conclude that our new tools are merely improvements on existing behaviors; this view is both right and wrong. The improvement is there, but it is an improvement so profound that it creates new effects. Philosophers sometimes make a distinction between a difference in degree (more of the same) and a difference in kind (something new). What

we are witnessing today is a difference in the degree of sharing so large that it becomes a difference in kind.

Shirky, *Here Comes Everybody*, 149.

53 Kirby Ferguson, *Everything Is Remix* (Remastered), September 12, 2015, https://vimeo.com; "Is Everything Remix?" *NPR TED Talk Radio Hour*, June 27, 2014, www.npr.org.

54 Linda Watson, *Wildly Affordable Organic: Eat Fabulous Food, Get Healthy, and Save the Planet—All on $5 Dollars a Day or Less* (Philadelphia: Da Capo Press, 2011).

55 See Linda Watson's greeter page on her blog, "Organic, Local, and Sustainable Cooking for Any Budget," *Cookforgood.com* (accessed February 26, 2019); and Andrea Weigl, "Frugal Meals Could Convert You," *Raleigh News and Observer*, June 3, 2009, https://www.newsobserver.com; Watson, *Fifty Weeks of Green: Romance and Recipes* (Raleigh, NC: Cook for Good, 2013), 20.

56 Linda Watson, "Why a Farmer Stars in My Response to *Fifty Shades of Grey* with a Recipe," *Mother Earth News*, February 17, 2015, http://www.motherearthnews.com.

57 Cole Mellino, "Climatarian Makes *New York Times* List of Top New Food Words for 2015," *EcoWatch*, December 22, 2015, http://ecowatch.com; Kate Yoder, "What on Earth Is a Climatarian?" *Grist*, December 16, 2015, http://grist.org. "Climatarians" eat food that has been produced locally and with few "fossil fuel miles," they strategize their eating habits in order to minimize any food waste, and they avoid meat from animals whose husbandry and processing consumes a lot of energy. Discussions and postings from and about climatarians can be followed on Twitter.com at #climatarian.

58 Watson, *Fifty Weeks of Green*, 21.

59 Watson, *Fifty Weeks of Green*, 153.

60 Watson, *Fifty Weeks of Green*, 154.

61 Watson, "Why a Farmer Stars."

62 Watson, "Why a Farmer Stars."

63 Michael Pollan's May 7, 2006, opinion piece, "Voting with Your Fork," was seminal in conversations about food system reform. See *New York Times*, http://pollan.blogs.nytimes.com. Pollan's book likewise presented the practice of small-scale food piety and the power of conscious consumerism as having the greatest efficacy in "changing the way we eat" and having a tangible impact. This stance is inflected by the Libertarian farmer Pollan features in the book. See *Omnivore's Dilemma: A Natural History of Four Meals* (New York: Penguin Press, 2006).

64 Pollan, "Voting with Your Fork."

65 Joe Fassler, "Michael Pollan's New Dilemma: Voting with Your Fork, He Says, Is So 2006; So What Comes Next?" *New Food Economy*, June 7, 2016, http://newfoodeconomy.com.

66 Ryan Hagan, "16 Sustainability Leaders Weigh In: How You Can Help to Reverse Global Warming," Crowdsourcing Sustainability, January 12, 2019, www.crowdsourcingsustainability.com (accessed February 26, 2019).

67 Watson, *Fifty Weeks of Green*, 148.
68 Watson, *Fifty Weeks of Green*, 124.
69 W. Somerset Maugham, "The Appointment in Samarra," in Alton Chester Morris, *Imaginative Literature; Fiction, Drama, Poetry* (San Diego: Harcourt, Brace, Javonovich, 1972), 8.
70 Gernot Wagner, "Going Green but Getting Nowhere," *New York Times*, September 7, 2011, www.nytimes.com.
71 See Gernot Wagner, *But Will the Planet Notice? How Smart Economics Can Save the World* (New York: Hill and Wang, 2011).
72 Wagner, "Going Green but Getting Nowhere."
73 Wagner, *But Will the Planet Notice*, 137–38.
74 Gernot Wagner and Martin Weitzman, "The Planet Won't Notice You Recycle and Your Vote Doesn't Count: But You Should Do It Anyway; The Definitive Guide to Screaming at, Coping with, and Profiting from Climate Change," *Salon*, March 29, 2015, http://www.salon.com.
75 Wagner and Weitzman, "The Planet Won't Notice."
76 J. Robert Hunter, *Simple Things Won't Save the Earth* (Austin: University of Texas Press, 1997), 2. In the "simple things" book genre, see Earthworks Group, *50 Simple Things You Can Do to Save the Earth* (New York: Hyperion, 1989); Earthworks Group, *50 Simple Things That Kids Can Do to Save the Earth* (Kansas City, KS: John Javna, 1990); and more recently, among others, Brangien Davis's humorous *Wake Up and Smell the Planet: The Non-Pompous, Non-Preachy Grist Guide to Greening Your Day* (Seattle: Skipstone, 2012). Gar Smith's "50 Difficult Things You Can Do to Save the Earth" can be found reprinted online from the original 1995 issue of *Earth Island Journal*: http://www.insular.com.
77 Derrick Jensen, *As the World Burns: 50 Simple Things You Can Do to Stay in Denial* (New York: Seven Stories Press, 2007).
78 As quoted from a November 5, 2008, *Newsweek* web exclusive in Wagner, *But Will the Planet Notice?* 12.
79 Sarah Fredericks, "Online Confessions of Eco-Guilt," *Journal for the Study of Religion, Nature, and Culture* 8, no. 1 (March 2014): 69.
80 Wagner, *But Will the Planet Notice?* 7.
81 Kristen Lopez, "How 'Fifty Shades Freed' Plays in the #MeToo Era," *Hollywood Reporter*, February 10, 2018, www.hollywoodreporter.com.
82 Lopez, "How 'Fifty Shades Freed' Plays."

CHAPTER 3. "I CAN'T! IT'S A PRIUS"

1 "Toyota Reveals Third-Generation Marketing Campaign: Harmony between Man, Nature, and Machine," Toyota USA Newsroom, May 11, 2009, http://pressroom.toyota.com. For a definition of what constitutes a "cultural icon" and the process by which a particular brand becomes one, see Douglas Holt, *How Brands Become Icons: The Principles of Cultural Branding* (Boston: Harvard Business School Press, 2004). Holt argues that certain brands generate "identity myths" that tap into

powerful symbolic resonances within the culture and soothe consumer anxieties about cultural change and uncertainty. These brands, such a Coca-Cola, Nike, Budweiser, or Harley-Davison, take on iconic status through a storying process that is often more intuitive than strategic in its deep cultural resonance.

2 Randy Kent, "Prius: Lover's Lane," YouTube.com, September 8, 2008, www.youtube.com/watch?v=NkWkd1zGGjo (accessed December 21, 2016); and see Zeropoint30 production company full commercial reel, including both "Lover's Lane" and "Heist" at www.zeropoint30productions.com (accessed December 21, 2016).

3 Kent, "Heist" (also posted at Vimeo.com, YouTube.com, and FunnyorDie.com).

4 See Josh Holt's "yuppie" characters Beatrice and Brody in his "Prius Lovers," and his viewers' comments in the "All Comments" section at www.youtube.com/watch?v=dufeipn8Lis. The "Prius Lovers" spot is also featured on DIY prosumer media sites such as DIY Factory, www.diyfactory.org, and AIOHOW.com, as well as standard media upload sites such as YouTube.com and Vimeo.com (accessed December 21, 2016).

5 The marketing web site LOHAS.com states that "Lifestyles of Health and Sustainability (LOHAS) describes an estimated $290 billion U.S. marketplace for goods and services focused on health, the environment, social justice, personal development and sustainable living. The consumers attracted to this market represent a sizable group in this country. Approximately 13–19% percent of the adults in the U.S. are currently considered LOHAS Consumers. This is based on surveys of the U.S. adult population estimated at 215 million." See the web site's "About" section on LOHAS background: www.lohas.com. The story of the Whole Foods video and its production can be found at the collective's web site, FogandSmog.com.

6 Delia Brown, "Whole Foods Parking Lot Rap Response: Revenge of the Black Prius," Youtube, July 5, 2011, www.youtube.com/watch?v=idG_Odfk9fI (accessed February 10, 2019). Brown is a painter and multimedia visual artist who specializes in representations of American wealth and entitlement. See www.deliabrown.net for further examples and descriptions of her work.

7 For a synthesis of the history of attitudes toward moral consumption since the 1700s, see Matthew Hilton, "The Legacy of Luxury: Moralities of Consumption since the 18th Century," *Journal of Consumer Culture* 4 (March 2004): 101–23.

8 See *South Park*, season 10, episode 2, "Smug Alert," www.youtube.com/watch?v=AnFAAdOBB1c (accessed December 15, 2016).

9 Mukherjee and Banet-Weiser, eds., *Commodity Activism*, 1–4.

10 "Through resonance," says Frye, "a particular statement in a particular context acquires a universal significance." As quoted in Postman, *Amusing Ourselves to Death*, 17.

11 Matthew Feinberg and Robb Willer, "The Moral Roots of Environmental Attitudes," *Psychological Science* 20, no. 10 (2012): 6. Feinberg and Willer's research builds on and extends social psychologist Jonathan Haidt's studies on liberal and conservative moral matrices. See Haidt, *The Righteous Mind*, particularly part 3 on how "morality binds and blinds," 219–319. The terms "liberal" and "conserva-

tive" in this context refer to political understandings of these terms within the United States.

12 Feinberg and Willer, "The Moral Roots of Environmental Attitudes," 6.

13 Feinberg and Willer, "The Moral Roots of Environmental Attitudes," 1.

14 See Pierre Bourdieu, *Distinction: A Social Critique of the Judgment of Taste* (Cambridge, MA: Harvard University Press, 1984), 188.

15 Krulik describes the ads on his web site, saying, "In a sea of virtually identical 'green' automobile campaigns, my goal was simply to do something different for the category. The result was an unexpectedly controversial viral campaign that spread across the Internet and overseas within a week. It was featured on many advertising and environmental blogs like Adfreak, TreeHugger, AutoblogGreen, and the *Huffington Post*." See Krulik's official site for the accompanying images, http://www.krulik.net.

16 Wagner, *But Will the Planet Notice*, 7.

17 Merritt et al., "Moral Self-Licensing," 349.

18 Mukherjee and Banet-Weiser, *Commodity Activism*, 1–3.

19 Wagner, *But Will the Planet Notice*, 181.

20 Wagner writes, "Owning a Prius makes you feel good about driving and also makes driving cheaper exactly because you use less fuel, so you may actually drive more. One effect—buying the more fuel-efficient car—may, in part, cancel out the other: driving less." As myself a driver of a plug-in hybrid (not a Prius), I am repeatedly "reality checking" would-be feelings of consumer virtue with Wagner's work in mind. Wagner, *But Will the Planet Notice*, 158.

21 Wagner, *But Will the Planet Notice*, 4.

22 Niclas Rolander et al., "The Dirt on Clean Electric Cars," *Bloomberg.com*, October 15, 2018.

23 Kendra Pierre-Louis, *Green Washed: Why We Can't Buy Our Way to a Green Planet* (New York: Ig Publishing, 2012), 78–79.

24 Willett Kempton, and coinvestigator cognitive anthropologists "mapped" many of these cultural resonances in their 1990s study of American attitudes and values about the environment. See Willett Kempton, James Boster, and Jennifer Hartley, *Environmental Values in American Culture* (Woburn, MA: MIT Press, 1996).

25 Marius Luedicke, Craig Thompson, and Markus Giesler, "Consumer Identity Work as Moral Protagonism: How Myth and Ideology Animate a Brand-Mediated Moral Conflict," *Journal of Consumer Research* 36 (April 2010): 1017–19. The researchers focus on the cult-like brand loyalty of Hummer owners and how this consumer brand loyalty is tied to culturally resonant American ideologies of the pioneer spirit and rugged individualism. Owners also position themselves against their environmentally activist adversaries and in so doing conceive of themselves as holding the moral high ground.

26 Luedicke et al., "Consumer Identity Work," 1016.

27 Stuart Hall, "Encoding, Decoding," in Simon During, ed., *The Cultural Studies Reader* (New York: Routledge, 1993), 98–100.

28 John Fiske, *Understanding Popular Culture*, 2nd edition (New York: Routledge, 2007), 83–85. For a more in-depth exploration of the strengths and weaknesses of Hall and Fiske's models, see Katherine Sender, "Selling Sexual Subjectivities: Audiences Respond to Gay Window Advertising," *Critical Studies in Mass Communication* 16 (1999): 175–76.

29 CarlockToyotaTupelo, "2010 Toyota Prius Harmony Commercial," YouTube, www.youtube.com/watch?v=k9LqWd3kkkM (accessed December 19, 2016).

30 CarlockToyotaTupelo, "The Making of the Prius Harmony TV Commercial," YouTube, www.youtube.com/watch?v=sAeIET4uufE (accessed December 19, 2016).

31 "Toyota Reveals Third-Generation Marketing Campaign."

32 Mukherjee and Banet-Weiser, *Commodity Activism*, 2.

33 "Toyota Prius 'Harmony,'" Green Washing Index, August 25, 2009, www.greenwashingindex.com; Sharon Silky Carty, "Greenwashed Car Ads Make People Feel Good about Polluting," *Huffington Post*, April 20, 2012, www.huffingtonpost.com. See also the Media Bites vlog critique of the "Harmony" commercial at www.youtube.com/watch?v=H7d7b4Hwf7M (accessed April 19, 2017); Jed Greer and Kenny Bruno's foundational exposé of greenwashing, *Greenwash: The Reality behind Corporate Environmentalism* (New York: Rowman and Littlefield, 1997); Frances Bowan, *After Greenwashing: Symbolic Corporate Environmentalism and Society* (Cambridge: Cambridge University Press, 2014); and Bruce Johansen, *Eco-Hustle: Global Warming, Greenwashing, and Sustainability* (Santa Barbara, CA: Praeger, 2015).

34 Paul Rauber, "Pollution Porn Is Now a Thing: Take That, Prius Drivers," *Sierra*, July 7, 2014, www.sierraclub.org (accessed January 21, 2017).

35 The public Facebook page for "Let the Coal Roll" lists 407,208 "likes" and 408,570 "followers," www.facebook.com (accessed November 27, 2018).

36 Tractor pulls, truck pulls, and monster truck rallies predate and constitute relevant antecedents to contemporary self-proclaimed "pollution porn" practices that are now recorded and circulated via the tools of new media and digital culture.

37 The following are nonrestricted: Pollution Porn Tumbler, http://pollutionporn.tumblr.com/; "Coal Rolling Rednecks," www.facebook.com; "Coal Roll'n Diesel," www.facebook.com; "Rolling Coal Community," www.facebook.com; "Coal Roller," www.facebook.com; and www.pollutionporn.com (accessed January 21, 2017).

38 Shirky, *Here Comes Everybody*, 54.

39 Sarah Banet-Weiser, *Empowered: Popular Feminism and Popular Misogyny* (Durham, NC: Duke University Press, 2018).

40 Howley, ed., *Media Interventions*, 1–7.

41 See the ongoing discussion, beginning with the 2016 "There's Nothing Wrong with Rolling Coal" thread and its responses in the "Automotive" forum in the collection of blogs and forums hosted at TheBeerBarrel.net, which defines itself as "a free speech community for even the most radical ideologue." Anything pretty much goes on this site, and it attracts mostly alt right sorts of topics and postings.

42 Outlaw (with Bottleneck), "Yuppie Folks Can't Survive," YouTube, October 15, 2018, www.youtube.com/watch?v=GVvgsV_NtFY (accessed July 20, 2018).
43 Chambers, as cited in Bennett and Royale, *An Introduction to Literary Criticism*, 59.
44 See a variety of "Hunting Pwius" Internet memes by typing "Hunting Pwius" in Google and then clicking on "Images for Hunting Pwius" (accessed February 8, 2019).
45 Not coincidentally, the kind of hypermasculinity asserted in Prius hunting/predation videos and memes are reminiscent of "gay bashing" images, especially as they deride the Prius for its effeminate, "sissy" qualities. One popular meme, for instance, shows an innocuous-looking Prius captioned by the following: "My friend said his Prius was leaking . . . I asked him if he wanted a tampon." To locate this image type "Hunting Pwius" in Google and then click on "Images for Hunting Pwius" (accessed December 21, 2016).
46 The online Urban Dictionary defines "nuffy" as a "retarded" or "stupid person." Some attribute the term to Australian slang in origin. See "Nuffy," *Urban Dictionary*, May 24, 2004, www.urbandictionary.com (accessed December 21, 2016). Religion and popular culture scholar Jeremy Biles provides a rich, in-depth analysis of the visual representations of class antagonism and diesel trucks versus conventional automobiles as performed in the spectacle of monster truck rallies in his ethnographic study, "Sunday! Sunday! Sunday! The Monster Trucks' Black Sabbath," University of Chicago Religion and Culture Web Forum, January 2005, https://divinity.uchicago.edu (accessed December 21, 2016).
47 Roaklin, "F U Self Righteous Prius Lovers," YouTube, June 18, 2012, www.youtube.com/watch?v=hZnhEKq01tA (accessed December 21, 2016).
48 Roaklin, "F U Self Righteous Prius Lovers."
49 Hall, "Encoding, Decoding," 93.
50 David Weigel, "Rolling Coal: Conservatives Who Show Their Annoyance with Liberals, Obama, and the EPA by Blowing Black Smoke from Their Trucks," *Slate*, July 3, 2014, www.slate.com (accessed December 21, 2016).
51 Haidt, *The Righteous Mind*, 204–5.
52 See, for example, Twitter, https://twitter.com/MrsAwwsum/status/646673266791198720; *ABC News Business Report* also featured a number of "guilty" Volkswagen owners, including Ari Levin, a New York–based diesel Jetta owner who filed a class-action suit against the auto manufacturer and reported "feeling guilty now about taking his Jetta out on the road." See Susanna Kim, "NY Man Sues Volkswagen over Emissions Scandal: 'Every Reason That I Bought the Car Was Based on a Lie,'" *ABC News*, September 22, 2015, http://abcnews.go.com; and Richard Conniff, "Me and My Jetta: How VW Broke My Heart," *New York Times*, September 25, 2015, www.nytimes.com.
53 Erica Shemper, "Sins of Emission: Playing My Part in the Volkwagen Scandal," ThinkChristian, September 28, 2015, http://thinkchristian.reframemedia.com.
54 Shemper, "Sins of Emission"; and Noe, "Clean Diesel: Too Good to Be True," National Public Radio, September 25, 2015, www.npr.org (accessed December 21, 2016).

55 For a guide to varieties and subgenres of pornography, see Shira Tarrant, *The Pornography Industry: What Everyone Needs to Know* (New York: Oxford University Press, 2016).
56 Andrew Ross, *No Respect: Intellectuals and Popular Culture* (New York: Routledge, 1989), 194, and chapter on "The Popularity of Pornography," 171–208.
57 Constance Penley, "Crackers and Whackers: The White Trashing of Porn," in Matt Wray and Anna Newitz, eds., *White Trash: Race and Class in America* (New York: Routledge, 1997), 99.
58 For further discussion of nature-dominating masculinity in truck and sport utility vehicle advertising, see Noel Sturgeon, *Environmentalism in Popular Culture: Gender, Race, Sexuality, and the Politics of the Natural* (Tucson: University of Arizona Press, 2009), 39–40.
59 "Raw Exhaust," Pollution Porn, September 21, 2015, http://pollutionporn.com (accessed December 15, 2016). Note that there are also protected links on PollutionPorn.com, not publicly available, that purport to be of "truck sluts" being shown coal roller pipes in action, having sex with the driver's "pipe," as it were. For the purposes of this project, I did not access any web sites or portions of web sites that were protected or required sign up or membership, nor did I review any content that was not free and published for an unrestricted public audience.
60 Pollution Porn Tumbler, http://pollutionporn.tumblr.com/page/2 (accessed January 20, 2017). Film scholar Linda Williams writes about the "visual domination" of the "money shot" in pornography and its raw power as a fetish: "If you don't have the come shots you don't have a porno picture" (126). In her chapter, "Fetishism and Hard Core: Marx, Freud, and the 'Money Shot,'" on the politics and aesthetics of the "money shot," she argues that the "money shot" does not exist in isolation and is instead part of an intricate matrix of enmeshed power dynamics. The "carbon cum shot," or diesel smoke stack "money shot," similarly, is suspended in a complex web of relations that include masculine sensibilities, working-class social critiques, antagonisms between "country folks" and city dwellers, notions of patriotism, anxieties about race and sexuality, and biblical notions of men legitimately, by God's decree, being granted "dominion" over the earth to use for their own purposes. See Williams's *Hard Core: Power, Pleasure, and the "Frenzy of the Visible,"* expanded edition (Berkeley: University of California Press, 1999), 89–112.
61 "Smoke Stack Girls 2," PollutionPorn, September 5, 2012, http://pollutionporn.com (accessed January 20, 2017).
62 "Chick Lickin' the Tip," PollutionPorn, June 28, 2015, http://pollutionporn.com (accessed December 15, 2016).
63 Pollution Porn Tumblr site posting, "If It Don't Blow Back, Take It Back," http://pollutionporn.tumblr.com/page/2 (accessed January 20, 2017).
64 Judy Anderson and Curtis Anderson, *Electric and Hybrid Cars: A History* (Jefferson, NC: McFarland, 2005), 24.
65 See, for instance, nature writer Noel Perrin's characterization of snowmobile culture and ecological conservationist culture in his home state of Vermont in *First*

Person Rural: Essays of a Sometime Farmer (Jaffrey, NH: Godine, 1990), 118–19; Tim Edensor and Sophia Richards, "Snowboarders vs. Skiiers: Contested Choreographies of the Slopes," *Leisure Studies* 26 (January 2007): 97–114; and for class differences in attitudes toward climate change, Thomas Laidley, "Climate, Class, and Culture: Political Issues as Cultural Signifiers in the U.S.," *Sociological Review* 61 (2013): 153–71.

66 Laura Kipnis, *Bound and Gagged: Pornography and the Politics of Fantasy in America* (Durham, NC: Duke University Press, 1998). On "class-bound" aesthetics, see p. 92. On class distinctions between pornography and art, see p. 85. On the problems of class in pornography, see also Linda Williams, *Hard Core*, 168.

67 See Daniel Bernardi, "Interracial Joysticks: Pornography's Web of Racist Attractions," in Peter Lehman, ed., *Pornography: Film and Culture* (New Brunswick, NJ: Rutgers University Press, 2006), 220–43; Jose Capino, "Asian College Girls and Oriental Men with Bamboo Poles: Reading Asian Pornography," in Lehman, ed., *Pornography*, 206–19; Jennifer Nash, *The Black Body in Ecstasy: Reading Race; Reading Pornography* (Durham, NC: Duke University Press, 2014).

68 Eric Lott, *Love and Theft: Blackface Minstrelsy and the American Working Class*, 20th edition (New York: Oxford University Press, 2013), xi, 4.

69 Show Me Diesels, "Rolling Coal on Protesters Compilation (Black Lives Matter, Trump Haters, and Tree Huggers)," YouTube, February 26, 2017, www.youtube.com/watch?v=rYPMbLO4pAY (accessed July 15, 2018).

70 Mukherjee and Banet-Weiser, *Commodity Activism*, 8.

71 Immigrant scapegoating, nativism, and political race baiting are far from new phenomena in U.S. history. See Ediberto Roman, *Those Damned Immigrants: America's Hysteria over Undocumented Immigration* (New York: NYU Press, 2013). On crumbling of the U.S. public school system and its infrastructure, see National Book Award winner and educator Jonathan Kozol's *Savage Inequalities: Children in America's Schools* (New York: Broadway Books, 2012); and *The Shame of the Nation: The Restoration of Apartheid Schooling in America* (New York: Random House, 2005). On the current lack of vocational schooling in the United States and the case for its reintroduction, see Valerie Strauss, "Why We Need Vocational Education," *Washington Post*, June 5, 2012, www.washingtonpost.com. For an excellent series that covers many of the issues I have enumerated, including the impact on the working class of free trade agreements and jobs shipped overseas, see the *Washington Post* series, "Liftoff and Letdown," particularly Jim Tankersley's feature, "The Devalued American Worker," *Washington Post*, December 14, 2014, http://www.washingtonpost.com. On the efficacy of "hate talk" in galvanizing radicalism, particularly among those from less educated, socioeconomically disadvantaged, and rural backgrounds, see David Neiwert, *The Eliminationists: How Hate Talk Radicalized the American Right* (New York: Routledge, 2009).

72 Jennifer Chu, "Study: Air Pollution Causes 200,000 Early Deaths Each Year in the U.S.: New MIT Study Finds Vehicle Emissions Are the Biggest Contributors to These Premature Deaths," *MIT News*, August 29, 2013, http://news.mit.edu; James

Ginda, "Adverse Health Effects of Diesel Particle Pollution" (Annual Cases in the U.S., 2010), www.dem.ri.gov (accessed December 22, 2016).

73 Andrew Ross, *No Respect*, 195.

74 Chafe, "Is There an American Narrative and What Is It?"

CHAPTER 4. GREEN IS THE NEW BLACK

1 Mike Monteiro, "How to Fight Fascism," Element Talks, August 23, 2017, www.elementtalks.com (accessed July 15, 2018).

2 See Carol Hanisch's essay, "The Personal Is Political," in Shulamith Firestone and Anne Koedt, eds., *Notes from the Second Year: Women's Liberation* (New York: Radical Feminism, 1970).

3 Adorno and Horkheimer, "The Culture Industry," 405–15.

4 Theodor Adorno, "The Stars Down to Earth: The *Los Angeles Times* Astrology Column," in *The Stars Down to Earth*, 2nd edition (London: Routledge, 2001), 46–171.

5 Robert Witkin, *Adorno on Popular Culture* (London: Routledge, 2002), 78.

6 See, for example, John Fiske, *Television Culture*, 2nd edition (New York: Routledge, 2011), especially chapter 5 on "Active Audiences" and "The Social Determinations of Meanings," 62–83. Fiske is influenced by Michel de Certeau's theorizing of agential "tactics" in reading culture in *The Practice of Everyday Life* (Berkeley: University of California Press, 1988). Fiske's student, media theorist Henry Jenkins, has in turn been influenced by this scholarly genealogy. Jenkins builds upon and extends both Fiske and de Certeau in his ethnographic studies: *Textual Poachers: Television Fans and Participatory Culture* (Routledge: New York, 1992); *Fans, Bloggers, Gamers: Exploring Participatory Culture* (New York: NYU Press, 2006); and *Convergence Culture*. See also Stuart Hall's essay, "Encoding, Decoding," 477–87. Some of the seminal reception research was conducted on audiences focused on the CBS network series *All in the Family*. Studies demonstrated that meaning making varied widely for audiences, who tended to see what they wanted to see in the character of Archie Bunker. What was gleaned by viewers of the program was often very different from the messages that the writers of the show and Norman Lear intended to communicate. For conservative audience viewers, Bunker's prejudices and resistance to social change served to confirm rather than challenge their own viewpoints. See Horace Newcomb and Paul Hirsh, "Television as a Cultural Forum: Implications for Research," *Quarterly Review of Film Studies* 8 (1983): 48–55; and David Thorburn, "Television as an Aesthetic Medium," *Critical Studies in Mass Communication* 2, no. 1 (June 1987): 161–73; Neil Vidmar and Milton Rokeach, "Archie Bunker's Bigotry: A Study in Selective Perception and Exposure," *Journal of Communication* 24 (Winter 1974): 36–47; Horace Newcomb, "The Television Artistry of Norman Lear," *Prospects* 2 (1976): 115–16. In James Chesebro and Caroline Hamsher's "Communication, Values, and Popular Television Series," in Horace Newcomb, ed., *Television: The Critical View*, 1st

edition (New York: Oxford University Press, 1976), 6–25, the authors conclude that "authorial messages in any given program, even when explicitly stated, are by no means automatically accepted by audiences, who tend to make their own meanings."

7 Bagdikian predicts that within a few short decades, five global media companies will control all media in the United States. See Bagdikian, *The New Media Monopoly*. On corporate surveillance, data mining, and privacy issues, see Fuchs, *Social Media*; Andrejevic, *iSpy*: and Shoshana Zuboff, *The Age of Surveillance Capitalism: The Fight for a Human Future at the New Frontier of Power* (New York: Hachette Book Group, 2019).

8 Paul Lewis, "Our Minds Can Be 'Hijacked': The Tech Insiders Who Fear a Smartphone Dystopia," *Guardian*, October 6, 2017, https://www.theguardian.com.

9 Lewis, "Our Minds Can Be 'Hijacked.'" See also Edward Herman and Noam Chomsky, *Manufacturing Consent*; and science-fiction writer William Gibson's portrayal of "wetware" and "jacking in" in his cyberpunk novel, *Neuromancer* (New York: Ace, 1984), 32–37.

10 Justin Harris, "How a Handful of Tech Companies Control Billions of Minds Every Day," TED, April 2017, https://www.ted.com (accessed January 24, 2018).

11 See, for example, Tim Wu's treatment of the attention economy in *The Attention Merchants: The Epic Scramble to Get inside Our Heads* (New York: Vintage, 2017); and Derek Thompson's *Hit Makers: The Science of Popularity in an Age of Distraction* (New York: Penguin, 2017).

12 Einstein, *Black Ops Advertising*.

13 See Little i Apps' Confession: A Roman Catholic App, iTunes Store, https://itunes.apple.com (accessed February 10, 2019); and https://apkpure.co/sin-tracker-confession (accessed February 27, 2019).

14 Rebel Box, "Sin Tracker: Confession," Google Playstore, https://play.google.com.

15 For the expanding trend in religious apps and their growing use as tools in daily religious practice, see Heidi Campbell, *When Religion Meets New Media* (New York: Routledge: 2010), especially chapters 2 and 3.

16 George Monbiot, "Paying for Our Sins," *Guardian*, October 18, 2006, http://www.theguardian.com; and *The Carbon Neutral Myth: Offset Indulgences for Your Climate Sins* (Carbon Trade Watch, 2007), www.carbontradewatch.org, which also criticizes the "business as usual" model that allows for our present levels of consumption without requiring fundamental social, political, and economic systemic changes.

17 For a discussion of moral arbitration, community confession, self-monitoring, and surveillance with regards to moral codes of good citizenship in American culture, see Jeremy Packer's *Mobility without Mayhem: Safety, Cars, and Citizenship* (Durham, NC: Duke University Press, 2008), 256–57.

18 See Rachel Wagner, *Godwired*, 204–5. For a concise guide to carbon "sin" tracking applications, see "The Green Guide to iPhone and Android Applications," Ask Green Irene, August 26, 2012, https://askgreenirene.tenderapp.com; and Jaymi

Heibuch, "More Than 100 iPhone Apps for Green Shopping, Eating, Travel, and Fun," October, 13, 2009, http://www.treehugger.com (accessed October 17, 2014).

19 "The Green Guide" and Heibuch, "More Than 100 iPhone Apps."

20 "The Green Guide" and Heibuch, "More Than 100 iPhone Apps."

21 Jaymi Heibuch, "ShopGreen Lets Online Retailers Incorporate Carbon Offsets into Purchases," Treehugger, April 15, 2009, www.treehugger.com (accessed January 8, 2016). DayFly Videos reviews the ShopGreen app and explains how the app cleverly "rewards you for ecofriendly activities" with shopping coupons for more consumption: www.veoh.com (accessed January 8, 2016).

22 Adele Peters, "This New App Turns Reducing Your Carbon Footprint into a Game," Fast Company, May 5, 2015, www.fastcompany.com.

23 See Oroeco's self-description on its main web page, www.oroeco.com (accessed January 8, 2016).

24 See David Harvey's *A Brief History of Neoliberalism* (New York: Oxford University Press, 2005). U.S. students, including doctoral students, and in my experience even business school students, when asked to define "neoliberalism," often assume this term is what it sounds like to American ears and thus logically refers to new forms of liberalism associated with the U.S. Democratic Party. The less confusing term "free-market fundamentalism" used in Harvey's book more clearly communicates the meaning of "neoliberalism" to a broader U.S. audience beyond simply the internal conversations of university professors. Throughout her scholarly writing on religion and media, Mara Einstein, a media ecologist who also holds an MBA, conscientiously defines this term to her audiences in accessible terms to avoid these confusions. See her clear explanation of the term, its history, and its philosophical context, in *Compassion, Inc.*, 114–16.

25 See, for example, Patricia Smith's critique of George Herbert Walker Bush's use of "1000 Points of Light" to champion the private charity model as superior to and more effective than the prospect of robust federal government involvement in the socioeconomic well-being of vulnerable U.S. populations. Smith makes the point that charity is indeed a good thing and should be encouraged, but it is simply not adequate to the task of dealing with a structurally unjust system. See Smith, "Intolerance and Exploitation: Civic Vice, Legal Norms, and Cooperative Individualism," in Christine Sistare, ed., *Civility and Its Discontents: Civic Virtue, Toleration, and Cultural Fragmentation* (Lawrence: University Press of Kansas, 2004), 69.

26 Robert Kunzig, "Germany Could Be a Model for How We'll Get Power in the Future," *National Geographic*, November 2015, https://www.nationalgeographic.com; Sören Amelang, "Renewables Cover about 100% of German Power Use for First Time Ever," Clean Energy Wire, January 5, 2018, www.cleanenergywire.org.

27 Kunzig, "Germany Could Be a Model"; Madeleine Cuff, "Renewables Meet Record-Breaking 78 Percent of German Electricity Demand," *Business Green*, July 29, 2015, http://www.businessgreen.com.

28 In this video interview and accompanying article, reporter Brian Merchant sits down to talk with Wagner about how "deeply uncomfortable" his book *But Will* *the Planet Notice?* has made some in the environmental community. In Merchant's "Recycle All You Want—the Planet Won't Notice," Treehugger, February 22, 2012, www.treehugger.com, he observes, "It's hard to accept the fact that our personal dedication to living in a planet-friendly way doesn't even begin to register on a global scale, and it's indeed depressing to consider that you could spend your life recycling, composting, eating organic food, taking public transportation, getting your power from solar, and biking, and essentially have no discernible impact on the planet. At all."

29 Gernot Wagner and Martin Weitzman, *Climate Shock: The Economic Consequences of a Hotter Planet* (New Brunswick, NJ: Princeton University Press, 2015), 131.

30 On the 50 percent statistic, see Wagner and Weitzman, *Climate Shock*, 132; and on the 84 percent statistic, see Elly Blue, *Bikenomics: How Bicycling Can Save the Economy* (Portland, OR: Microcosm Publishing, 2013), 14.

31 Wagner and Weitzman, *Climate Shock*, 131. Emphasis is mine.

32 See Robert Shiller, "How Idealism Expressed in Concrete Steps Can Fight Climate Change," *New York Times*, March 27, 2015, www.nytimes.com.

33 See a comprehensive timeline and review of Copenhagen's "bicycle heaven" in "Bike City Copenhagen: This Is the Ultimate Bicycle Friendly City," Icebike, August 28, 2015, http://www.icebike.org; Ben Shiller, "How Copenhagen Became a Cycling Paradise by Considering the Full Cost of Cars," Fast Company, June 10, 2015, www.fastcompany.com; Street Films, "Cycling Copenhagen, through North American Eyes," documentary film short, https://vimeo.com. See also Feargus O'Sullivan, "Copenhagen's Bike Lane in the Sky," *Atlantic*, November 23, 2015, http://www.theatlantic.com; Diane Cardwell, "Copenhagen Lighting the Way to Greener, More Efficient Cities," *New York Times*, December 8, 2014, www.nytimes.com.

34 Charles Komanoff, "Climate Idealism Can't Hold a Candle to Collective Action," Carbon Tax Center, March 30, 2015, www.carbontax.org.

35 Monteiro, "How to Fight Fascism."

36 Shiller, "How Idealism Expressed in Concrete Steps." I remind readers that I do not approach these comments as any sort of quantitative data but as primary source material for qualitative content and rhetorical analysis.

37 Patricia Smith, "Intolerance and Exploitation: Civic Vice, Legal Norms, and Cooperative Individualism," in Christine Sistare, ed., *Civility and Its Discontents* (Lawrence: University Press of Kansas, 2004), 69.

38 Wagner, *But Will the Planet Notice*, 11.

39 Wagner, *But Will the Planet Notice*, 8.

40 This phrase comes from Wagner's, *But Will the Planet Notice.*

41 For the original article on the Oroeco application in *Grist*, see Samantha Larson, "The New App That Tracks Your Carbon Footprint—and Lords It over Your Friends," *Grist*, May 7, 2014, http://grist.org. The Facebook reader responses follow the *Grist* Facebook posting, "This app wants you to beat your Facebook friends at a new game: Who has the lowest carbon footprint?" www.facebook.com (accessed, January 8, 2016).

42 Derrick Jensen, "Forget Shorter Showers: Why Personal Change Does Not Equal Political Change," *Orion*, July 2009, https://orionmagazine.org.

43 *Grist* Facebook posting on Oroeco application, https://www.facebook.com. Postings all took place during May of 2014.

44 *Grist* Facebook posting; and *Orion* magazine comments section for Jensen, "Forget Shorter Showers."

45 Haidt, *The Righteous Mind*, 55.

46 In the realm of uncivil, malicious exchanges, the interjection of computerized "chatbots" or "talkbots" into online political and/or campaign discussion spaces has cast a dark shadow on some of the hopes and possibilities for productive civic conversations online. These computer programs are designed to simulate human conversation and, although often used to spam chat spaces, can be programmed with phrases associated with particular political bents in order to give the appearance of there being more support or opposition toward an issue discussed in a chat space than actually humanly exists. See Andrew Zaleski, "How Bots, Twitter, and Hackers Pushed Trump over the Finish Line," *Wired*, November 10, 2016, https://www.wired.com. On the limits of participatory social media, see chapter 5, "The Power and Political Economy of Social Media," in Fuchs, *Social Media*, 97–125; and for a review of corporate control of media, see Croteau and Hoynes, *Media/Society*, 32–71. On the power and potentialities of networked conversations in the digital mediasphere, see especially Shirky's chapter, "It Takes a Village to Find a Phone," in *Here Comes Everybody*, 1–24; and Nickolas Chirstakis and James Fowler's *Connected: The Surprising Power of Our Social Networks and How They Shape Our Lives* (New York: Little, Brown, 2011). Jenkins's aforementioned research on civic engagement, digital communities, and political action, including the actions of the Harry Potter Alliance, among other groups, are also germane to the consideration of the expanding scope and scale of digital forums for moral arbitration of contemporary issues.

47 See Klein, *This Changes Everything*; McKibben, *Eaarth*; Pulitzer Prize–winning author Elizabeth Kolbert's, *Field Notes from a Catastrophe: Man, Nature, and Climate Change* (London: Bloomsbury, 2006). See also Mike Tidwell's essay, "To Really Save the Earth, Stop Going Green," in Bill McKibben, ed., *The Global Warming Reader* (New York: Penguin, 2012), 233–38.

48 Tidwell, "To Really Save the Earth, Stop Going Green"; and Jensen, "Forget Shorter Showers."

49 See Jan Semenza et al., "Public Perceptions of Climate Change," *American Journal of Preventative Medicine* 35, no. 5 (2008): 479. Note that this percentage may be much higher since the 2017 devastation of Hurricane Harvey.

50 Semenza et al., "Public Perceptions of Climate Change," 483.

51 Semenza et al., "Public Perceptions of Climate Change," 486.

52 Semenza et al., "Public Perceptions of Climate Change," 483.

53 George W. Bush, "At O'Hare, President Says 'Get on Board,'" September 27, 2001, White House Press Release, http://georgewbush-whitehouse.archives.gov; Andrew Bacevich, "He Told Us to Go Shopping: Now the Bill Is Due," *Washington Post*, October 5, 2008, http://www.washingtonpost.com; Robert Shiller, "Spend, Spend, Spend: It's the American Way," *New York Times*, January 14, 2012, www.nytimes.com.
54 Bush, "At O'Hare"; Bacevich, "He Told Us to Go Shopping"; Shiller, "Spend, Spend Spend."
55 Bush, "At O'Hare"; Bacevich, "He Told Us to Go Shopping"; Shiller, "Spend, Spend Spend."
56 Andrew Weigert, *Mixed Emotions: Certain Steps toward Understanding Ambivalence* (Albany: State University of New York Press, 1991), 111.
57 Weigert, *Mixed Emotions.*
58 Sarah McFarland Taylor, "Shopping and Consumption," in Eric Mazur and John Lyden, eds., *The Routledge Companion to Religion and Popular Culture* (New York: Routledge, 2015), 317–35.
59 For a critical reading and analysis of the voyeuristic aspects of reality TV, see Lemi Baruh's "Publicized Intimacies on Reality Television: An Analysis of Voyeuristic Content and Its Contributions to the Appeal of Reality Programming," *Journal of Broadcasting and Electronic Media* 53, no. 2 (2009): 190–210. On the intimacy and immediacy of "reality" television, see Robin Nabi et al., "Reality-Based Television Programming and the Psychology of Its Appeal," *Media Psychology* 5, no. 4 (2003): 303–30; Steven Reiss and James Wiltz, "Why People Watch Reality TV," *Media Psychology* 6, no. 4 (2004): 363–78.
60 Season 1 appeared on HGTV in 2007, and season 2 appeared in 2009 on the Planet Green Channel, a relaunching of the Discovery Home Channel with environmentally themed content.
61 "Toyota Prius Pop Culture Icon," https://pressroom.toyota.com/releases/2013+thwt+prius+pop+culture.download (accessed March 22, 2019); Kelly Carter, "Hybrid Cars Were Oscars' Politically Correct Rides," *USA Today*, March 30, 2003, http://usatoday30.usatoday.com; Jim Henry, "Oscar Night Star Power Provides Priceless Publicity for Prius," *Automotive News*, October 29, 2007, www.autonews.com.
62 See the following episodes: "A Model's Model Home," *Living with Ed*, season 2, episode 1, HGTV, August 26, 2007; "Talking Green with Larry Hagman," *Living with Ed*, season 2, episode 2, HGTV, August 27, 2007; "Jay Leno's in Hot Water!" *Living with Ed*, season 2, episode 2, HGTV, September 2, 2007; "Mr. Green and Mr. Browne," *Living with Ed*, season 2, episode 2, HGTV, September 3, 2007.
63 "A Model's Model Home."
64 "Talking Green with Larry Hagman."
65 "Jay Leno's in Hot Water!"
66 "Green Tips from Ed Begley, Jr.: Help Your Community, Tip #5," *(it) magazine*, May 31, 2014, http://itmagazine.org; and posted at www.BegleyLiving.com (accessed March 29, 2019).

67 *See* https://begleyliving.com and BegleyBest.com (accessed February 28, 2019). Note that Pierre-Louis argues that the illusion that we can simply "buy our way to a green planet" is gravely harmful and counterproductive. Although she acknowledges that the "green consumer movement" has multiple virtues, she contends that it masks the reality that it is the volume of our consumption that has the greatest negative environmental impact on the planet. See Pierre-Louis, *Green-Washed.*

68 See Peter Ward's discussion of celebrity veneration in *Gods Behaving Badly: Media, Religion, and Celebrity Culture* (Waco, TX: Baylor University Press, 2011); and also the sections on Kim Kardashian and Honey Boo Boo Child in McFarland Taylor, "Shopping and Consumption," 327–30.

69 See George Lipsitz, "The Meaning of Memory: Family, Class, and Ethnicity in Early Network Television Programs," in Lynn Spigel and Denise Mann, eds., *Private Screenings* (Minneapolis: University of Minnesota Press, 1992), 71.

70 Lipsitz, "The Meaning of Memory," 73.

71 Lipsitz, "The Meaning of Memory," 74–76.

72 Lipsitz, "The Meaning of Memory," 78. Some notable exceptions would be television programming shown on premium cable channels, such as Home Box Office and Showtime, where arguably the goal of the programming is not to sell specific products but to sell premium cable package subscriptions by providing high-quality programming.

73 Joseph Henrich and Francisco Gil-White, "The Evolution of Prestige: Freely Conferred Deference as a Mechanism for Enhancing Benefits of Cultural Transmission," *Evolution and Human Behavior* 22 (2001): 165.

74 Henrich and Gil-White, "The Evolution of Prestige," 176.

75 Henrich and Gil-White, "The Evolution of Prestige," 184.

76 Other key qualities celebrities impart to brands, as long as their reputations remain unbesmirched, include trustworthiness, attractiveness, and expertise. See Abdullah Malik and Bushan Sudhakar's review of celebrity endorsement research, "Brand Positioning through Celebrity Endorsement," *International Review of Management and Marketing* 4, no. 4 (2014): 259–75; Anita Elberse and Jeroen Verleun, "The Economic Value of Celebrity Endorsements," *Journal of Advertising Research* 52, no. 2 (June 2012): 149–65; B. Zafer Erdogan, "Celebrity Endorsement: A Literature Review," *Journal of Marketing Management* 15, no. 4 (1999): 291–314.

77 Ed Keller and Jon Berry, *The Influentials* (New York: Fress Press, 2003), 1.

78 Keller and Berry, *The Influentials*; and Ward, *Gods Behaving Badly*, 9–34.

79 See more generally, Philip Maciak, "Why Do So Many People Watch HGTV?" *Pacific Standard*, September 29, 2014, http://www.psmag.com; and more specifically, Mimi White, "House Hunters, Real Estate Television, and Everyday Cosmopolitanism," in Laurie Ouellette, ed., *A Companion to Reality Television* (Malden, MA: Wiley Blackwell, 2014), 386–401; Mimi White, "A House Divided," *European Journal of Cultural Studies* 20, no. 5 (October 2017): 575–91; and Mimi White,

"Gender Territories: House Hunting on American Real Estate TV," *Television and New Media*, 14, no. 3 (May 2013): 228–43.

80 Tamsin Blanchard, *Green Is the New Black: How to Change the World with Style* (London: Hodder and Stoughton, 2007), 14.

81 Blanchard, *Green Is the New Black*, 41.

82 Tom Crompton, "Weathercocks and Signposts," *World Wildlife Fund Report*, April 2008, http://assets.wwf.org.uk (accessed October 14, 2014).

83 See P. Wesley Schlutz et al., "The Constructive, Destructive, and Reconstructive Power of Social Norms," *Psychological Science* 18, no. 5 (May 2007): 429–34. Here I am indebted to Ezra Markowitz and Azim Shariff's thorough and insightful review article, "Climate Change and Moral Judgment," *Nature Climate Change* 2 (April 2012): 243–47.

84 Molly Goodson, "America's Top Model Rundown: The Models Go Green," Pop-Sugar, September 27, 2007, www.popsugar.com (accessed March, 29, 2019).

85 John Steve, "ANTM 9, Episode 2, the Models Go Green," *Jonnasphere*, October 14, 2007, http://jonnathanst.blogspot.com.

86 "The Models Go Green, Episode Reviews," TV.com, September 27, 2007, www.tv.com.

87 "The Models Go Green."

88 "The Models Go Green."

89 Celebrity media blogger Molly Goodson also advances this theory in Goodson, "America's Top Model Rundown," asserting that restricting the models from smoking is the perfect way for the producers to "up the ratings with testy catfights."

90 See reader comments for Amelie Gillette's review, "America's Next Top Model: The Models Go Green," A.V.Club, September 26, 2007, http://www.avclub.com.

91 See Thomas McGlaughlin's *Street Smarts and Critical Theory: Listening to the Vernacular* (Madison: University of Wisconsin Press, 1996). Henry Jenkins's multiple ethnographic studies of fans, bloggers, and gamers also points to the kind of vernacular theorizing produced by these communities. See Jenkins, *Fans, Bloggers, and Gamers*.

92 Croteau and Hoynes, *Media/Society*, chapter 8 on "Active Audiences and the Construction of Meaning," 260–92.

93 Croteau and Hoynes, *Media/Society*, 261.

94 Croteau and Hoynes, *Media/Society*, 293.

95 For an analysis of the appeal of the "makeover show," see the special issue, "Television Transformations: Revealing the Makeover Show," *Journal of Media and Cultural Studies* 22, no. 4 (August 2008), especially, Beverly Skeggs and Helen Wood, "The Labour of Transformation and Circuits of Value 'around' Reality Television," 559–72.

96 Find episodes of *Greenburg* posted here: https://www.youtube.com/playlist?list=PLBADC311A00CAD435 (accessed January 8, 2016).

97 "About the Nautical Tribe," Facebook, www.facebook.com (accessed December 15, 2017).
98 The project's web site, which features photos of the boat, pictures of prospective student contestants, would-be host Fabian Cousteau, and even professors who were lined up to be judges, as of 2016, was taken down. The defunct site was located at www.nauticaltribe.com. Some of these images survive on the Twitter and Facebook sites, and in blogs of those who wrote about the project and shared its trailer videos. See, for instance, Lynn Hasselberger, "I Approve This Reality TV Show," WordPress.com, April 26, 2011, https://icountformyearth.wordpress.com (accessed January 3, 2018). The series' promotional trailers, like Rizor's "What If?" video, can be found at Hasselberger's blog and also on YouTube at www.youtube.com/watch?v=nIuor4bcNew (accessed January 3, 2018).
99 The Nautical Tribe Facebook page: www.facebook.com (accessed January 3, 2018).
100 See Mark Andrejevic's critique of the uses of "tribalism" and "totemism" in the series *Survivor* and similar reality TV programs that mimic tribal cultures as treating "neotribalism" as a commodity to be sold. Mark Andrejevic, *Reality TV: The Work of Being Watched* (Oxford, UK: Rowman and Littlefield, 2004), 197–98.
101 Jim Rizor, "What If?" (video trailer), YouTube, www.youtube.com/watch?v=nIuor4bcNew&list=UUnWlqWVQ49WZ3Bai574vNgQ (accessed January 20, 2016).
102 Barry King, "Training Camps of the Modular: Reality TV as a Form of Life," in David S. Escoffery, ed., *How Real Is Reality TV* (Jefferson, NC: McFarland, 2006), 43.
103 See Elizabeth Franko, "Democracy at Work? The Lessons of Donald Trump and *The Apprentice*," in Escoffrey, *How Real Is Reality TV*, 247–58.
104 For further discussion of capitalism and "reality TV," see Andrejevic, *Reality TV*, 195–228.
105 Postman, *Amusing Ourselves to Death*, 83–98.
106 See Michael Wolff, *Television Is the New Television: The Unexpected Triumph of Old Media in the Digital Age* (New York: Portfolio, Penguin Books, 2015); Michael Strangelove, *Post-TV: Piracy, Cord-Cutting, and the Future of Television* (Toronto: University of Toronto Press, 2015).

CHAPTER 5. VEGETARIAN VAMPIRES

1 See Lynn Messina's *Little Vampire Women* (New York: HarperCollins, 2010), 2–3.
2 Amos Bronson Alcott and transcendentalist abolitionist Charles Lane established the Fruitlands community in Harvard, Massachusetts, in 1843.
3 Richard Francis, *Fruitlands: The Alcott Family and Their Search for Utopia* (New Haven, CT: Yale University Press, 2010), 7.
4 Francis, *Fruitlands*, 23.
5 Francis, *Fruitlands*, 7.
6 Messina, *Little Vampire Women*, 12.
7 Maureen Ogle, *In Meat We Trust: An Unexpected History of Carnivore America* (New York: Houghton Mifflin, 2013); Josh Ozersky, *The Hamburger: A History* (New Haven, CT: Yale University Press, 2009), 11–12, 63.

8 On the topic of "extractivist capitalism" and economies based on the principle of "forcible draining," see Henry Veltmayer and James Petras, *The New Extractivism: a Post-Neoliberal Development Model or Imperialism of the Twenty-first Century?* (London: Zed Books, 2014). See also David Korten's discussion of what he terms the "extractive Money Economy" in *Change the Story, Change the Future*, 108. Naomi Klein's chapter, "Beyond Extractivism," is devoted to critical examination of extractivist economics and its consequences in *This Changes Everything*, 161–90.

9 *This Changes Everything*, directed and produced by Lewis Cuarón, Hossain Barnes, Naomi Klein, and Louverture Hector's production company (Message Productions, LLC, 2015). In Klein's book, she talks of the historical embeddedness of what she calls "core civilization myths" in Western culture that have espoused and promoted "humanity's duty to dominate a natural world that is believed to be at once limitless and entirely controllable" (159). Klein's characterization of this story echoes historian Carolyn Merchant's work in *The Death of Nature: Women, Ecology, and the Scientific Revolution* (San Francisco: Harper and Row, 1980).

10 See Ellen Moore's second chapter, "Cradle to Crave: The Commodification of the Environment in Family Film," in *Landscape and Environment in Hollywood Film* (New York: Praeger, 2017), 31–55.

11 See psychoanalytic semiotician Mary Ann Doane's *Desire to Desire: The Women's Film of the 1940s* (Bloomington: Indiana University Press, 1987), in which she explores the problematic relationship of women to mass culture, particularly how Hollywood paradoxically avows and denies female desire; and Roland Marchand, *Advertising the American Dream: Making Way for Modernity, 1920–1940* (Berkeley: University of California Press, 1985).

12 Judith Williamson, *Consuming Passions: The Dynamics of Popular Culture* (New York: Marion Boyars), 15.

13 Nina Auerbach, *Our Vampires, Ourselves* (Chicago: University of Chicago Press, 1995), 1.

14 Auerbach, *Our Vampires, Ourselves*, 1.

15 On the theoretical concept of "floating signifers" and the contributions of Ferdinand de Saussure, Claude Lévi-Strauss, Jacques Lacan, and Stuart Hall to the eventual articulation of this concept, see Irene Rima Makaryk, ed., *Encyclopedia of Contemporary Literary Theory: Approaches, Scholars, Terms* (Toronto: University of Toronto Press, 1993), 546–47.

16 Milly Williamson, *The Lure of the Vampire: Gender, Fiction, and Fandom from Bram Stoker to Buffy* (New York: Columbia University, Wallflower Press, 2005), 183.

17 Williamson, *The Lure of the Vampire*.

18 Josh Ozersky's *Hamburger: A History* (New Haven, CT: Yale University Press, 2009), especially chapter 1 on "Sizzle and Symbolism" and chapter 4, "Have It Your Way."

19 Ogle, *In Meat We Trust*, p. 82.

20 Commissioned by the "The Beef Industry Council" from their Beef Check Program, the "Beef. It's What's for Dinner" campaign was launched in 1992 and narrated by Robert Mitchum. The campaign won both Sappi and Effie advertising industry awards. The original commercial can be viewed here: https://www.youtube.com/watch?v=rUxkizsmi70 (accessed February 22, 2018).

21 "The Best Advertising Slogans of All Time according to Digg Users," Fast Company, September 3, 2008, https://www.fastcompany.com (accessed February 24, 2018).

22 Richard Greene and Silem Mohammad, *Vampires, Zombies, and Philosophy* (Chicago: Open Court, 2010), viii. This is also a reference, I presume, to *The Lost Boys*. In the 1987 vampire cult classic *The Lost Boys*, one of the film's characters learns all the "ins and outs" of vampire hunting by reading a 1950s (pre–Comics Code) comic book called *Vampires Everywhere!* "Vampires Everywhere!" is also the name of a California-based Goth punk band.

23 See, for example, Vampire Wine, vampire gourmet products, and Bloody Mary mix at www.vampire.com (accessed February 2, 2018); and Desi Jedeikin, "14 Weirdest Pieces of *Twilight* Merchandise," Smosh, November 18, 2011, www.smosh.com. Store.VampireFreaks.com is one-stop shopping for vampire condoms, jewelry, cosmetics, clothes, and myriad accessories for the vampire enthusiast. For a scholarly overview of the intersections of youth culture and enthrallment with vampire media franchise, see the introduction to Melissa Click, Jennifer Aubrey, and Elizabeth Behm-Morawitz, eds., *Bitten by* Twilight*: Youth Culture, Media, and the Vampire Franchise* (New York: Peter Lang, 2010), 1–17.

24 "Vampires," Environmental Protection Agency, http://www.epa.gov (accessed February 29, 2016). This audio file was removed from the EPA web site in 2017.

25 "Reducing Electricity Use and Costs," U.S. Department of Energy, https://energy.gov (accessed February 2, 2018).

26 "Vampires Sucking You Dry," U.S. Department of Energy, http://energy.gov (accessed February 23, 2016). This image has since been replaced with a page entitled "4 Ways to Slay Vampire Energy" and features an animated short appealing to children and promoting "Energyween," the Department of Energy's Halloween-themed campaign to get kids to be more aware of wasted energy. See "4 Ways to Slay Energy Vampires This Halloween," U.S. Department of Energy, https://energy.gov (accessed February 2, 2018). The original U.S. Department of Energy article, its full text, and the image of "Nosferatu" survive at the news service BreakingEnergy.com, which has archived it: John Schueler, "Are Energy Vampires Sucking You Dry," https://breakingenergy.com (accessed February 2, 2018).

27 Schueler, "Are Energy Vampires Sucking You Dry?"

28 Department of Energy, *Intrusive Residential Standby Service Report*, 2005, www.ourbreathingplanet.com. And another example can be found here: "Vampire Power, Electronics Turned Off Still Cost You Money," Apparently Apparell, www.apparentlyapparel.com (accessed February 24, 2018).

29 See the U.S. Department of Energy's "Energy Saver" page, which warns the public, "Energy Vampires Are Attacking Your Home—Here's How to Stop Them," October 20, 2014, https://energy.gov (accessed February 2, 2018). Energy utility companies like "ComEd" also use vampire language and imagery in their Internet energy-saving campaigns, in which villainous home-invading vampires "steal" and "suck" energy from unsuspecting consumers. See ComEd's "Beware the Energy Vampires Lurking in Your Home This Halloween," October 31, 2017, https://poweringlives.comed.com (accessed February 2, 2018).

30 "Happy Halloween from Energy Star!" EnergyStar.gov, www.energystar.gov (accessed February 29, 2016).

31 Ernesta Jones, "Is Your Energy Bill Scary? Slaying Energy Vampires Can Save Americans Millions," Environmental Protection Agency, October 27, 2008, http://yosemite.epa.gov (accessed February 26, 2018). Document was removed in 2017 but can still be found in the EPA online archives: https://archive.epa.gov (accessed February 2, 2018).

32 See Cynthia Hendershot, *I Was a Cold War Monster: Horror Films, Eroticism, and the Cold War Imagination* (Madison: Popular Press, University of Wisconsin, 2001), especially chapter 6 on the homefront and home invasion; Stacey Abbott, *Celluloid Vampires: Life after Death in the Modern World* (Austin: University of Texas Press, 2007), 71–72; Aspasin Stefanon, *Reading Vampire Gothic through Blood: Bloodlines* (New York: Palgrave Macmillan, 2014), chapter 2. Note that here, I refer to the second Red Scare (1947–1957), as opposed to the first Red Scare (1919–1920). For further delineation of this distinction, see Griffin Fariello, *Red Scare: Memories of an American Inquisition* (New York: Norton, 2008).

33 See for example, Ali Behdad, *A Forgetful Nation: On Immigration and Cultural Identity in the United States* (Durham, NC: Duke University Press, 2008); and Leonard Dinnerstein and David Reimers, *Ethnic Americans: A History of Immigration*, 5th edition (New York: Columbia University Press, 2009). Note that xenophobia and fear of Eastern European immigrants, particularly Eastern European Jews, are prevalent in Bram Stoker's *Dracula* and constitute much of what makes the dark Count Dracula so terrifying to a British audience witnessing an influx of dark-haired, darker-skinned immigrants. See the British Library curator of contemporary literary archives Greg Buzwell's "Dracula: Vampires, Perversity, and Victorian Anxieties," British Library (online articles on archives and collections), www.bl.uk (accessed February 2, 2018).

34 Each year in the United States, which uses 28 percent of the world's energy, enough energy is "sucked" and wasted by "electronic vampires" to power the entire country of Italy for a year. Julie Carr Smyth, "Electronic 'Vampires' Suck Energy, Not Blood," *USA Today*, October 30, 2007, https://usatoday30.usatoday.com.

35 Corvis Nocturnum, *Allure of the Vampire: Our Sexual Attraction to the Undead* (Fort Wayne, IN: Dark Moon Publishing, 2006), 174–75.

36 Nocturnum, *Allure of the Vampire*, 175. There is also a Daoist phenomenon of "sexual vampirism," in which one partner sucks vital energy out of the other during sexual acts and in which a consort can function as a kind of "energy pod." I am indebted to Asian religions scholar and colleague Sarah Jacoby for bringing this to my attention.

37 Nocturnum, *Allure of the Vampire*, 198.

38 Nocturnum, *Allure of the Vampire*, 201. For a fuller ethnographic and historic treatment of the "Real Vampire Community," see Rosemary Ellen Guiley, *Vampires among Us: The Surprising Story of Real-Life Vampires in the 20th Century* (New York: Pocket Books, 1991) and Katherine Ramsland, *Piercing the Darkness: Undercover with Vampires in America Today* (New York: Harper Voyager, 1998).

39 Nocturnum, *Allure of the Vampire*, 202.

40 Nocturnum, *Allure of the Vampire*, 205–6.

41 See, for example, Susan Power Bratton and Shawn Hinz, "Ethical Responses to Commercial Fisheries Decline in the Republic of Ireland," *Ethics and the Environment* 7, no. 1 (2002): 54–91.

42 Keith Peterman, "Climate Vampire: Frankenstorm Foreboding," Global Hot Topic, October 30, 2012, www.yorkblog.com (accessed February 24, 2018).

43 Justin Hovland, "Renewables vs. Vampires," *San Francisco Examiner*, June 10, 2011, www.examiner.com (accessed July 31, 2013).

44 Dupré has also posted videos of testimony from families suffering the toxic consequences of the explosion and its cleanup to her Youtube.com channel: www.youtube.com/watch?v=bQf2NWbmb34 (accessed March 18, 2014). See also Dupré's "The Story behind Vampire of Mocondo," January 24, 2013, https://thestorybehindthebook.wordpress.com (accessed February 24, 2018).

45 Although the phrase "not all men" has been most recently popularized by its deployment in social media to address issues of generalization when discussing sexual assault, Charles Dickens used the phrase "but not all men" more than a century earlier in his 1836 novel, *The Pickwick Papers*, reissue edition (New York: Penguin Classics, 2000), 108. Articles in *Time* and on *Slate* both criticized the phrase "Not all men" for deflecting conversations away from serious considerations of the links among gender, power, and violence. See Jess Zimmerman, "Not All Men: A Brief History of Every Dude's Favorite Argument," *Time*, April 28, 2014, http://time.com; and Phil Plait, "#YesAllWomen," *Slate*, May 27, 2014, www.slate.com.

46 The series refers to the actual historical Hungarian political society, the Order of the Dragon, which reportedly had ties to the father of Vlad the Impaler (Count Dracula), drawing links between the Latin "draco," the image of the dragon, and the name "Dracula." See Radu Florescu and Raymond McNally, *Dracula: Prince of Many Faces, His Life and His Times* (Boston: Back Bay Books, 1990), xviii, 28–29.

47 Sean Daly, "Downton Stabby? NBC's 'Dracula' Borrows Blood, Bodices from Hot Shows," *Tampa Bay Times*. October 13, 2013, www.tampabay.com (accessed February 24, 2018).

48 W. Scott Poole, *Monsters in America: Our Historical Obsession with the Hideous and the Haunting* (Waco, TX: Baylor University Press, 2011), 15.

49 Sara Hackenberg, associate professor in English at San Francisco State University, quoted in Nocturnum, *Allure of the Vampire*, 57.

50 Tim Surrette, "Dracula Series Premier Review: I Want to Suck Away Your Energy Costs," TV.com, October 26, 2013, www.tv.com (accessed February 24, 2018). See also Saler, *As If*.

51 Gün R. Semin and Kenneth Gergen, *Everyday Understanding: Social and Scientific Implications* (Thousand Oaks, CA: Sage, 1990); and Kenneth Gergen, *The Saturated Self: Dilemmas of Identity in Contemporary Life* (New York: Basic Books, 2000).

52 Tom Standage, *Writing on the Wall: Social Media—The First 2,000 Years* (New York: Bloomsbury, 2014), 21–47.

53 Peters, *The Marvelous Clouds*, 19.

54 Poole, *Monsters in America*, 208.

55 Alan Rusbridger, "Climate Change: Why the *Guardian* is Putting Threat to the Earth Front and Center," *Guardian*, March 6, 2015, www.theguardian.com; Adam Vaughan, "Earth Day: Climate Scientists Say That 75% of Known Fossil Fuel Reserves Must Stay in Ground," *Guardian*, April 22, 2015, www.theguardian.com.

56 Alan Rusbridger, "Dear Bill Gates: Will You Lead the Fight against Climate Change?" *Guardian*, April 30, 2015, www.theguardian.com; Mrill Ingram, "It's the 'Biggest Story in the World': A Direct Appeal to One of the World's Most Generous Givers," *Upworthy*, May 14, 2015, www.upworthy.com. In 2015, the Bill and Melinda Gates Foundation divested $825 U.S. from Exxon Mobil, and in 2016, the foundation divested $187 million dollars from its investments with BP. Damian Carrington, "Bill and Melinda Gates Foundation Divests Entire Holding in BP," *Guardian*, May 12, 2016, www.theguardian.com. The Wellcome trust increased its investment in fossil fuel after the challenge was issued.

57 Bill McKibben, "Climate Fight Won't Wait for Paris, Vive La Résistance," *Guardian*, March 9, 2015, www.theguardian.com.

58 Some examples of phallic-suggestive keep-it-in-the-ground memes can be found here: www.merkley.senate.gov; "In the Ground Jagger," http://axisoflogic.com; "Keep It in the Ground," http://greenhousepr.co.uk; "Keep Fossil Fuels in the Ground," https://fromthestyx.files.wordpress.com (accessed March 2, 2019).

59 Malthus's work has been controversial, especially as, in some instances, its content has been deployed by the eugenics movement in support of its aims. See editor Geoffrey Gilbert's introduction to Thomas Malthus, *An Essay on the Principle of Population* (New York: Oxford University Press, 1999), xvii. Mathus's book was first published in 1798.

60 See historian Robert Mayhew's *Malthus: The Life and Legacies of an Untimely Prophet* (Boston: Belknap, 2014); and Ahmed Husser, *Principles of Environmental Economics: An Integrated Economic and Ecological Approach*, 3rd edition (New York: Routledge, 2012), 249–52.

61 William Catton, *Overshoot: The Ecological Basis of Revolutionary Change* (Urbana: University of Illinois Press, 1980).
62 On "using sex to sell abstinence," see Christine Gardner, *Making Chastity Sexy: The Rhetoric of Evangelical Abstinence Campaigns* (Berkeley: University of California Press, 2011), especially 43–61.
63 See Sara Moslener, *Virgin Nation: Sexual Purity and American Adolescence* (New York: Oxford University Press, 2015).
64 See, for instance Sumathi Ramaswamy's treatment of ecstatic piety in *Passions of the Tongue: Language Devotion in Tamil India, 1891–1970* (Berkeley: University of California Press, 1997), 95–97.
65 P. Wesley Schultz and Lynnette Zelezny, "Reframing Environmental Messages to Be Congruent with American Values," *Research in Human Ecology* 10, no. 2 (2003): 126, 134.
66 Schultz and Zelezny, "Reframing Environmental Messages," 134.
67 Robert Cox, *Environmental Communication and the Public Sphere*, 3rd edition (Los Angeles: Sage, 2012), 235.
68 Cox, *Environmental Communication*, 235. See also Bluestem, https://www.bluestem.org (accessed March 22, 2019).
69 Poole, *Monsters in America*, 15.
70 Frederick Dahlstand, *Amos Bronson Alcott: An Intellectual Biography* (Rutherford, NJ: Fairleigh Dickinson University Press, 1983), 197.
71 See Samantha Amber Oakley, "'I Could Kill You Quite Easily, Bella, Simply by Accident': Violence and Romance in Stephenie Meyer's *Twilight* Saga," *Theses, Dissertations, and Other Capstone Projects* (Mankato: Minnesota State University, 2012), and Kristina Deffenbacher, "Rape Myths, *Twilight*, and Women's Paranormal Revenge in Romantic and Urban Fantasy Fiction," *Journal of Popular Culture* 45, no. 5 (October 2014): 923–36. Deffenbacher looks at revenge narratives that counter the hackneyed trope of romance novels that the hero rapes the heroine "until she loves him."
72 Leah Lamb, "The Earth Is Our Host and We Are One Helluva Mega Coven of Vampires," Current Green Blog, September 21, 2009, http://blogs.current.com (accessed April 6, 2010).
73 Lamb, "The Earth Is Our Host."
74 Lamb, "The Earth Is Our Host."
75 And seduce they have, as evidenced by a proliferation of vampire-themed fan and enthusiast Internet sites, consumer products, and works of mediated popular culture. In addition to Click et al., *Bitten by* Twilight and Auerbach's *Our Vampires, Ourselves*, see Giselle Liza Anatol, ed., *Bringing Light to* Twilight*: Perspectives on the Pop Culture Phenomenon* (New York: Palgrave Macmillan, 2011); Mary Hallab, *Vampire God: The Allure of the Undead in Western Culture* (New York: SUNY Press, 2009); Richard Greene and K. Silem Mohammad, *Zombies, Vampires and Philosophy: New Life for the Undead* (Chicago: Open Court, 2010).

76 See the Nature Conservancy *Twilight* press release here: "*Twilight* Vampire Habitat Protected by the Nature Conservancy," November 19, 2012, Cision PR Newswire, www.prnewswire.com (accessed February 24, 2018).

77 "Washington Creating more *Twilight*," Nature Conservancy, www.nature.org (accessed December 18, 2013); and "*Twilight* Vampire Habitat Protected by the Nature Conservancy," November 12, 2012, http://payload161.cargocollective.com (accessed March 2, 2019).

78 See Gabriel Peters-Lazaro, "Civic Imagination at the 2016 Salzburg Academy on Media and Global Change," Medium, September 28, 2016, https://medium.com.

79 Jenkins et al., *By Any Media Necessary*.

80 Marah Eakin, "Pop Pilgrims: How *Twilight* Helped Keep a Struggling Logging Town on Washington's Olympic Peninsula Afloat," A.V. Club, October 31, 2013, https://music.avclub.com (accessed February 24, 2018); "*Twilight* Star Robert Pattinson Increases Tourism by 1000 Percent in Washington State Town Where New Moon Is Set," *Daily Mail*, November 13, 2009, http://www.dailymail.co.uk (accessed February 24, 2018); Lois Farrow Parhsley, "Travels in *Twilight* Territory," *Atlantic*, November 17, 2011, https://www.theatlantic.com; and Cynthia Wills-Chun, "Touring the *Twilight* Zone: Cultural Tourism and Commodification on the Olympic Peninsula," in Click et al., *Bitten by* Twilight, 261–80.

81 See, for instance, the "Forever *Twilight*" Forks Festival list of events, Discover Forks, Washington, http://www.forkswa.com (accessed February 20, 2019).

82 Nicole Feenstra, "A *Twilight* Trip in Seattle and Forks, Washington," Canoe, May 22, 2013, http://blogs.canoe.ca (accessed May 30, 2013).

83 This was a part of PETA's larger campaign in which they had beautiful actresses and models disrobe as an activist statement against wearing fur. See PETA video, "I'd Rather Go Naked Than Wear Fur," which tells the history of the "Go Naked" campaign: www.peta.org (accessed February 24, 2018). Christian Serratos's photo in the PETA campaign can be found here: "Christian Serratos Is Rosita on the Walking Dead," I-95, http://i95rocks.com/ (accessed February 24, 2018).

84 "*Twilight* Star Hits Cemetery in Sexy, Dark PETA Ad," PETA, November 12, 2012, www.peta.org.uk (accessed February 24, 2018).

85 Laura Wright, "Vegan and Vegetarian Vampires," Vegan Body Project, February 27, 2011, http://veganbodyproject.blogspot.com (accessed February 24, 2018).

86 See SilverCloud234, "Green Bella," FanFiction.net, May 10, 2010, https://m.fanfiction.net (accessed February 5, 2018).

87 "Volvo Target Tweens in *Twilight: New Moon*," TheCarConnection, October 28, 2009, www.thecarconnection.com (accessed February 24, 2018).

88 Natalie Wilson, "Can You Buy That *Twilight* Feeling?" Seduced by *Twilight*, October 8, 2009, https://seducedbytwilight.wordpress.com (accessed February 24, 2018).

89 Christine Mathias, "The Ten Most Baffling *Twilight* Products," *Salon*, June 28, 2010, https://www.salon.com (accessed February 24, 2018).

90 Wilson, "Can You Buy That *Twilight* Feeling?"; and Jedeikin, "14 Weirdest Pieces of *Twilight* Merchandise."

91 "More Disturbing *Twilight* Products," Monster Scholar, December 2009, https://monsterscholar.blogspot.com (accessed February 24, 2018).

92 J. Turner Masland, "Stephenie Meyer Is the Worst Ecological Catastrophe of the 21st Century," *Dewey's Not Dead*, September 22, 2009, http://deweysnotdead.blogspot.com (accessed February 24, 2018).

93 Ibid.

94 De Certeau, *The Practice of Everyday Life*; Fiske, *Understanding Popular Culture*; Hall, "Encoding, Decoding"; and Jenkins, *Convergence Culture*.

95 Matt Richenthal, "Ian Somerhalder Makes Plans for *The Vampire Diaries* to Go Green," TV Fanatic, July 7, 2010, www.tvfanatic.com (accessed February 24, 2018); Maranda Pleasant, "Ian Somerhalder Interview," *Origin*, April 8, 2013, 20.

96 *The Vampire Diaries*, "Bad Moon Rising," season 2, episode 3.

97 See Antoine Faivre, *Theosophy, Imagination, Tradition: Studies in Western Esotericism*, trans. Christine Rhone (Albany: SUNY Press, 2000), xvi, xxi, 23, 50–52.

98 See the fan feed comments in response to the "Nature" entry at the *Vampire Diaries* wiki, http://vampirediaries.wikia.com (accessed February 24, 2018).

99 "Do You Believe the Balance of Nature Line?" Vampire-Diaries.net Forums, July 9, 2013 (accessed March 5, 2015). On FanFiction.net, there are also a number of fan fiction postings that take up this thorny issue of the "balance of nature" in the series and expand upon it.

100 *True Blood*, season 2, episode 1, "Nothing But the Blood," aired June 14, 2009.

101 See, for example, Nana Rivera, "Vampire Bill," www.pinterest.com; "Even Vampires Recycle!" So Fresh and Green, June 19, 2009, https://sofreshsogreen.wordpress.com; Transcript, "True Blood in Dallas" chat room, https://docuri.com (accessed February 24, 2018).

102 TrueBlood@TrueBloodHBO (Twitter account), April 22, 2011, https://twitter.com/truebloodhbo/status/61417347772186624; True Blood Official Facebook page, "Happy Earth Day," Facebook, www.facebook.com (accessed February 24, 2018).

103 See, for example, Gerri Miller, "*True Blood* Couple Obsessively Recycles," Mother Nature Network, August 3, 2011, https://www.mnn.com (accessed February 24, 2018).

104 Rick Perlstein, "It's Mourning in America," *Village Voice*, October 26, 2004, https://www.villagevoice.com (accessed February 2, 2018). This image first appeared in *True Blood* in season 1, episode 4, "Escape from Dragon House," which first aired on September 29, 2008.

105 Matt Haber, "2004 *Village Voice* Cover Makes a Cameo on HBO Vampire Series," *Observer*, September 30, 2008, http://observer.com; Shadaliza, "Vampire George," The Vault (True Blood Fan Site), December 23, 2008, www.trueblood-online.com (accessed February 2, 2018).

106 Shayne Pepper, "Public Service Entertainment: HBO's Interventions in Politics and Culture," in Howley, ed., *Media Interventions*, 128.

107 Pepper, "Public Service Entertainment," 131.

108 *True Blood*, season 3, episode 9, "Everything Is Broken," first aired August 15, 2010.

109 Andrea Chalupa, "*True Blood* Is Making Me Want to Be a Vegan," *Huffington Post*, August 16, 2010, https://www.huffingtonpost.com (accessed February 24, 2018).

110 On the characterization of nonexpert participatory audience theorizing of media content as, in effect, "lay theorizing," see Patrick McCurdy, "Theorizing 'Lay Theories of Media': A Case Study of the Dissent! Network at the 2005 Gleneagles G8 Summit," *International Journal of Communication* 5 (2011): 619–38.

111 Marina Hanes, "*True Blood* Vampire's Bite: Show's Strong Enough to Make You Go Vegan," VeganMainstream, August 23, 2010, https://www.veganmainstream.com (February 24, 2018).

112 Pepper, "Public Service Entertainment," 141.

113 Christine Patton, "Vampire Coalition Unveils 'Save the Humans' Program," reprinted on Resilience.org, December 12, 2012, and originally on Patton's blog at http://peakoilhausfrau.blogspot.com (accessed February 24, 2018).

114 Jean-Paul Sartre uses the word "trouble" to denote disturbance or agitation. See *Being and Nothingness: A Phenomenological Essay on Ontology* (New York: First Washington Square Press, 1993), 503. Sartre compares consciousness "troubled" by desire to "troubled" water that is no longer transparent due to unseen disturbance. In this sense, the vegetarian vampire disturbs or agitates notions of desire and their connection to notions of environmental virtue.

115 See Couldry's "Afterword," in Howley, ed., *Media Interventions*, 402.

CHAPTER 6. COMPOSTING A LIFE

1 Mary Catherine Bateson, *Composing a Life* (New York: Grove Press, 2001).

2 Bateson, *Composing a Life*.

3 Sunya Vatomsky, "Thinking about a Green Funeral? Here's What to Know," *New York Times*, March 22, 2018, https://www.nytimes.com.

4 Vatomsky, "Thinking about a Green Funeral?"

5 Vatomsky, "Thinking about a Green Funeral?"

6 Jesicca Dickler, "Most Americans Live Paycheck to Paycheck," *NBC News*, August 24, 2017, https://www.cnbc.com.

7 Stephen Christy Jr., "The Final Stop for Land Trusts: It's Time for Land Trusts to Enter the Land of the Dead," Land Trust Alliance, http://www.greenburialcouncil.org (accessed July 8, 2018).

8 Harold Schechter, *Whole Death Catalog: A Lively Guide to the Bitter End* (New York: Ballantine Books, 2009), 209.

9 "FAQ," Glendale Memorial Nature Preserve web site, http://glendalenaturepreserve.org (accessed July 20, 2018).

10 Gary Laderman, *The Sacred Remains: American Attitudes toward Death, 1799–1883* (New Haven, CT: Yale University Press, 1996), 152–54; Jessica Mitford, *The American Way of Death Revisited* (New York: Vintage Books, 1998), 55, 142–44.

11 See Gary Laderman's account of Lincoln's death and the public exhibition of his embalmed corpse in *The Sacred Remains*, 157–63. It is also important to note that media, particularly the then new technology of photography, played a critical role in processing the Civil War dead, as photographs were used to "capture" the dead while they were still "fresh," making it possible for relatives to identify the dead before heat and decay made them unrecognizable. See Ross Kelbaugh, *Introduction to Civil War Photography* (New York: Thomas Publishing, 1991). In terms of decades of near ubiquity of embalming in the funerary trade, note that traditional Jewish and Muslim burials disallow this practice.
12 Robert Pogue Harrison, *The Dominion of the Dead* (Chicago: University of Chicago Press, 2003), xi.
13 Mary Douglas's *Purity and Danger: An Analysis of the Concepts of Pollution and Taboo* (1966), as quoted in Mary Bradbury, *Representations of Death: A Social Psychological Perspective* (New York: Routledge, 1999), 116.
14 Here, I refer to the work of cultural anthropologist Ernest Becker and psychologists who have theorized the "denial of death" to be a potent motivating force shaping human behavior. Whether one agrees with Becker's assessment or not, Becker's best-selling book nonetheless brought the notion of "death denial" into popular cultural consciousness and discourse. His concept of "death denial" as a motivating factor in human behavior continues to shape the world of marketing and advertising. See Becker's classic *The Denial of Death* (New York: Simon and Schuster, 1973). For an explanation of actor-network theory (ANT) and "actants," as they are used in material semiotic social theory as a way to signal and describe interacting networks of human and nonhuman combinations of materialities, human agency, ideas, cultural practices, and social processes, see Bruno Latour, *Science in Action: How to Follow Scientists and Engineers through Society*, reprint edition (Cambridge, MA: Harvard University Press, 1988), 84–91.
15 Woodward, *Understanding Material Culture*, 172.
16 In the field of marketing, the "consumer journey" or "consumer decision journey" (CDJ) refers to the processes, pathways, and "funnels" a consumer goes through in order to decide on what products to purchase. This is nearly always a "hero's journey," as comparative mythologist Joseph Campbell termed it, and so is predicated on story that is focused on the individual actor.
17 Peters, *The Marvelous Clouds*, 383.
18 Peters, *The Marvelous Clouds*, 2–3.
19 Peters, *The Marvelous Clouds*, 266. On bodies as media, see also Bernadette Wegenstein, *Getting under the Skin: Body and Media Theory* (Cambridge: MA: MIT Press, 2006), in which Wegenstein asserts the body to be the fundamental *basis* of media.
20 See Kenneth Kramer, *The Sacred Art of Death and Dying: How the World's Religions Understand Death* (Mahwah, NJ: Paulist Press, 1988), 5.
21 This theme is taken up, for instance, in Mark Harris, *Grave Matters: A Journey through the Modern Funeral Industry to a Natural Way of Burial* (New York:

Scribners, 2007); Suzanne Kelly, *Greening Death: Reclaiming Burial Practices and Restoring Our Tie to the Earth* (Lanham, MD: Rowman and Littlefield, 2015); Amy Brown, Tony Hale, Jeremy Kaplan, and Brian Wilson's environmental film documentary *A Will for the Woods* (Overwhelming Umbrella Productions, 2014); and Bob Butz, *Going Out Green: One Man's Adventure Planning His Own Funeral* (Traverse City, MI: Spirituality and Health Books, 2009), among numerous titles now addressing the greening of death and burial.

22 Peters, *The Marvelous Clouds*, 145.

23 Peters, *The Marvelous Clouds*, 145–46.

24 Stephanie Wienrich and Josephine Speyer of the Natural Death Centre in London, an advocacy and resource group for woodland/green burial in the United Kingdom, feature multiple accounts of "getting back to nature" in death as in life in their edited volume, *The Natural Death Handbook* (London: Rider, 2003).

25 See, for instance, Verplank Enterprises' "Centurion 38," fabricated of solid bronze, www.verplank.com (accessed December 13, 2017). The Hecox Goodwin company also carries a "Citadel" casket, as does "Casket Depot," and "Citadel" tends to be a popular name in casket products and merchandising.

26 See "Stainless steel by the numbers. Why Clark's 304 stainless steel offers the best protection," www.clarkgravevault.com (accessed December 13, 2017).

27 See "Solid Copper 12 Gauge," www.clarkgravevault.com (accessed December 13, 2017).

28 See "How a Clark Vault Protects," www.clarkgravevault.com (accessed December 13, 2017).

29 David Chidester writes from a religious studies perspective about the mystical and utopian promise of Tupperware, indeed the "cult" of Tupperware, in *Authentic Fakes*, chapter 3 on "Plastic Religion." Marketing for "sealer technology" caskets bears striking resemblance to the Tupperware cult and its utopian promise of eternal freshness in a world blissfully and magically free of spoilage.

30 Josh Slocum, "What You Should Know about Exploding Caskets," *Washington Post*, August 11, 2014, https://www.washingtonpost.com; Kieron Monks, "Exploding Caskets: 'Tupperware for the Dead' Blamed for Freakish Phenomena," *Metro World News*, July 8, 2013, https://www.metro.us; Mark Harris, *Grave Matters*, 35–36.

31 A felt crafting company named Bellacouche in Devon, UK, makes and markets their Leafcocoon as an "eco-coffin alternative." Adorned with natural-looking leaves, the Leafcocoon looks like the cross between a shroud and a "hot pocket" for the dead.

32 See, for instance, http://www.naturalburialcompany.com; Nature's Casket, www.naturescasket.com; Kinkaraco Green Funeral Products, https://kinkaraco.com; *Passages: The Natural Choice Product Catalog* (for green funeral products), www.passagesinternational.com (accessed March 5, 2019).

33 See "Slate Banana Green Casket," Memorials.com, https://www.memorials.com (accessed July 8, 2018).

34 Schechter, *The Whole Death Catalog*, 265.
35 "*Six Feet Under* Kills in the Ratings," *Chicago Tribune*, August 24, 2005, http://articles.chicagotribune.com.
36 Marc Graser, "Toyota Prius Scores a Perfect Product Placement: Why 'Six Feet Under' Was an Uplifting Brand Experience," *Ad Age*, September 7, 2005, http://adage.com.
37 Susan Chumsky, "The Rise of Back-to-the-Basic Funerals: Baby Boomers Are Drawn to Green and Eco-Friendly Funerals," *New York Times*, March 12, 2014, https://www.nytimes.com; Dana Tim, "Death Goes Green with Eco-Funeral Burials," *Oregonian*, November 21, 2009, http://www.oregonlive.com.
38 "Singing for Our Lives," *Six Feet Under*, season 5, episode 8.
39 Wendy MacNaughton, "Green Burials: At the End of Life, Thinking outside the Coffin," *New York Times*, November 15, 2018, www.nytimes.com.
40 In providing a "monkey wrench" to open land development, the eco-pious dead take on both symbolic and legal agency as active "wards" or talismans protecting and defending the land.
41 A classic of literature on death and American culture, muckraking journalist Jessica Mitford's 1963 *American Way of Death* succeeded in opening a previously taboo discussion about death, consumerism, profit, and value. Mitford is credited for launching a wave of consumer consciousness about funerary products and services. The character Nate in *Six Feet Under* is prophesying a similar shift in funerary awareness ushered in by green burial advocates.
42 "About Us," Kinkaraco Funeral Products, https://kinkaraco.com (accessed August 1, 2017).
43 Kinkaraco products' web site, https://kinkaraco.com/pages/earth (accessed March 3, 2019).
44 Kinkaraco products' web site.
45 Esmerelda Kent, "Practicing Buddhist Economics in the Funeral Industry," Kinkaraco Funeral Products Blog, https://kinkaraco.com (accessed July 23, 2018).
46 "About Us: The Founder's Story," Kinkaraco web site, https://kinkaraco.com (accessed March 6, 2019); Aurora Wells, "What Is a Green Funeral? Interview with Esmerelda Kent," *Seven Ponds*, May 3, 2014, http://blog.sevenponds.com (accessed March 6, 2019). On the conversion to Buddhism of Alan Ball, who was raised a Southern Baptist, see Thomas Fahy's *Alan Ball: Conversations* (Jackson: University Press of Mississippi, 2013), 42–45.
47 Derek Beres, "Reconsidering the Business of Death," *Big Think*, April 22, 2014, http://bigthink.com.
48 Available on DVDs and YouTube; see also the series web site at "Gail Rubin: A Good Goodbye," http://agoodgoodbye.com (accessed February 11, 2019). The series can also be viewed on Channel 26 and 27 in Albuquerque's Comcast cable system in New Mexico (as of December 2017). On Youtube.com, go to https://www.youtube.com/watch?v=yIkoLxOjVVk (accessed December 15, 2017).

49 A go-to resource in this genre is Rabbi Arnold Goodman's *Plain Pine Box: A Return to Simple Jewish Funeral and Eternal Traditions* (New York: KTAV Publishing, 1981, 2003). The Jewish burial society Kavod v'Nichum also maintains and updates sources for Jews on green burial resources and practices: http://jewish-funerals.org (accessed February 11, 2019).

50 Monica Emerich, *The Gospel of Sustainability: Media, Market, and LOHAS* (Champagne: University of Illinois Press, 2011). The automotive industry has also conducted intensive, focused research into the appeal of alternative-fuel vehicles (AFVs) to consumers heavily identified with LOHAS marketing. See, for instance, Maximillian Heiler, *Consumer Behavior and the Decision-Making Process of the LOHAS Target Group in the Automotive Industry* (Munich: Grin, 2015).

51 Trish Sheffield, "Advertising," in Gary Laderman and Luis Leon, eds., *Religion and American Cultures*, volume 2 (Santa Barbara, CA: ABC-CLIO, 2003), 445–46; and Sheffield's *Religious Dimensions of Advertising* (Basingstoke, UK: Palgrave Macmillan, 2006).

52 See United Parcel Services' online brochure entitled, "Your Partner in Sustainability," in which the company lays out how consumers can secure carbon-neutral shipping for most anything. This includes relocating deceased loved ones who are packed in dry ice for the journey: www.completeview.ups.com (accessed February 11, 2019). This method of shipping is also featured in the environmental documentary, *A Will for the Woods* (2014).

53 This referential spectrum of "light green" to "dark green" was first formally outlined and theorized in an ethical context by Patrick Curry in *Ecological Ethics: An Introduction* (Cambridge, UK: Polity, 2005). Curry's model has been highly generative to ecological ethical conversations since, and judging by Crouch's comments, is also influential on green burial product and service marketers.

54 See AGreenerFuneral.org for explanations and gradations of funerary ecopiety.

55 Glennys Howarth, *Death and Dying: A Sociological Introduction* (Cambridge, UK: Polity, 2007), chapter 5, "Death and the Media," 102.

56 Note that Rubin's show is also sponsored by funeral product suppliers, and her interview with Crouch could be construed as essentially a free "infomercial" for the company's green burial line of products.

57 See Amos Tversky and Daniel Kahneman's classic essay, "Judgment under Uncertainty: Heuristics and Biases," *Science*, September 27, 1974, 1124–31; Kahneman's *Thinking, Fast and Slow* (New York: Farrar, Straus, and Giroux [Macmillan], 2011), section 4 on "Choices," 269–376; and Derek Thompson, "The Irrational Consumer: Why Economics Is Dead Wrong about How We Make Choices," *Atlantic*, January 16, 2013, www.theatlantic.com.

58 Historian David Shi writes about just how caught up romantic notions of "the simple life" and ideals of "living simply" are with American notions of moral virtue. From Puritan piety to hippie back-to-the-land commune piety, the American virtue of "simplicity" is chronically morally fraught and highly complicated. See

Shi's *Simple Life: Plain Living and High Thinking in American Culture* (Athens: University of Georgia Press, 2007). In the practice of green burial, as in other realms of ecopiety, antimaterialistic sentiments of "simplicity" ironically often require a whole new set of elaborate consumer acquisitions.

59 See, for instance, "New York Green Burial Meetup" on Meetup.com (accessed February 11, 2019); "Green Burials and DIY Funerals," Radio Times, WHYY.org, January 23, 2017; Mark Harris, "Taking on Alabama's Restrictive Funeral Laws," *Progressive*, August 3, 2015, http://progressive.org.

60 Ann Hoffner, "Why Does Grave Depth Matter for Green Burial?" GreenBurial-Naturally, March 2, 2017, www.greenburialnaturally.org (accessed March 6, 2019).

61 Joel Banner Baird, "Green Burial Plans Elude Burlington-Area Hospice Patient," *Burlington Free Press*, January 25, 2018, www.burlingtonfreepress.com.

62 "Advocate with Action," National Home Funeral Alliance, www.homefuneralalliance.org (accessed July 8, 2018).

63 Joshua Slocum and Lisa Carlson, *Final Rights: Reclaiming the American Way of Death* (Hinesburg, VT: Upper Access Press, 2011), 143.

64 Slocum and Carlson, *Final Rights*, 141.

65 Slocum and Carlson, *Final Rights*, 143.

66 Slocum and Carlson, *Final Rights*.

67 Slocum and Carlson, *Final Rights*, 142–43.

68 Slocum and Carlson, *Final Rights*, 145.

69 Wendii Miller, "A Very Natural D.I.Y. (Dig It Yourself) Burial," https://www.youtube.com/watch?v=BOXMa1oWEXY (accessed December 5, 2017).

70 "Everything You Know about Funerals Is Wrong," YouTube.com, September 9, 2013, https://www.youtube.com/watch?v=wIBkKGPfJ9Q (accessed June 30, 2018).

71 Joe Dominguez and Vicki Robin, *Your Money or Your Life* (New York: Penguin, 1993).

72 The broader trend toward eco-spiritual sensibilities is chronicled in Roger Gottlieb's *Greener Faith*.

73 Lisa Carlson, *Caring for the Dead: Your Final Act of Love* (Hinesburg, VT: Upper Access, 1998), 13–14.

74 Carlson, *Caring for the Dead*, 36.

75 Carlson, *Caring for the Dead*, 45. Here, Carlson is quoting Thomas Lynch's *Undertaking: Life Studies from the Dismal Trade* (New York: Penguin Books, 1997).

76 Lynch quoted in Carlson, *Caring for the Dead*, 45.

77 Charles Landman, "Green Burial: My Body, My Self," Conscious Elders Network, www.consciouselders.org (accessed July 22, 2018).

78 Jim Harper, "My Big Fat Green Funeral," *Biscayne Times*, February 2012, www.biscaynetimes.com.

79 Shannon Palus, "How to Be Eco-Friendly When You're Dead," *Atlantic*, October 30, 2014, www.theatlantic.com.

80 The 2017 environmental documentary film *Here After*, directed and produced by Sarah Friedland and Esy Casey, provides a stunning entrée into the "Neptune Memorial Reef" located 3.25 miles of Miami's coast. Built from human cremains and concrete, the reef now spans sixteen miles. See www.hereafterdoc.us (accessed December 15, 2017).

81 "What to Expect," Eternal Reefs web site, www.eternalreefs.com (accessed December 5, 2017).

82 "Eternal Reefs Golden Beach Reef, a Tour of New and Mature Reef Balls," https://www.youtube.com/watch?v=ohm9kxmGfSg; Eternal Reefs promotional DVD; Eternal Reefs promotional brochure, www.eternalreefs.com; "Eternal Reefs Let You Rebuild an Ecosystem after You're Gone" (video), https://vimeo.com (accessed December 12, 2017).

83 Innis as quoted in Peters, *Marvelous Clouds*, 19.

84 Eternal Reefs web site, www.eternalreefs.com (accessed December 12, 2017).

85 Eternal Reefs Video Brochure, www.youtube.com/watch?v=W_4jU1psvLI (accessed February 11, 2019).

86 "About Reef Balls," Eternal Reefs, www.eternalreefs.com (accessed December 16, 2017).

87 "The Eternal Reefs Story," http://www.eternalreefs.com (accessed December 3, 2018).

88 "The Eternal Reefs Story."

89 "The Eternal Reefs Story."

90 "Eternal Reefs Final Expense Planning," http://www.eternalreefs.com (accessed December 3, 2018).

91 Greensprings Natural Cemetery Preserve, Newfield, NY, www.naturalburial.org; "Walk in the Park: Greensprings Natural Cemetery Preserve" (video), www.youtube.com/watch?v=zRo8nB5hEEc (accessed July 31, 2017).

92 Mary Woodsen, Greensprings "Education" page, www.naturalburial.org (accessed December 15, 2017).

93 Political philosopher Jane Bennett offers a third way of "thing power" between the conventional binary of "subject" and "object," illuminating the vitality of both human and nonhuman "actants" to produce effects as part of a complex web of relationships. Bennett's use of the notion of "actants" in relationship to the work of Bruno Latour will be discussed later in this book. See Jane Bennett, *Vibrant Matter: A Political Economy of Things* (Durham, NC: Duke University Press, 2010), 8–10.

94 Kittler, as described by Peters, *Marvelous Clouds*, 25.

95 C. A. Beal, "Be a Tree: The Natural Burial Guide for Turning Yourself into a Forest," http://www.beatree.com (accessed March 6, 2019).

96 Beal, "Be a Tree."

97 Beal, "Be a Tree."

98 Beal, "Be a Tree."

99 See Mircea Eliade's discussion of the concept of *axis mundi*, including the world tree, in *Images and Symbols*, trans. Philip Mairet (Princeton, NJ: Princeton University Press, 1991), 48–51.

100 The Capsula Mundi Team, "The Burying of the Future: An Interview with Capsula Mundi Designers," official Capsula Mundi web site, September 22, 2017, www.capsulamundi.it (accessed March 12, 2019).

101 "Welcome to the Capsula Mundi Shop," www.capsulamundi.it (accessed March 12, 2019).

102 See Katie Rogers, "Mushroom Suits, Biodegradable Urns, and Death's Green Frontier," *New York Times*, April 22, 2016, www.nytimes.com.

103 Woodward, *Understanding Material Culture*, vi–vii.

104 "Back to Nature, Back to Life," Bios Urn web site, https://urnabios.com (accessed December 10, 2017).

105 "Stories of the Bios Urn" (video), www.youtube.com/watch?v=stnFgIusJuo (accessed December 15, 2017).

106 "Bios Incube," Kickstarter.com campaign, www.kickstarter.com (accessed December 8, 2017).

107 The term "the Internet of things" refers to the ways in which everyday consumer items are increasingly linked to the Internet, sharing our consumer information with manufacturers, marketers, and consumer databases, often without our cognizance of where and how our data will be used. See Merced Bunz and Graham Meikle, *The Internet of Things* (Cambridge, UK: Polity, 2017).

108 Elizabeth Stinson, "Turn Your Dead Grandma into a Tree with This Smart Planter," *Wired*, July 29, 2017, www.wired.com; Lulu Chang, "Grow a Tree from a Loved One's Ashes, and the Bios Incube Will Keep It Alive," *Digital Trends*, July 7, 2017, www.digitaltrends.com.

109 See Billy Campbell's comments in Nancy Rommelmann, "Crying and Digging: Reclaiming the Realities and Rituals of Death," *L.A. Times*, February 6, 2005, http://articles.latimes.com.

110 Carlson, *Caring for the Dead*, 9–14.

111 Ann Braude, *Radical Spirits: Spiritualism and Women's Rights in Nineteenth-Century America* (Bloomington: Indiana University Press, 2001), 14.

112 Jeremy Sconce, in *Haunted Media: Electronic Presence from Telegraphy to Television* (Durham, NC: Duke University Press, 2000), details the ways in which supernatural "presence" has historically infused electronic media technologies. Sconce's chapter on "Static and Stasis" probes radio and television technologies and their culturally imagined connections with supernatural communication.

113 Patty Khuly, "When Your Pet Is Gone, What to Do with Those Ashes?" *Pet MD*, July 2, 2009, www.petmd.com; Bess Lovejoy, "Cremation Is on the Rise, but Where to Put the Ashes?" *Time*, June 13, 2013, http://ideas.time.com.

114 "Bios Urn: Back to Nature, Back to Life," https://urnabios.com/urn (accessed December 3, 2018).

115 On the "move," see Robert Orsi, *The Cambridge Companion to Religious Studies* (Cambridge: Cambridge University Press, 2012), 99. Adam Vaughan, "Data Centre Emissions Rival Air Travel as Digital Demand Soars: How Viral Cat Videos Are Warming the Planet," *Guardian*, September 25, 2015, www.theguardian.com;

Ingrid Burrington, "The Environmental Toll of a Netflix Binge: The Data Centers That Support the Internet Use a Huge Amount of Energy," *Atlantic*, December 15, 2015, www.theatlantic.com. Burrington's article disabuses readers of the notion that there are such things as "clean clouds" and that using the Internet is somehow more environmentally virtuous than old-fashioned paper-based communication.

116 Ernest Becker, *The Denial of Death* (New York: Simon and Schuster, 1973).

117 Phillippe Ariès, *Western Attitudes toward Death from the Middle Ages to the Present* (Baltimore, MD: Johns Hopkins University Press, 1974), 13–14.

118 These acts, as outlined in Becker's work, are well summarized by Roger Gottlieb in *A Spirituality of Resistance: Finding a Peaceful Heart and Protecting the Earth* (New York: Rowman Littlefield, 1999), 20.

119 See, for instance, Scranton's *Learning to Die in the Anthropocene*, in which Scranton implicates the widespread Western pathology of death denial for human failure to face climate crisis head on and to deal with it honestly and practically. Scranton, a practicing Buddhist, finds the Buddhist acceptance of the naturalness and inevitability of impermanence to be a much better social-spiritual/cognitive tool for humans to employ as we inhabit the Anthropocene.

120 See Julie Beck, "What Good Is Thinking about Death?" *Atlantic*, May 28, 2015, www.theatlantic.com; and Sheldon Solomon, *The Worm at the Core: On the Role of Death in Life* (New York: Random House, 2015).

121 William Irvine, *A Guide to a Good Life: The Ancient Art of Stoic Joy* (New York: Oxford University Press, 2008), 198. See also Julie Beck's discussion of Irvine's book in "What Good Is Thinking about Death?"

122 Justin McDaniel, *The Lovelorn Ghost and the Magical Monk: Practicing Buddhism in Modern Thailand* (New York: Columbia University Press, 2011), 125. I am indebted to Buddhologist Sarah Jacoby for consulting on this concept. Jacoby adds that this kind of meditation on objects of repulsion also can be a contemplative tool for dissuading lust.

123 See Dutch consumer researchers Naomi Mandel and Dirk Smeesters, "The Sweet Escape: The Effect of Mortality Salience on Consumption Quantities for High and Low Self-Esteem Consumers," *Journal of Consumer Research* 35 (August 2008): 309–23; and "Morbid Thoughts Whet the Appetite," *Science Daily*, June 25, 2008, www.sciencedaily.com (accessed March 6, 2019). See also Stephanie O'Donohoe and Darach Turley's treatment of Benetton clothing company's death and disaster campaigns and the company's successful use of "catastrophe aesthetics," including bloody soldier shirts, grave sites, and images of death row, in "Dealing with Death: Art, Mortality, and the Marketplace," in Stephen Brown and Anthony Patterson, eds., *Imagining Marketing* (New York: Routledge, 2007), 97–98.

124 "Morbid Thoughts Whet Appetite."

125 "Morbid Thoughts Whet Appetite."

126 Sylvain-Jacques Desjardins, "Shopping with the Grim Reaper in Mind: Concordia Study Examines Links between Fear of Death and Shopping Behaviour," *Concordia University News*, February 22, 2010, www.concordia.ca.

127 Stephanie O'Donohue and Darach Turley, "Dealing with Death: Art, Mortality, and the Marketplace," in Stephen Brown and Anthony Patterson, eds., *Imagining Marketing* (New York: Routledge, 2007), 95.

128 Scranton, *Learning to Die in the Anthropocene*, 27.

CHAPTER 7. EXPANDING THE SCOPE OF JUSTICE

1 See social psychologist and justice researcher Susan Opotow's seminal definition in "The Scope of Justice, Intergroup Conflict, and Peace," in Linda R. Tropp, ed., *The Oxford Handbook of Intergroup Conflict* (New York: Oxford University Press, 2012), 72–88.

2 For a detailing of the marketing of Christian "witnessing" products at the International Christian Retailing Show, see Stefanie Simon, "What Would Jesus Sell?" *Los Angeles Times*, July 21, 2006, http://articles.latimes.com; R. Laurence Moore, *Selling God: American Religion in the Marketplace of Culture* (New York: Oxford University Press, 1995), especially chapter 8 on religious marketing; Colleen McDannell, *Material Christianity: Religion and Popular Culture in America* (New Haven, CT: Yale University Press, 1995); and on purity witnessing tools, Sara Moslener's "Don't Act Now! Selling Abstinence in the Christian Marketplace," in Eric Mazur and Kate McCarthy, ed., *God in the Details: American Religion in Popular Culture* (New York: Routledge, 2010), 197–219.

3 For a more detailed sense of what this term encompasses and "media making" as a movement, see Lawrence Grossberg, Ellen Wartella, D. Charles Whitney, and J. MacGregor Wise, *MediaMaking: Mass Media in Popular Culture*, 2nd edition (Thousand Oaks, CA: Sage, 2005).

4 Amanda Baugh, *God and the Green Divide: Religious Environmentalism in Black and White* (Oakland: University of California Press, 2017), 33.

5 Baugh, *God and the Green Divide.*

6 Rural communities are also, of course, plagued by ecoracism and environmental injustices, but Green For All's media making focuses on health and economic challenges in urban spaces.

7 Susan Opotow, "From Moral Exclusion to Moral Inclusion: Theory for Teaching Peace," in Susan Opotow, Janet Gerson, and Sarah Woodside, *Theory into Practice* 44, no. 4 (Autumn 2005): 303–18.

8 For insights into Green For All cofounder Van Jones's vision of "lifting all boats" via an economic "green wave," see Anna Fahey, "A Green Wave Shall Lift All Boats, Says Van Jones," The Sightline Institute, November 9, 2007, www.sightline.org. Green For All's initiatives and community programs work to support the creation of a "more just, peaceful, and verdant world." This phrase also happens to be the motto of the MacArthur Foundation, and Green For All cofounder Majora Carter is one of the foundation's celebrated fellows for her work as an urban revitalization strategist.

9 "Sacrifice zones" are low-income communities located next to, within, or close to polluting industrial sites and exposed disproportionately to high levels of

hazardous chemicals. See Stephen Lerner, *Sacrifice Zones: The Front Lines of Toxic Chemical Exposure in the U.S.* (Cambridge, MA: MIT Press, 2010).

10 Baugh, *God and the Green Divide*, 2.

11 Baugh, *God and the Green Divide*, 30–33.

12 Video of this public "live tattooing" is posted on Tatzoo's open-access, public Facebook page and is designed for outreach and activist circulation. Even so, here I have respectfully omitted this fellow's name and the names of other fellows mentioned in this chapter.

13 Dr. Seuss, *The Lorax* (New York: Random House, 1971).

14 Sarah McFarland Taylor, "Religion and the Environment in Popular Culture," in Gary Laderman and Luis Leon, eds., *Religion and American Cultures.* Volume 3, *Tradition, Diversity, and Popular Expression*, 2nd edition (Santa Barbara, CA: ABC-CLIO, 2014), 1147–48.

15 Tatzoo official web site, http://tatzoo.org (accessed June 1, 2018).

16 Victor Turner, "Betwixt and Between: The Liminal Period in Rites de Passage," in *Forest of Symbols: Aspects of the Ndembu Ritual* (Ithaca, NY: Cornell University Press, 1967), 23–59; Arnold Van Gennep, *The Rites of Passage*, trans. Monica Vizedom and Gabriel Caffee (Chicago: University of Chicago Press, 1961).

17 John Berger, *Why Look at Animals* (New York: Penguin, 2009), 37.

18 This dynamic of "speaking for" has traditionally been part of the proceedings of the Bioregional Congress, which began in the 1980s to include "animal representatives" at its national meeting councils. Animal discernment and "speaking for" other species is also used in deep ecologists John Seed and Johanna Macy's co-created "Council of All Beings" activist process, as portrayed in John Seed, Joanna Macy, Pat Fleming and Arne Naess, *Thinking like a Mountain: Toward a Council of All Beings* (Gabriola Island, B.C.: New Society Publishers, 1998).

19 Tatzoo official video on Facebook.com, https://www.facebook.com/tatzoo/videos/10201279550853715/ (accessed June 1, 2018).

20 Marshall McLuhan, *Understanding Media: The Extensions of Man* (New York: Signet, 1964).

21 Lorraine Chow, "Costa Rica Wants to Become the First Country to Ban Single-Use Plastics," EcoWatch, April 7, 2017, www.ecowatch.com.

22 Sophie Eastaugh, "France Becomes Frist Country to Ban Plastic Cups and Plates," *CNN News*, September 20, 2016, www.cnn.com.

23 Imogen Calderwood, "16 Times Countries and Cities Have Banned Single-Use Plastics," *Global Citizen*, April 25, 2018, www.globalcitizen.org (accessed July 8, 2018).

24 Brynna Strand and Charlie Ann Kerr, "10 Countries and Cities Confronting Plastic Bag Pollution Head-On," EarthDay.org, April 20, 2018, www.earthday.org (accessed July 8, 2018).

25 Marilyn Brewer and Wendi Gardner, "Who Is This We? Levels of Collective Identity and Self Representation," *Journal of Personality and Social Psychology* (July 1996): 84.

26 Margo DeMello, *Animals and Society: An Introduction to Human-Animal Studies* (New York: Columbia University Press, 2012), 34.

27 Peters, *The Marvelous Clouds*, 266.

28 Howley, ed., *Media Interventions*, 5.

29 Howley, ed., *Media Interventions*, 397.

30 Mark Fischetti, "Endangered Tattoos: Volunteers Get Inked to Help Save Species," *Scientific American*, October, 19, 2012, www.scientificamerican.com.

31 extInked Project, www.extinked.org.uk (accessed April 4, 2018).

32 extInked Project.

33 Xavier, "Skinvertising: the World of Human Billboards" Tattoodo, 2015, www.tattoodo.com (accessed April 4, 2018).

34 Sam Walker, "On Sports: This Skin for Rent," *Wall Street Journal*, November 9, 2001, W10; John Vukelej, "Post No Bills: Can the NBA Prohibit Its Players from Wearing Tattoo Advertisements?" *Fordham Intellectual Property, Media, and Entertainment Law Journal* 15, no. 2 (2005): 508–35.

35 Andrew Adam Newman, "The Body as Billboard: Your Ad Here," *New York Times*, February 17, 2019, www.nytimes.com; Vi-An Nguyen, "Would You Get a Tattoo for a 15 Percent Raise?" *Parade*, May 2, 2013, https://parade.com; Rob Garver's article on subsidized corporate logo tattoos, "How Fintech Firms Get Peak Performance from Employees," *American Banker*, March 5, 2018, https://www.americanbanker.com; Abha Bhattarai, "This Local Pizza Chain Pays for Employee Tattoos of the Company's Logo," *Washington Post*, July 9, 2016, https://www.washingtonpost.com.

36 On gift economies, see David Cheal, *The Gift Economy* (New York: Routledge, 2017); Charles Eisenstein, *Sacred Economics: Money, Gift, and Society in the Age of Transition* (Berkeley, CA: North Atlantic Books, 2011).

37 Rhino and skulls, www.bushwarriors.org (accessed June 1, 2018).

38 Dori Gurwitz, "About," Bush Warriors Ink, https://bushwarriors.wordpress.com (accessed June 1, 2018).

39 "21 Responses to Tattoo of the Day," Bush Warriors Ink, June 28, 2010, https://bushwarriors.wordpress.com (accessed April 10, 2018).

40 Bron Taylor, "Earth First! From Primal Spirituality to Ecological Resistance," in Roger Gotlieb, ed., *This Sacred Earth: Religion, Nature, and Environment*, 1st edition (New York: Routledge, 1996), 545; and *Dark Green Religion: Nature Spirituality and the Planetary Future* (Berkeley: University of California Press, 2009).

41 Julia Butterfly Hill, *The Legacy of Luna: The Story of a Woman, a Tree, and the Struggle to Save the Redwoods* (San Francisco: Harper One, 2001), 112–15.

42 Sociologist Émile Durkheim famously argued that "totems" are the simplest and earliest from of religion and that totems collectively represent different aspects of society. The idea of the totem took on an obsessive quality for scholars who were fascinated by them and projected onto them evolutionary cultural theories. See, for instance, Émile Durkheim, *The Elementary Forms of the Religious Life: A Study* *in Religious Sociology* (London: Allen & Unwin, 1915); Sigmund Freud, *Totem and Taboo*, trans. A. A. Brill (New York: Moffat, Yard, 2018).

43 Margo DeMello, *Bodies of Inscription: A Cultural History of the Modern Tattoo Community* (Durham, NC: Duke University Press, 2000), 46–47; and Peter Gathercole, "Contexts of Maori Moko," in Arnold Rubin, ed., *Marks of Civilization* (Los Angeles: Museum of Cultural History, University of California, 1988). I am indebted to Greg Kares for bringing the Maori traditions to my attention.

44 Martha Bebinger, "Extra-Permanent Ink: Preserving Your Tattoos after Death," *National Public Radio*, October 11, 2014, https://www.npr.org; Aaron Smith, "New Company Preserves Tattoos When You Die," *CNN*, October 1, 2017, http://money.cnn.com.

45 Margaret Badore, "Extinction Empathy Tattoos Commemorate Less Charismatic Species," TreeHugger, October 3, 2013, https://www.treehugger.com.

46 Screech Owl looks on during "live tattoo," https://www.facebook.com/tatzoo/photos/a.136669789694393.20125.120452987982740/1345877298773630/?type=3&theater (accessed June 1, 2018).

47 Tatzoo Facebook videos, "Tatzoo Was Live: Live Tattooing at Wild Center in Adirondacks," July 30, 2016, https://www.facebook.com (accessed June 1, 2018).

48 See, for instance, Catholic Worker cofounder Dorothy Day's autobiographical *From Union Square to Rome* (Maryknoll, NY: Orbis, 2006 [1938]), 7–8. Day writes of her extended self, suffering with the poor as part of the mystical body of Christ.

49 For an ethnography that provides further such examples of suffering, piety, and devotion, see Robert Orsi, *Thank You, St. Jude: Women's Devotion to the Patron Saint of Hopeless Causes* (New Haven, CT: Yale University Press, 1996).

50 Testimonials: http://tatzoo.org (accessed June 1, 2018).

51 As an example of contemporary mediated ecstatic experience of tattoo pain in popular culture, see Spencer Drew, "Church of Pain: Religion, Ritual, and the Body in the New Series Spin-Off, *S-Town*," *Religion Dispatches*, April 11, 2017, http://religiondispatches.org.

52 See ethnic studies scholar Alberto Pulido's discussion of the brotherhood of flagellants in *The Sacred World of the Penitentes* (Washington, DC: Smithsonian Institution Press, 2000); and sociologist Michael Carroll's *Penitente Brotherhood: Patriarchy and Hispano-Catholicism in New Mexico* (Baltimore, MD: Johns Hopkins University Press, 2002), 76–89.

53 Victoria Pitts, *In the Flesh: The Cultural Politics of Body Modification* (New York: Palgrave Macmillan, 2003), 46–48.

54 Testimonials: http://tatzoo.org (accessed June 1, 2018).

55 Martha Henriques, "Tattoo Ink Nanoparticles Pose Long-Term Risk by Clogging Lymph Nodes," *International Business Times*, reposted to *Environmental Health News*, September 14, 2017; Jason Daley, "Tattoo Ink May Stain Your Lymph Nodes," reposted from Smithsonian.com to Environmental Health News, September 14, 2017; Dan Garisto, "Inked Mice Hint at How Tattoos Persist in People,"

reposted from *Science News* to Environmental Health News, March 16, 2018, www.ehn.org (accessed June 15, 2018).

56 Brett Israel, "Inkling of Concern: Chemicals in Tattoo Inks Face Scrutiny," *Center for Public Integrity*, August 3, 2011, https://www.publicintegrity.org.

57 See Debbie McCulliss's synopsis of the medical literature on tattoo health risks, including environmental pollutants, in "The Risk of Getting Inked," *Healthy Living*, March 2015, https://healthylivingmagazine.us; and World Health Organization, *Environmental Health Criteria: Dermal Exposure* (Geneva, Switzerland, 2014), 300–315, www.who.int.

58 See anthropologist and cultural theorist Mary Douglas's, *Purity and Danger: An Analysis of Concepts of Pollution and Taboo* (New York: Routledge, 1966).

59 Lerner, *Sacrifice Zones*, 2–3.

60 See Sarah Alisabeth Fox, *Downwind: A People's History of the Nuclear West* (Lincoln: University of Nebraska Press, 2014).

61 These are referred to by the government technically as "exclusion zones," which means that people are restricted from actually inhabiting these zones. "Sacrifice zones," on the other hand, are inhabited by those whose health and lives are deemed to be acceptable sacrifices by planners and officials who permit and even provide incentives for polluting industries to locate in these areas.

62 Traci Brynne Voyles, *Wastelanding: Legacies of Uranium Mining in Navajo Country* (Minneapolis: University of Minnesota Press, 2015); Michael Amundson, *Yellowcake Towns: Uranium Mining Communities in the American West* (Boulder: University of Colorado Press, 2004).

63 For a history of NIMBY politics and its impact on low-income and mostly minority communities, see Carl Zimring, *Clean and White: A History of Environmental Racism in the United States* (New York: NYU Press, 2016); and Dorceta Taylor, *Toxic Communities: Environmental Racism, Industrial Pollution, and Residential Mobility* (New York: NYU Press, 2014).

64 Mark Winne, *Closing the Food Gap: Resetting the Table in the Land of Plenty* (Boston: Beacon, 2009); and Alison Hope Alkon, *Cultivating Food Justice: Race, Class, and Sustainability* (Cambridge, MA: MIT Press, 2011).

65 On the high rates of correlation between urban minority populations in the United States and skyrocketing asthma rates, see Jon Wallace, "Study Links Unhealthy Segregated Neighborhoods to Childhood Asthma," Princeton University Woodrow Wilson School of Public and International Affairs, August 4, 2017, wws.princeton.edu. The study finds that the biggest health risk for asthma among African American children comes from being trapped in unhealthy neighborhoods located near the output of hazardous industries. These children are made unwilling "sacrifices" to companies' bottom lines.

66 Miller, *Religion and Hip Hop*, 71.

67 Markese Bryant, "Behind the Video: Dream Reborn," YouTube, December 14, 2009, www.youtube.com/watch?v=wem2MC63Ufs.

68 Markese Bryant, "The Dream Reborn (My President Is Green)," YouTube, December 14, 2009, www.youtube.com/watch?v=I52vlPByrUc.
69 Note that when this video was first produced in 2009 and funded by the organization Green For All, cofounded by then-Obama administration environmental advisor Van Jones, it reflected a great optimism about what the Obama administration would be able to accomplish in terms of environmental policy.
70 Green For All, "A New Sound: Green For All," www.greenforall.org. See also a video about a family dealing with their daughter's environmental health problems, including emergency room visits for asthma, in Green For All's "The Cost of Delay," May 22, 2012, www.youtube.com/watch?v=mBb3ZSi-cIs (accessed June 15, 2018).
71 Seasunz and J. Bless with "stic.man," "Global WarNing," YouTube, June 15, 2012, www.youtube.com/watch?v=aV3C-e2ukWw (accessed June 5, 2018).
72 On the trope of nature as moral "safety valve" for the corruption and volatility of the city, see American historian Peter Schmitt, *Back to Nature: The Arcadian Myth in Urban America*, 2nd edition (Baltimore, MD: Johns Hopkins University Press, 1990). For an argument for cities as "solutions" rather than "problems," see B. D. Wortham Galvin et al., eds., *Sustainable Solutions: Let Knowledge Serve the City* (New York: Routledge, 2016).
73 Anthony Pinn, *Why Lord? Suffering and Evil in Black Theology* (New York: Continuum, 1995).
74 This is not to say that polar bears and whales are not important; they are. However, when their seemingly far-away plight is utilized as motivation for action, it is a tough sell when minority children are ingesting lead, struggling for basic nutrition, dealing with asthma crises while lacking health care coverage, facing the general health risks of poverty, and being shot at.
75 Christopher Weber, "Speak Your Mind, That's Wind Power!" *E*, July 25, 2011, https://emagazine.com.
76 On rap music as an important tool for moral critique and social resistance, see Theresa Martinez, "Popular Culture as Oppositional Culture: Rap as Resistance," *Sociological Perspectives* 40, no. 2 (1997): 265. On popular culture as oppositional culture, see Sterling Stuckey, *Slave Culture* (New York: Oxford University Press, 1987); James Scott, *Domination and the Arts of Resistance* (New Haven, CT: Yale University Press, 1990); Bonnie Mitchell and Joe Feagin, "America's Racial-Ethnic Cultures: Opposition within a Mythical Melting Pot," in Benjamin Bowser, ed., *Toward the Multicultural University* (Westport, CT: Praeger, 1995); Tricia Rose, *Black Noise: Rap Music and Black Culture in Contemporary America* (Hanover, NH: Wesleyan University Press, 1994). For a more complete discussion of rap music and moral panics, see Cathy Cohen's chapter, "Gangsta Rap Made Me Do It: Moral Panics," in *Democracy Remixed: Black Youth and the Future of American Politics* (New York: Oxford University Press, 2012), 18–48.

77 See Miller, *Religion and Hip Hop*; Monica Miller, Anthony Pinn, and Bernard Freeman, eds., *Religion in Hip Hop: Mapping the New Terrain in the U.S.* (New York: Bloomsbury, 2015); Murray Forman and Mark Anthony Neal, eds., *That's the Joint! The Hip Hop Studies Reader*, 2nd edition (New York: Routledge, 2011); Cathy Cohen, *Democracy Remixed: Black Youth and the Future of American Politics* (New York: Oxford University Press, 2012).

78 Brian Cross, *It's Not about Salary: Rap, Race, and Resistance in Los Angeles* (New York: Verso, 1993).

79 On NWA, see Rose, *Black Noise*, 101–2. The quotation on bridging the gap comes from Martinez, "Popular Culture as Oppositional Culture," 279.

80 See Houston Baker, *Black Studies, Rap, and the Academy* (Chicago: University of Chicago Press, 1993). Martinez acknowledges that she, along with "the work of Lears (1985), Clarke et al. (1976), Hall (1981), Williams (1991), Frith (1981) and Gilroy (1993) would assert, popular culture may be embedded within and even contribute to a dominant hegemonic framework, but it is still capable of resisting that framework. (See also Chandra Mukerji and Michael Schudson, *Rethinking Popular Culture* [Berkeley: University of California Press, 1991]). Rap music is, perhaps, once of the most intriguing examples of such resistance." This quotation comes from Martinez, "Popular Culture as Oppositional Culture," 272.

81 Legally, environmental racism or eco-racism is defined as the intentional or unintentional targeting of low-income and/or racially homogenous communities in planning or policy decisions such that those communities are faced with more negative environmental effects than are more affluent communities. See Luke Cole and Sheila Foster, *From the Ground Up: Environmental Racism and the Rise of the Environmental Justice Movement* (New York: NYU Press, 2001).

82 See Beyoncé's, "Formation" (Official Music Video), YouTube, September 29, 2016, www.youtube.com/watch?v=hG2D1fvEwIo (accessed June 1, 2018).

83 Brenton Mock, "After Harvey and Irma, People of Color Face Displacement," *Grist*, September 14, 2017, https://grist.org. On the historic impact of discrimination, segregation, pollution, and economic injustice in Houston's traditionally African American Third Ward, see Robert Bullard, *Invisible Houston: The Black Experience in Boom and Bust* (College Station, TX: A&M Press, 2000).

84 Tatiana Cirisano, "'Hand in Hand' Telethon: Vic Mensa Wants Climate Change Deniers Removed from Gov't," *Billboard*, September 12, 2013, https://www.billboard.com.

85 Green For All, "Common Joins Green For All's #FixThePipes Campaign," YouTube, September 11, 2017, www.youtube.com/watch?v=mchBmwRykeg.

86 Lynn Feinerman, producer/director, *Eco Rap: Voices from the Hood* (San Francisco: Green Planet Films, 1993). Some researchers have argued that this intertwining of lead and violence in exposed communities is literal, as lead contamination has been found to be causally linked to higher rates of violence in young adults who have been exposed to elevated levels of contamination as children. See, for instance, Paul Stetesky and Michael Lynch, "The Relationship between

Lead and Crime," *Journal of Health and Social Behavior* (June 2004): 214–29, and Herbert Needleman et al., "Bone Lead Levels and Delinquent Behavior," *Journal of the American Medical Association* 275, no. 5 (1996): 363–69.

87 See Harvard sociologists Robert Sampson and Alix Winter's "Racial Ecology of Lead Poisoning: Toxic Inequality in Chicago's Neighborhoods, 1995–2013," *DuBois Review* (Cambridge, MA: Hutchins Center for African and African American Research, 2016), 1–23, https://scholar.harvard.edu; Emily Weyrauch, "Black Kids Are 10 Times More Likely Than White Kids to Die from Guns, Study Says," *Time*, June 20, 2017, http://time.com.

88 Olga Khazan, "Being Black in America Can Be Hazardous to Your Health," *Atlantic*, July/August 2018, www.theatlantic.com.

89 "Campus Consciousness Tour, Fall Tour October 2010," Reverb, www.reverb.org (accessed February 8, 2019); Cristina Ramos, "Drake Announces Eco-Friendly College Tour," MTV News, February 8, 2010, http://www.mtv.com/news.

90 William Banfield, "The Rub: Markets, Morals, and the 'Theologizing' of Music," in Anthony Pinn, ed., *Noise and Spirit: The Religious and Spiritual Sensibilities of Rap Music* (New York: NYU Press, 2003), 174.

91 Banfield, "The Rub."

92 Roderick Nash, *Wilderness and the American Mind* (New Haven, CT: Yale University Press, 1967); Leo Marx, *The Machine in the Garden: Technology and the Pastoral Ideal in America* (New York: Oxford University Press, 1964); Douglas Strong, *Dreamers and Defenders: American Conservationists* (Lincoln: University of Nebraska Press, 1971).

93 Feinerman, *Eco Rap*.

94 In 1989, Kimberlé Crenshaw coined the term "intersectional" to talk about intersections of race and sex. Since then "intersectional" has been used to talk about politics that attend to multiple areas of activism and identity simultaneously. "Intersectional environmentalism" refers to environmental activism that does not separate out environmentalist causes from struggles against discrimination based upon race, class, sex, and so forth. See Kimberlé Crenshaw, "Demarginalizing the Intersection of Race and Sex: A Black Feminist Critique of Antidiscrimination Doctrine, Feminist Theory, and Antiracist Politics," *University of Chicago Legal Forum* 1989 (1989): https://chicagounbound.uchicago.edu.

95 Bryant, "Dream Reborn."

96 Van Jones, *The Green Collar Economy: How One Solution Can Fix Two of Our Biggest Problems* (New York: Harper One, 2008), chapter 4, "The New Green Deal," 85–120; Van Jones, "Working Together for a New Green Deal," *Nation*, November 17, 2008, www.thenation.com. The activist slogan "I Can't Breathe" became iconic in the Black Lives Matter movement as a rallying cry referencing Eric Garner, an African American man who died in Staten Island, New York, after police officers held him in an illegal chokehold. Eco-justice advocates also draw attention to African Americans not being able to breathe in polluted neighborhoods and the resulting disease and death.

97 Jones, *The Green Collar Economy.*

98 Xiuhtezcatl Martinez, "Hip-Hop Environmental Activism," TEDxYouth@ MileHile, May 14, 2014, https://www.youtube.com/watch?v=02V2yVkedtM; and Lorraine Chow, "Pending Youth Climate Case Inspires Nationwide Movement," Ecowatch, October 29, 2018, www.ecowatch.com.

99 For a more detailed description of this suit and a copy of the complaint, see "Details of Proceedings, Juliana v. United States," Our Children's Trust, 2018, www.ourchildrenstrust.org (accessed July 1, 2018).

100 Laura Parker, "Support Is Surging for Teens' Climate Change Suit," *National Geographic*, March 5, 2019, www.nationalgeographic.com.

101 Xiuhtezcatl Martinez, *We Rise: The Earth Guardians Guide to Building a Movement That Restores the Planet* (Emmaus, PA: Rodale, 2017).

102 Coco McPherson, "Environmental Activist Xiuhtezcatl Martinez: A Teen on the Front Lines," *Rolling Stone*, July 19, 2017, www.rollingstone.com.

103 Claire Elise Thompson, "Youth-led Climate Protests Sweep across Europe," *Grist*, February 15, 2019, https://grist.org. The European student climate marches were inspired by the example set by sixteen-year-old Swedish student Greta Thunberg, who had engaged in a weekly climate strike since August 2018 and whose protests had gone viral on social media.

104 Martinez, *We Rise*, 29–30.

105 On the rhetorical power of short form or "tout court" historically and in the digital era, see Robert Hariman, "New Wines in Old Wine Bottles: Quotations and the Rhetoric of Fiction," *Quarterly Journal of Speech* 99, no. 2 (May 2013): 233–41. Not coincidently, these are also the short forms of storytelling that make consumer marketing, including stealth of "black ops" marketing, so effective in digital spaces. See Einstein, *Black Ops Advertising.*

106 Martinez, *We Rise*, 200–201.

107 Elizabeth Royte, *Garbage Land: On the Secret Trail of Trash* (New York: Little, Brown, 2004); Edward Humes, *Garbology: Our Love Affair with Trash* (New York: Avery, Penguin Group, 2013); Sally French, "The Huge Floating Island of Trash in the Pacific Is Now Twice the Size of Texas," MarketWatch, March 23, 2018, www.marketwatch.com.

108 Stephen Lasko, "Resuscitating Two Birds with One Breath," *Gaia*, Spring 2009, https://environmentalmediaproject.files.wordpress.com (accessed July 1, 2018).

109 See the transcript of Michel Martin's interview with Van Jones, "White House Advisor Pushes 'Green Collar Jobs,'" National Public Radio, April 22, 2009, www.npr.org.

110 Lauren McCauley, "To Change Everything, We Need Everyone," Eco-Watch, January 4, 2017, www.ecowatch.com.

CONCLUSION

1 See Brené Brown, *Rising Strong: How the Ability to Reset Transforms the Way We Live, Love, Parent, and Lead* (New York: Random House, 2015), 6; and Paul Zak, *The Moral Molecule: The Source of Love and Prosperity* (New York: Dutton, 2012),

28–50; and Giovanni Rodridguez, "This Is Your Brain on Story," *Forbes*, July 21, 2017, www.Forbes.com.

2 Brown, *Rising Strong*, 9.

3 See Theodor Adorno, "Freudian Theory and the Pattern of Fascist Propaganda," in Geza Róheim, ed., *Psychoanalysis and the Social Sciences*, vol. 3 (Oxford: International Universities Press), 279–300; and Max Horkheimer and Theodor Adorno, *Dialectic of Enlightenment* (Palo Alto, CA: Stanford University Press, 2007), especially chapter 4 on "The Culture Industry: Enlightenment as Mass Deception," 94–136.

4 Lawrence Grossman et al., *Mediamaking: Mass Media in Popular Culture* (Thousand Oaks, CA: Sage, 2005), xxii.

5 See chapter 6, "Rules for the Human Park: A Response to 'Heidegger's Letter on Humanism,'" in Peter Sloterdijk, *Not Saved: Essays after Heidegger* (Cambridge, UK: Polity, 2017), 211.

6 Peter Sloterdijk, *Neither Sun nor Death* (Los Angeles: Semiotext(e) [MIT Press], 2011), 84–85. See also Roy Scranton's analysis of media loops in his chapter on "The Compulsion of Strife" and his detailing of the ways in which we feed social media "vibrations" of fear, social energetics of distrust and despair, when we pass them on. This constant stream of threats and the fear they generate "short circuit our own autonomous desires, divert us from our goals," distract us, and interject static into our cognitive processing. Like Sloterdijk, Scranton advocates interrupting these loops and not feeding certain "social energetics," or at the very least, transforming them. Roy Scranton, *Learning to Die in the Anthropocene*, 78. I am indebted to my professor Elizabeth Ellsworth for steering me toward this literature in her "Media in the Anthropocene" course at the New School.

7 Chafe, "Is There an American Narrative and What Is It?"

8 John Ciardi, *How Does a Poem Mean: A Question Posed by the Distinguished Poet* (Boston: Houghton Mifflin, 1959), 4–5.

9 Ciardi, *How Does a Poem Mean*, 6.

10 Brian Sutton-Smith, *The Ambiguity of Play* (Cambridge, MA: Harvard University Press, 1997), 198; see also discussion of Dr. Stuart Brown's National Institute of Play in *The Gifts of Imperfection* (Center City, MN: Hazelden, 2010), 101.

11 Scranton, *Learning to Die in the Anthropocene*, 55.

12 To close her book, *Vibrant Matter*, Bennett even issues a "Nicene Creed for would-be materialists" (122). In interviews, Bennett has also spoken about how her foundational notions about justice were shaped early on by the liberation-theology-inflected Catholicism of her Italian/Irish Roman Catholic upbringing. See Michael James, "Agency, Nature, and Emergent Properties: An Interview with Jane Bennett," Synthetic Zero, May 24, 2013, https://syntheticzero.net (accessed July 3, 2018). Matthew Fox is a former Dominican priest who has since left Roman Catholic orders and has been received as a priest into the Episcopal Church.

13 Matthew Fox, *Wrestling with the Prophets: Essays on Creation Spirituality and Everyday Life* (San Francisco: Harper San Francisco, 1995), 39.

14 See Bennett, *Vibrant Matter*, xi. On "delight and disturbance," see Bennett, *The Enchantment of Modern Life: Attachments, Crossings, Ethics* (Princeton, NJ: Princeton University Press, 2001), 128. Without enchantment, a strange combination of delight and disturbance, says Bennett, we lack the reservoir of energy needed to challenge injustice.

15 Bennett, *Vibrant Matter*, 122, 127.

16 Sam Wang and Sandra Aamodt, "Play, Stress, and the Learning Brain," *Cerebrum*, September 24, 2012, https://www.ncbi.nlm.nih.gov.

17 See part 4, "The Serious Business of Play," and chapter 14, on "Play in Adult Life" in Sandra Aamodt and Sam Wang, *Welcome to Your Child's Brain: How the Mind Grows from Conception to College* (New York: Bloomsbury, 2011).

18 Wang and Aamodt, "Play, Stress, and the Learning Brain."

19 Wang and Aamodt, "Play, Stress, and the Learning Brain."

20 Wang and Aamodt, "Play, Stress, and the Learning Brain."

21 Elizabeth Kolbert, *The Sixth Extinction*, 249.

22 Catherine Keller, *Apocalypse Then and Now: A Feminist Guide to the End of the World* (Boston: Beacon, 1996), xiii.

23 On the images streamed of "Starman," Naomi Klein tweeted, "About Elon's Big Day: This is a car commercial in space. Everyone: Please stop participating," February 6, 2018, https://twitter.com/naomiaklein/status/961003065636806656?lang=en (accessed February 15, 2018).

24 David Mosher and Kelly Dickerson, "Elon Musk: We Need to Leave Earth as Soon as Possible," *Business Insider*, October 10, 2015; "Elon Musk: We Need to Leave Earth as Soon as Possible for One Critical Reason," Physics-Astronomy, October 11, 2015, www.physics-astronomy.com.

25 Paul Harris, "Interview with Elon Musk: 'I'm Planning to Retire to Mars," *Guardian*, July 31, 2010, www.theguardian.com. On Musk's prophecies of AI robots "killing us all," see James Cook, "Elon Musk Thinks Robots Could Start Killing Us in Five Years," *Inc.*, November 7, 2014, www.inc.com/business-insider. Unclear is why the killer robots would not simply follow us to Mars.

26 "You back up your hard drive . . . maybe we should back up life, too," Musk has said, repeatedly making the earth/computer hardware-storage-system analogy. See Chris Heath, "How Elon Musk Plans on Inventing the World (and Mars)," *GQ*, December 12, 2015, https://www.gq.com.

27 For American studies scholarship on the "next great place" frontier mentality and the allure of virgin territory, see Annette Kolodny, *The Land before Her: Fantasy and Experience of American Frontiers, 1630–1860* (Chapel Hill: University of North Carolina Press, 1984); and the earlier classic in this genre, Henry Nash Smith's *Virgin Land: The American West as Symbol and Myth* (New York: Vintage, 1950). For a more in-depth history of the cultural wallpaper that is "unmarked Protestantism" in the United States, see Tracy Fessenden, *Culture and Redemption*.

28 See, more generally, Christian Davenport's *Space Barons: Elon Musk, Jeff Bezos, and the Quest to Colonize the Cosmos* (New York: Hatchett, 2018), especially

chapter 7, "The Risk," and quotations from Musk on 177; Matt Kalpin, "Elon Musk: Manifest Destiny in the New Space Frontier," *Northeastern University Science*, March 19, 2017, https://nuscimag.com (accessed July 15, 2018).

29 See Nicole Karlis and Keith Spencer, "How Taxpayer Money Could End Up Paying for Rich People to Go to Space," *Salon*, January 21, 2018, www.salon.com.

30 Jerry Hirsch, "Elon Musk's Growing Empire Is Fueled by $4.9 Billion in Government Subsidies," *Los Angeles Times*, May 30, 2015, www.latimes.com.

31 See Shop SpaceX Online Catalog, https://shop.spacex.com (accessed July 15, 2018).

32 Ten different varieties of SpaceX flags, including one with an "occupy" message, can be found here: "Flags," https://shopspacex.com (accessed July 15, 2018).

33 "Cosplay" is short for "costume play" and is a popular practice within various fandoms, in which fans dress up as favorite characters from films, novels, television series, comic books, or video games, and then gather and role play, sometimes engaging in reenactments of storylines or creating new ones.

34 Some examples would be Andy Weir's best-selling novel, *The Martian*, originally self-published and then republished by Crown Publishers in 2014; Haemimont Games' sci-fi builder simulation game, *Surviving Mars* (2018); *Mars: War Logs* (2013); *Waking Mars* (2012), a Mars terraforming game, and many other examples.

35 See Keith Spencer, "Against Mars-a-Lago: Why SpaceX's Mars Colonization Should Terrify You," *Salon*, October 8, 2017, www.salon.com; Tim Hewitt-Coleman, "Elon Musk Is Dead Wrong about Mars," Permaculture Research Institute, September 11, 2017, https://permaculturenews.org.

36 See Georgia Frances King, "Should We Leave Earth to Colonize Mars? A NASA Astronaut Says 'Nope,'" *Quartz*, February 10, 2017, https://qz.com; and astrobiologist David Warmflash's, "Forget Mars: Here's Where We Should Build Our First Off-World Colonies," *Discover*, September 8, 2014, http://blogs.discovermagazine.com; Adam Ozinek, "Sorry Nerds, but Colonizing Mars Is Not a Good Plan," *Forbes*, May 6, 2017, www.forbes.com.

37 Alan Boyle, "Life, Liberty, and the Pursuit of Spaceflight? Jeff Bezos Links Blue Origin to Saving Earth," *CNN News*, November 15, 2017, www.geekwire.com.

38 Warmflash, "Forget Mars."

39 See Richard Callahan Jr., Kathryn Lofton, and Chad Seales, "Allegories of Progress: Industrial Religion in the United States," *Journal of the American Academy of Religion* 78, no. 1 (March 2010): 1–39; and Tricia Sheffield, *The Religious Dimensions of Advertising* (New York: Palgrave, 2006), 53–99.

40 "Elon Musk Named 2007 Entrepreneur of the Year," *Inc.*, December 1, 2007, www.inc.com.

41 As demonstrated in the wildfire phenomenon of *Fortnite* in its *Save the World* and *Battle Royale* game modes, profits for which resulted in game maker Epic Games grossing nearly $300 million in a single month (April 2018) alone. David Thier, "Fortnite's Massive Prize Pool Just Made It the Biggest Game in Esports," *Forbes*, May 21, 2018, www.forbes.com.

42 This wording comes from a game review quoted on Ken Eklund's web site and blog Writerguy.com (accessed July 3, 2018).
43 As cited in "Welcome to a World without Oil," a seven-minute video featured on the greeter page for the World Without Oil archive, http://writerguy.com (accessed February 15, 2018).
44 Ken Eklund, "Lesson Ten: World without Oil," Writerguy, May 2007, http://writerguy.com (accessed July 8, 2018). On levers for social change, see "Ethan Zuckerman and the Levers of Change," MIT Civic Media Project, December 12, 2012, https://civic.mit.edu.
45 "Welcome to a World without Oil"; see also McGonigal chronicling of her ARG game designs and her aim in creating these games on her web site: https://janemcgonigal.com (accessed July 15, 2018).
46 Belinda Acosta, "See Jane Game: Interview with Jane McGonigal," *Austin Chronicle*, February 9, 2008, www.austinchronicle.com.
47 "World without Oil, the Texts," http://wwotext.blogspot.com (accessed July 3, 2018).
48 Saler, *As If*.
49 Acosta, "See Jane Game."
50 "Staying in Touch and Keeping This Journal Going," LiveJournal, June 8, 2007, http://fallingintosin.livejournal.com (accessed July 15, 2018).
51 Naomi Alderman, "The First Great Works of Digital Literature Are Already Being Written," *Guardian*, October 13, 2015, www.theguardian.com.
52 See Paul Hawken, ed., *Drawdown: The Most Comprehensive Plan Ever Proposed to Reverse Global Warming* (New York: Penguin, 2017).
53 Catherine Keller, "Women against Wasting the World: Notes on Eschatology and Ecology," in Irene Diamond and Gloria Orenstein, eds., *Reweaving the World: The Emergence of Ecofeminism* (San Francisco: Sierra Club Books, 1990), 263.
54 Starhawk, *The Fifth Sacred Thing* (New York: Bantam Random House, 1994), 3.
55 "About—Sustainable Human," Susatainable Human, https://sustainablehuman.tv and "By Changing the Story, We Change Our World," posting on Facebook, July 24, 2014, www.facebook.com (accessed July 15, 2018).
56 Sustainable Human, "To Change Our World, We Must Change Our Story: A New Story of the People," Facebppl. November 2017, www.facebook.com (accessed July 15, 2018).
57 Gottschall, *The Storytelling Animal*, xiv.
58 Liana Gamber-Thompson and Arely Zimmerman document such use of media as "bridge to social action" in their study of undocumented youth "DREAMer" media activism. See "DREAMing Citizenship: Undocumented Youth, Coming Out, and Pathways to Participation," in Jenkins et al., *By Any Media Necessary*, 207.
59 See, for example, the work of Yvette Alberdingk Thijm, of the organization Witness, who advocates "more eyes and more justice"—the use of citizen video and photography by "live-streaming activists" to expose injustices. This grassroots movement has now extended in the United States from not only documenting

police brutality and shootings but to posting videos of white people who call the police on African Americans who are simply shopping, waiting for a friend at a coffee house, and even knocking on constituents' doors as a legislative representative. See Thijm's TEDxStoll talk, "The Power of Citizen Video to Create Undeniable Truths," TED, April 2017, www.ted.com (accessed July 15, 2018). Thijm rightly cautions that this culture of surveillance is a double-edged sword and not without its serious drawbacks, if not violent intrusions into private life.

60 Harris, "How a Handful of Tech Companies Control Billions of Minds Every Day."

61 Jenkins et al., *By Any Media Necessary*, 20. Here, Jenkins details the arguments about the thorny politics of voice and what voices do and do not make it into public life found in Nick Couldry's *Why Voice Matters: Culture and Politics after Neoliberalism* (London: Sage, 2010).

62 Jenkins et al., *By Any Media Necessary*, 15–17.

63 Larry McCaffery and Sinda Gregory, "Everything on the Verge of Becoming Something Else: An Interview with Ted Mooney," The Write Stuff, 2012, www.altx.com (accessed July 15, 2018).

64 Gottschall, *The Storytelling Animal*, 197.

INDEX

ABOUT THE AUTHOR

Sarah McFarland Taylor is Associate Professor in the Department of Religious Studies and in the Program in Environmental Policy and Culture at Northwestern University. She is the award-winning author of *Green Sisters: A Spiritual Ecology*.